Tasks for vegetation science 41

SERIES EDITORS

The titles published in this series are listed at the end of this volume.

Plants and Climate Change

Edited by

JELTE ROZEMA, RIEN AERTS and HANS CORNELISSEN
Vrije Universiteit, Amsterdam, The Netherlands

Springer

A C.I.P. Catalogue record for this book is available from the library of Congress.

ISBN-1-4020-4442-9

Published by Springer,
P.O. Box 17, 3300 AA Dordrecht, The Netherlands

www.springeronline.com

Printed on acid-free paper

Printed in the Netherlands

Plant Ecology

An international journal Formerly Vegetatio

CONTENTS • VOLUME 182 • NO. 1–2 • 2006

Special issue on:
PLANTS AND CLIMATE CHANGE

Edited by:
Jelte Rozema, Rien Aerts and Hans Cornelissen

Plant Ecology (2006)
DOI 10.1007/s11258-005-9063-6

Preface

Special issue – *Plants and Climate Change*

April 23, 2004 a symposium was held at the Vrije Universiteit, Amsterdam, The Netherlands, entitled Plants and (Present and Past) Climate Change on the occasion of the new Chair on Climate–Biosphere Interactions for Jelte Rozema. The aim of that Chair is to promote research on how climate affects or affected the biosphere and *vice versa* both in the present and past.

Contributors were invited to write review-like papers providing the State-of-the-Art of topics relating to plant–climate interactions, but also new research data are presented here.

This Special Issue of the International Journal Plant Ecology on Plants and Climate Change covers 14 peer-reviewed papers highlighting plant responses to atmospheric CO_2 increase, to global warming and to increased ultraviolet-B radiation as a result of stratospheric ozone depletion.

Dependent on how and how well plant responses to increased temperature, atmospheric CO_2 and ultraviolet-B have been preserved in the (sub)-fossil record, past climates and past atmospheric chemistry may be reconstructed. Pollen and tree-ring data reflect plant species composition and variation of temperature and precipitation over long or shorter time intervals. In addition to well preserved morphological and chemical plant properties, new analytical techniques such as stable isotopes are becoming increasingly important in this respect. The development and validation of such biotic climate and environment proxies builds a bridge between biological and geological research. This highlights that Plant–Climate Change research is becoming a multi- and transdisciplinary field of relevant research.

The guest editors acknowledge the opportunity provided by Springer Publishers and the Editor-in-Chief of Plant Ecology, Prof. Arnold van der Valk, to prepare and edit this Special Issue. We are also indebted to the many referees who helped to review the submitted papers.

JELTE ROZEMA, RIEN AERTS, HANS CORNELISSEN
Department of Systems Ecology
Institute of Ecological Science
Vrije Universiteit
De Boelelaan 1087
1081 HV
Amsterdam
The Netherlands

Photo. Geothermally influenced ground, such as illustrated here from Bellingshausen Island in the South Sandwich Islands, provides protection from some of the most extreme conditions of the Antarctic, and is often colonised by exceptional plant communities otherwise only known from lower latitudes, illustrating the separate controls on processes involved in long distance transport to (colonisation) and subsequent establishment at an Antarctic location.

Plant Ecology (2006) 182:1–10
DOI 10.1007/s11258-005-9022-2

Responses of terrestrial Antarctic ecosystems to climate change

P. Convey* and R. I. L. Smith
*British Antarctic Survey, Natural Environment Research Council, High Cross, Madingley Road, Cambridge, CB3 0ET, UK; *Author for correspondence (e-mail: p.convey@bas.ac.uk; phone: +44-1223-221588; fax: +44-1223-221259)*

Received 1 September 2004; accepted in revised form 15 December 2004

Key words: Alien species, Antarctic, Bryophyte, Climate change, Colonisation, Microbiota, Phanerogam, Terrestrial ecosystem, UV radiation

Abstract

Antarctic terrestrial biota are generally limited by the inexorably linked environmental factors of low summer temperature and lack of available water. However, in parts of the Antarctic, both these factors are changing rapidly on contemporary timescales. Terrestrial biota have concurrently been faced with changes in the timing of UV-B maxima associated with spring ozone depletion. The region of the Antarctic Peninsula and Scotia Arc has experienced one of the most rapid rates of environmental warming seen worldwide over the last 30–50 years. Together with local changes in precipitation, this has resulted in a rapid reduction in extent and thinning of many ice-fields and glaciers, exposing new terrain for colonisation while, at the same time, altering patterns of water availability in terrestrial habitats. The rapid development of communities on newly-exposed ground is also facilitated by the existence of soil propagule banks, which contain propagules of both local and exotic origin. In this paper we collate and review evidence from a range of observational and manipulative studies that investigate the effect of climate change, especially increased temperature, on the processes of colonisation and subsequent community development by plants in the Antarctic. Biological changes that have been associated with climate change are visible in the form of expansions in range and local population numbers amongst elements of the flora. Environmental manipulation experiments further demonstrate the possibility of large and rapid species and community responses to climate amelioration, with many resident biota responding positively, at least in the absence of increased competition from exotic colonists. Manipulation studies are also starting to elucidate more subtle responses to climate changes, at levels ranging from cell biochemistry to habitat and food web structure. Integrating such subtle responses is vital to improving our ability to understand the consequences of climate change, as these may lead to much greater consequential impacts on communities and ecosystems.

Introduction

Antarctic terrestrial ecosystems

This paper focuses on two of the three biogeographical zones that are conventionally recognised in Antarctica, the sub- and maritime Antarctic (Smith 1984), which include the west coast of the Antarctic Peninsula northwards from Alexander Island, through the island archipelagos of the Scotia Arc (South Shetland, South Orkney and South Sandwich Is.), to South Georgia.

The maritime Antarctic climate is markedly seasonal, with a strong maritime influence especially during summer that becomes reduced during winter through seasonal sea ice formation. The maritime influence on sub-Antarctic climates is present year-round, resulting in a damping of seasonal temperature variation relative to other polar regions (Convey 1996a; Danks 1999). Faunal and floral diversity are very restricted in comparison with lower latitude locations (Block 1984; Smith 1984; Convey 2001a), generally regarded as a consequence of low summer temperatures, geographical isolation from neighbouring temperate landmasses, and the relatively short period available for colonisation since the commencement of the most recent phase of glacial retreat. For instance, only two higher insects and two flowering plants are present in the maritime Antarctic, where the fauna otherwise comprises simple communities of soil arthropods (Acari, Collembola) and other invertebrates (Nematoda, Tardigrada, Rotifera), and the flora is cryptogamic (bryophytes, lichens). Contemporary communities typically have low species richness and complexity, with few trophic links and interactions. Communities of the most extreme Antarctic environments appear to be at the first stages of colonisation or succession, and are limited purely by the extreme conditions rather than biotic interactions. They are expected to be very sensitive to changes in climate or consequential processes (Callaghan and Jonasson 1995; Freckman and Virginia 1997; Frenot et al. 2005).

Antarctic climate change

In the context of climate change, three elements are fundamental to the biology of Antarctic terrestrial organisms, temperature, water and solar irradiance. Low thermal energy input is a key defining characteristic of Antarctic terrestrial habitats, even when compared with the Arctic. Mean air temperatures even during the warmest summer months are low (below 0 °C in the continental, 0–2 °C in the maritime and 5–10 °C in the sub-Antarctic; Walton 1984; Convey 1996a; regional definitions after Smith 1984), a factor that is particularly significant in the context of climate change, being near to lower threshold temperatures for many biological functions. In this situation, a small temperature increment has a potentially greater biological impact than one of similar scale in a less extreme environment.

Recent air temperature increases are well documented along the Antarctic Peninsula and islands of the Scotia Arc (Smith 1990; Fowbert and Smith 1994; King and Harangozo 1998; Skvarca et al. 1998; King et al. 2003), with those along the west coast of the Antarctic Peninsula being the most rapid in the Southern Hemisphere. At several locations, increases in annual air temperatures of more than 1 °C have occurred over the last 30–50 years, with rates of warming during the winter months being even greater, up to 0.1 °C per year. A feature of Antarctic Peninsula warming is that much stronger trends are seen in the winter months (King and Harangozo 1998; King et al. 2003), which has a potentially important biological impact, by effectively shortening the winter season. Even greater rates of warming have been reported in some Antarctic freshwater ecosystems (Quayle et al. 2002, 2003), which appear to magnify the signal seen in air temperatures.

The availability of liquid water may be even more important than temperature to biological activity in Antarctic terrestrial habitats (Kennedy 1993; Sømme 1995; Block 1996). Water availability is influenced directly by precipitation events (especially in summer), and also by release from seasonal snow cover and glaciers (i.e. the timing can be separated from the original precipitation event). Increased precipitation is predicted in this region (Budd and Simmonds 1991), with some documentary evidence published (Turner et al. 1997). There are also reports of decreasing trends in precipitation, more particularly in the sub-Antarctic (Frenot et al. 1997; Bergstrom and Chown 1999), but also from the maritime Antarctic South Orkney Is. (Noon et al. 2001), highlighting the importance of understanding trends at the appropriate (local) scale. Rapid decreases in glacial extent, and loss of smaller 'permanent' snow banks, may both in the short-term release increased quantities of liquid water into terrestrial habitats. Such decreases are documented along the Antarctic Peninsula and Scotia Arc (Smith 1990; Gordon and Timmis 1992; Fowbert and Smith 1994; Pugh and Davenport 1997; Fox and Cooper 1998; Vaughan et al. 2001; Quayle et al. 2003). In the longer term, loss or seasonal exhaustion of previously permanent sources of water supply may itself provide another limitation to terrestrial habitats (see Convey et al. 2003).

Finally changes in radiation climate have been predicted to impact terrestrial biota. The 'ozone hole' that forms annually in the austral spring over Antarctica has occurred, at least in the recent era, only since the early 1980s (Farman et al. 1985). When present, increased penetration of biologically damaging shorter wavelength UV-B radiation is possible. While maximum intensities of radiation received under the ozone are similar to normal summer maxima, they differ in two important features in that maxima now occur earlier in the season (in particular, when biota may not be fully physiologically active and able to respond) and that lower wavelengths penetrate to ground level than is normal at this time of year.

Predictions

It has been recognised since the late 1980s that the climate changes being experienced in parts of Antarctica are likely to lead to clear responses in these simple terrestrial ecosystems (Roberts 1989; Smith and Steenkamp 1990; Voytek 1990). As has happened globally, this recognition has encouraged the development of a predictive literature (e.g. Adamson and Adamson 1992; Wynn-Williams 1994, 1996; Kennedy 1995a; Convey 1997; Walton et al. 1997; Bergstrom and Chown 1999; Smith 2003; Frenot et al. 2005), addressing potential consequences for microbial groups, invertebrate fauna, flora, colonisation and ecosystem level processes. As a very broad generalisation, changes in two of the three major environmental variables (temperature, water) lead to predictions of positive responses in indigenous Antarctic biota, essentially through relaxation of current abiotic constraints on biological activity (Kennedy 1995a; Convey 2003). However, they are also predicted to lead to an increase in colonisation by exotic species (Kennedy 1995a; Smith 1996; Convey 1997, 2003; Frenot et al. 2005), generating increased diversity and trophic complexity and altering the physical structure of habitats. In the longer term, the latter process may lead to the loss of Antarctic species and communities through increased competition, as defining features of contemporary Antarctic biota and ecosystems include poor competitive abilities and general insignificance of competition (Convey 1996b). In contrast, changes in the radiation environment experienced as a result of exposure to increased UV-B under the ozone hole are likely to result in negative consequences for biota, as the costs of mitigation strategies are increased.

Over the last decade, a range of Antarctic terrestrial biological studies have been designed to test predictions associated with climate change, and there is now a literature pertaining to the real consequences of change in the Antarctic [see reviews by Convey (2001b, 2003) and Smith (2001); Walther et al. (2002) provide a global review of such evidence]. These can be separated into (i) descriptive and observational studies, and (ii) manipulative studies applied to more or less realistic natural habitats. In the remainder of this paper, we aim to summarise and collate the evidence obtained from such studies relating to the consequences of climate change for botanical elements of Antarctic terrestrial ecosystems.

Evidence from field observations

While the recent period of rapid regional climate change in the Antarctic is, justifiably, receiving much attention, it is also the case that the region has experienced change on a scale of 100s–1000s of years, during and subsequent to the cycles of Pleistocene glacial advance and recession (Convey 2003). During the early 1960s, within the first two decades of the current warming phase, rapid ice recession was already apparent at a number of maritime Antarctic locations, such as Signy I. and the Argentine Is. (R.I.L. Smith, personal observations). At this time, lichen trim-lines visible adjacent to glaciers, and the re-exposure of vegetation buried under ice, indicated that a previous cold period (possibly equating to the 'mini Ice Age' of Europe) had allowed greater extent of glaciers and icefields (Smith 1972, 1990; Corner and Smith 1973; Fenton 1982).

The most frequently quoted and striking proposed consequence of regional warming in the maritime Antarctic are the recent increases (1–2 orders of magnitude; Table 1) in populations of the two native Antarctic flowering plants (*Deschampsia antarctica* and *Colobanthus quitensis*) (Fowbert and Smith 1994; Smith 1994, 2001, 2003; Grobe et al. 1997). These increases do not include

Table 1. Increase in numbers of plants of *Deschampsia antarctica* and *Colobanthus quitensis* recorded between 1964 and 1990 at three sites in the Argentine Islands, western Antarctic Peninsula (extracted from Fowbert and Smith 1994).

Year	*Deschampsia antarctica*	*Colobanthus quitensis*
1964	610	62
1990	c. 17,000	377

any southwards extension of the species' geographical ranges, which remain defined by a lack of suitable ice free habitat south of the current limit in the Terra Firma Islands of southern Marguerite Bay. The most important factor behind the expansions is likely to be through more frequent success in the maturation of seeds (cf. Edwards 1974; Convey 1996c), which may have a further impact through their being able to remain dormant in soil propagule banks (McGraw and Day 1997).

Rapid population increases have been seen amongst bryophytes and microbiota in the maritime Antarctic (Smith 1993, 2001; Wynn-Williams 1996). These are again assisted by the characteristic of many of these groups of having propagules which are capable of remaining dormant in the soil propagule bank for many years (Table 2) (Smith and Coupar 1987; Smith, 1987, 1993). Antarctic soils have been manipulated *in situ* and subjected to laboratory culture in order to examine the diversity of (culturable) propagules present, with results indicating that propagule banks contain locally-occurring and, rarely, exotic species (Smith 1987, 1993, 2000; Smith and Coupar 1987). Aerobiological and palynological studies further demonstrate infrequent transfer of exotic biological material into the region, primarily from southern South America (Barrow 1978; Marshall 1996). While pollen of South American origin is frequently a component of such sampling studies, to date no spores of bryophyte or lichen species of confirmed non-Antarctic origin have been found (Marshall and Convey 1997). However, the presence of lower latitude bryophytes at ephemeral geothermally active sites in the maritime and continental Antarctic (Smith 1991; Bargagli et al.

Table 2. The contribution of soil propagule banks to the establishment of plants on Antarctic soils under simulated warming conditions.

	Age (years exposure) of moraine					
	5	—	15	25	35	45
(a)						
Total plant cover (%)	11		26	43	53	75
Number of species	2		5	8	8	10
Location	Biogeographical zone		Number of species			
(b)						
Husdal, South Georgia, 54 °S	Sub-Antarctic		18			
Signy Island, South Orkney Islands, 60 °S	Maritime Antarctic (north)		8			
			15 (calcareous ground)			
Cierva Point, northern Antarctic Peninsula, 64 °S	Maritime Antarctic (central)		6			
Rothera Point, Adelaide Island, 67 °S	Maritime Antarctic (central)		5			
Clarke Peninsula, Wilkes Land, 67 °S	Continental Antarctic (coast)		2			
Mars Oasis, Alexander Island, 72 °S	Maritime Antarctic (southern)		5			
Edmondson Point, Victoria Land, 74 °S	Continental Antarctic (coast)		4			

(a) Plant species diversity present after three years under screens placed on a moraine chronosequence on sub-Antarctic South Georgia.
(b) Numbers of bryophyte species cultured from unvegetated surface soil collected at recently deglaciated sites along a latitudinal gradient from sub-Antarctic South Georgia to continental Antarctic Victoria Land. Data extracted from Smith (2001).

1996; Convey et al. 2000) demonstrates that such transfer must occur.

It is clear from transplant experiments carried out during the 1960s (Edwards and Greene 1973; Edwards 1980), and from the many accidental introductions to sub-Antarctic islands (Frenot et al. 2005) that many non-Antarctic flowering plants are capable of establishment if the problems of dispersal can be overcome. Some of these species, such as *Poa annua* on South Georgia are invasive, and have rapidly occupied areas where indigenous plant communities have been removed by vertebrate activity, particularly grazing by introduced reindeer. Human activity is already particularly important in this context – more than 50% of the vascular flora of South Georgia is accounted for by persistent anthropogenically-introduced species (Smith 1996) – with successful introductions expected to increase, and extend to more southerly locations as climate amelioration continues (Frenot et al. 2005).

There have been few attempts at the biochemical or ecophysiological level to quantify responses of Antarctic plants *in situ* to changing climatic conditions. In the Antarctic, the studies of Newsham et al. (2002) and Newsham (2003) provide the only examples of measurement of a direct biochemical response to increased UV-B radiation during episodes of ozone depletion in non-manipulated field populations of three bryophyte taxa (*Andreaea*, *Cephaloziella*, *Sanionia*). These show patterns of protective pigment synthesis and loss that are most strongly correlated with their recent (natural) radiation exposure history, indicating a rapid and dynamic biochemical response to this environmental stress. Rousseaux et al. (1999) provide data linking foliar DNA damage with ozone depletion in a southern South American herb.

Evidence from field manipulations

The use of simple field manipulation techniques is widespread in Antarctic studies. Most are based around the use of some form of chamber or screen (greenhouse methodologies), which is placed over an area of habitat in order to alter aspects of the thermal and radiation climates. In their simplest form, these can be left in place in remote locations year-round. With more regular access for researchers, and availability of a power source, methodolologies can be adopted which include water or nutrient addition, and the use of lamps to alter light and UV radiation climates. However, in detail, the environmental changes achieved are often more complex and inter-related than is widely appreciated (Caldwell and Flint 1994; Kennedy 1995b, c), and care is required in both experimental design and interpretation of the data obtained.

Greenhouse methodologies have generated very rapid population responses in studies of Antarctic microbes (Wynn-Williams 1993, 1996), and bryophytes and phanerogams (Smith 1990, 1993, 1994, 1999), with these responses also linked with changes in invertebrate populations (Kennedy 1994; Convey and Wynn-Williams 2002; Convey et al. 2002). The most rapid or largest responses appear to be obtained in manipulations of more extreme habitats, for instance at higher altitude (Kennedy 1994) or latitude (Convey and Wynn-Williams 2002). Plant species typically achieve greater coverage, success in establishment, lusher growth, and increased population densities, reproductive output, enhanced sexual reproduction and juvenile survival rates. These types of manipulation are known to cause changes in vegetation growth form (Smith 1990, 2001; Day et al. 1999, 2001; Sullivan and Rozema 1999; Ruhland and Day 2000) and, it is hypothesised, subtle alterations in habitat structure and microclimate. Such changes, possibly linked with changes in diet quality (see below), are thought to underlie or facilitate some of the responses seen in faunal communities (Convey et al. 2002). The synthesis across trophic levels of responses to climate change is a complex but important task as, while subtle and apparently insignificant at one level, they may combine to create much greater effects elsewhere in the ecosystem (Day 2001; Searles et al. 2001; Johnson et al. 2002).

Studies of the effects of environmental change on plant physiology and biochemistry in the Antarctic have focused on the consequences of radiative changes experienced as a result of ozone hole formation. The potential negative effects of UV-B radiation on cell function are well-known, as are the responses available to plants and other organisms to mitigate the damage (e.g. Vincent and Quesada 1994; Wynn-Williams 1994; Cockell and Knowland 1999; Rozema 1999). In laboratory experiments the confirmation of damage predictions is relatively straightforward (e.g. Quesada et al. 1995). However, effects seen in the field (through natural exposure or manipulation) are far smaller in magnitude or even apparently non-existent (Fiscus and Booker 1995; Allen et al. 1998), which may suggest that many experimental results are in reality artefacts of the manipulation technique used. Where identified, the biochemical impacts of exposure to UV-B follow the predicted changes in reaction pathways, particularly those involved in protective pigment synthesis (e.g. Rozema 1999; Paul 2001). One major ecological implication of these lies in the requirement for changes in resource allocation strategies, while consequences, through changes in diet quality, may again be felt much more widely through the ecosystem.

Different research groups have used both amendment (supplementary lamps) and screening (selective absorbtion of incoming wavelengths) techniques to identify the effects of changes in UV radiation exposure on Antarctic autotrophs (Wynn-Williams 1996; Quesada et al. 1998; Huiskes et al. 1999; Montiel et al. 1999; Smith 1999; George et al. 2001; Lud et al. 2003), although note the caution (above) that all these techniques introduce potentially confounding artefacts. The groups targeted include algae, cyanobacteria, bryophytes, phanerogams and lichens, while the processes studied range through photosynthetic performance and mitigation strategies (protective pigments) to DNA damage.

Organisms that are fully physiologically active may be able to use a range of mitigation mechanisms whenever they are exposed to damage from UV-B. However, the effectiveness of some repair mechanisms depends on the UV-B:PAR ratio, which is much higher during spring ozone depletion than in midsummer. It is also possible that sensitivity to damage may also be influenced by other environmental factors or change through developmental processes.

Photosynthetic performance in some species appears to be remarkably robust under exposure to UV-B. The two phanerogams, *Deschampsia antarctica* and *Colobanthus quitensis*, have been studied by several groups and show no reduction of photosynthetic parameters when exposed to a range of realistic irradiance stresses (Montiel et al. 1999; Lud et al. 2001), although UV screening causes changes in concentrations of UV-B absorbing pigments, growth and production (Day et al. 1999; Ruhland and Day 2000; Xiong and Day 2001). Similarly, the field measurements of Newsham et al. (2002) on two Antarctic bryophytes (*Cephaloziella varians*, *Sanionia uncinata*) showed no change in photosynthetic yield at the same time as increased concentrations of UV-B screening pigments and carotenoids correlated with the level of recent exposure to natural levels of UV-B. Lud et al. (2001) also reported no change in yield in a lichen (*Usnea antarctica*) at different levels of UV-B exposure. Other experimental studies of bryophytes and cyanobacteria produce equivocal results, with exposure to increased levels of UV-B leading to increases in protective pigments in some, but not all, species (Quesada et al. 1998; Huiskes et al. 1999; Montiel et al. 1999; George et al. 2001).

Taken together, these studies broadly indicate that many Antarctic autotrophs already possess considerable abilities to resist or mitigate the consequences of irradiance stress on their photosynthetic apparatus, and that the recent advent of the Antarctic ozone hole does not directly compromise these abilities. In a related study of resistance to DNA damage in Antarctic terrestrial microbiota, George et al. (2002) found that levels of damage were very low relative to than those reported in microbiota from the shallow marine environment, proposing this to relate to much greater development of screening systems in terrestrial organisms over evolutionary time. Lud et al. (2002) found no increase in DNA damage (measured by cyclobutyl pyrimidine dimer formation) in the moss *Sanionia uncinata* under natural levels of UV-B radiation in the Antarctic and, although damage was caused by experimental UV-B enhancement, concluded that repair processes were sufficient on a daily timescale. As mentioned above, dynamic switches in utilisation of biochemical pathways relating to pigment production and metabolism, and repair processes, may carry both direct resource costs to the organism concerned and indirect costs elsewhere in the food web (ecosystem) through consequential changes in trophic linkages.

Conclusions

Rapid changes in three major environmental variables are being experienced in the region of the

Antarctic Peninsula and Scotia Arc, these being temperature, liquid water availability and radiation climate.

In the short term, many Antarctic terrestrial biota are likely to benefit from generally reduced environmental stresses, as they are already well-adapted to cope with the existing, rapidly fluctuating, stresses of their highly variable environment. The direct consequences of increases in UV-B exposure are likely to be negative, though subtle, as well as being expressed through consequential impacts elsewhere in the foodweb. In the longer term, colonisation of the region by lower latitude species with greater competitive ability, most likely with inadvertent human assistance, will become increasingly important, and could lead to large-scale change in the biological composition and possibly trophic complexity in some existing Antarctic terrestrial ecosystems.

Botanical responses to recent climate amelioration are already visible in the maritime and sub-Antarctic, in the form of rapid local expansion of populations of flowering plants, comparable changes in bryophytes, and rapid colonisation of ground recently exposed by snow and ice recession. These changes facilitate and are rapidly followed by the development of typical terrestrial invertebrate communities. The results of field manipulations mimicking predicted levels of thermal amelioration generally confirm these patterns of response, as well as highlighting the importance of propagule banks in the soil in accelerating the processes of establishment and community development.

Acknowledgements

This paper contributes to the British Antarctic Survey BIRESA (Biological Responses to Environmental Stress in Antarctica) project and to the SCAR RiSCC (Regional Sensitivity to Climate Change in Antarctica) Program.

References

Adamson, H. and Adamson, E. 1992. Possible effects of global climate change on Antarctic terrestrial vegetation. In: Impact of Climate Change on Antarctica–Australia. Australian Government Publishing Service, Canberra, pp. 52–62.

Allen D. J., Nogués S. and Baker N.R. 1998. Ozone depletion and increased UV-B radiation: is there a real threat to photosynthesis?. J. Exp. Bot. 49: 1775–1788.

Bargagli R., Broady P.A. and Walton D.W.H. 1996. Preliminary investigation of the thermal biosystem of Mount Rittmann fumaroles (northern Victoria Land, Antarctica). Antarct. Sci. 8: 121–126.

Barrow C.J. 1978. Postglacial pollen diagrams from South Georgia (sub-Antarctic) and West Falkland Island (South Atlantic). J. Biogeog. 5: 251–274.

Bergstrom D.M. and Chown S.L. 1999. Life at the front: history, ecology and change on southern ocean islands. Trends Ecol. Evol. 14: 472–476.

Block W. 1984. Terrestrial microbiology, invertebrates and ecosystems. In: Laws R.M. (ed.), Antarctic Ecology vol 1., Academic Press, London, pp. 163–236.

Block W. 1996. Cold or drought – the lesser of two evils for terrestrial arthropods? Eur. J. Entomol. 93: 325–339.

Budd W.F. and Simmonds I. 1991. The impact of global warming on the Antarctic mass balance and global sea level. In: Weller G., Wilson C.L. and Severin B.A.B. (eds), Proceedings of the International Conference on the Role of Polar regions in Global Change.

Caldwell M.M. and Flint S.D. 1994. Stratospheric ozone reduction, solar UV-B radiation and terrestrial ecosystems. Clim. Change 28: 375–394.

Callaghan T.V. and Jonasson S. 1995. Arctic terrestrial ecosystems and environmental change. Phil. Trans. R. Soc. Lond., A 352: 259–276.

Cockell C.S. and Knowland J. 1999. Ultraviolet radiation screening compounds. Biol. Rev. 74: 311–345.

Convey P. 1996a. Overwintering strategies of terrestrial invertebrates from Antarctica – the significance of flexibility in extremely seasonal environments. Eur. J. Entomol. 93: 489–505.

Convey P. 1996b. The influence of environmental characteristics on life history attributes of Antarctic terrestrial biota. Biol. Rev. 71: 191–225.

Convey P. 1996c. Reproduction of Antarctic flowering plants. Antarct. Sci. 8: 127–134.

Convey P. 1997. Environmental change: possible consequences for life histories of Antarctic terrestrial biota. Korean J. Polar Res. 8: 127–144.

Convey P. 2001a. Antarctic Ecosystems. In: Levin S.A. (ed.), Encyclopedia of Biodiversity vol 1., Academic Press, San Diego, pp. 171–184.

Convey P. 2001b. Terrestrial ecosystem response to climate changes in the Antarctic. In: Walther G.-R., Burga C.A. and Edwards P.J. (eds), 'Fingerprints' of climate change – adapted behaviour and shifting species ranges, Kluwer, New York, pp. 17–42.

Convey P. 2003. Maritime Antarctic climate change: signals from terrestrial biology. In: Domack E., Burnett A., Leventer A., Convey P., Kirby M. and Bindschadler R. (eds), Antarctic Peninsula Climate Variability: Historical and Palaeoenvironmental Perspectives Antarctic Research Series, vol. 79., American Geophysical Union, Washington D.C, pp. 145–158.

Convey P. and Wynn-Williams D.D. 2002. Antarctic soil nematode response to artificial environmental manipulation. Eur. J. Soil Biol. 38: 255–259.

Convey P., Block W. and Peat H.J. 2003. Soil arthropods as indicators of water stress in Antarctic terrestrial habitats?. Global Change Biol. 9: 1718–1730.

Convey P., Smith R.I.L., Hodgson D.A. and Peat H.J. 2000. The flora of the South Sandwich Islands, with particular reference to the influence of geothermal heating. J. Biogeog. 27: 1279–1295.

Convey , P. , Pugh P.J.A., Jackson C., Murray A.W., Ruhland C.T., Xiong F.S. and Day T.A. 2002. Response of Antarctic terrestrial arthropods to multifactorial climate manipulation over a four year period. Ecology 83: 3130–3140.

Corner R.W.M. and Smith R.I.L. 1973. Botanical evidence of ice recession in the Argentine Islands. Br. Antarct. Surv. Bull. 35: 83–86.

Danks H.V. 1999. Life cycles in polar arthropods – flexible or programmed? Eur. J. Entomol. 96: 83–102.

Day T.A. 2001. Multiple trophic levels in UV-B assessments – completing the ecosystem. New Phytol. 152: 183–186.

Day T.A., Ruhland C.T. and Xiong F. 2001. Influence of solar UV-B radiation on Antarctic terrestrial plants: results from a 4-year field study. J. Photochem. Photobiol., B Biol. 62: 78–87.

Day T.A., Ruhland C.T., Grobe C.W. and Xiong F. 1999. Growth and reproduction of Antarctic vascular plants in response to warming and UV radiation reductions in the field. Oecologia 119: 24–35.

Edwards J.A. 1974. Studies in *Colobanthus quitensis* (Kunth) Bartl and *Deschampsia antarctica* Desv.: VI: reproductive performance on Signy Island. Br. Antarct. Surv. Bull. 39: 67–86.

Edwards J.A. 1980. An experimental introduction of vascular plants from South Georgia to the maritime Antarctic. Br. Antarct. Surv. Bull. 49: 73–80.

Edwards J.A. and Greene D.M. 1973. The survival of Falklands Island transplants at South Georgia and Signy Island, South Orkney Islands. Br. Antarct. Surv. Bull. 33-34: 33–45.

Farman J.C., Gardiner B.G. and Shanklin J.D. 1985. Large losses of total ozone in Antarctica reveal seasonal ClO_x/No_x interaction. Nature 315: 207–210.

Fenton J.H.C. 1982. Vegetation re-exposed after burial by ice and its relationship to changing climate in the South Orkney Islands. Br. Antarct. Surv. Bull. 51: 247–255.

Fiscus E.L. and Booker F.L. 1995. Is increased UV-B a threat to crop photosynthesis and productivity? Photosynth. Res. 43: 81–92.

Fowbert J.A. and Smith R.I.L. 1994. Rapid population increase in native vascular plants in the Argentine Islands, Antarctic Peninsula. Arct. Alpine Res. 26: 290–296.

Fox A.J. and Cooper A.P.R. 1998. Climate-change indicators from archival aerial photography of the Antarctic Peninsula. Ann. Glaciol. 27: 636–642.

Freckman D.W. and Virginia R.A. 1997. Low-diversity Antarctic soil nematode communities: distribution and response to disturbance. Ecology 78: 363–369.

Frenot Y., Gloaguen J.-C. and Trehen P. 1997. Climate change in Kerguelen Islands and colonization of recently deglaciated areas by *Poa kerguelensis* and *P. annua*. In: Battaglia B., Valencia J. and Walton D.W.H. (eds), Antarctic Communities: Species, Structure and Survival., Cambridge University Press, Cambridge.

Frenot Y., Chown S.L., Whinam J., Selkirk P., Convey P., Skotnicki M. and Bergstrom D. 2005. Biological invasions in the Antarctic: extent, impacts and implications. Biol. Rev. 80: 45–72.

George A.L., Murray A.W. and Montiel P.O. 2001. Tolerance of Antarctic cyanobacterial mats to enhanced UV radiation. FEMS Microbiol. Ecol. 37: 91–101.

George A.L., Peat H.J. and Buma A.G.J. 2002. Evaluation of DNA dosimetry to assess ozone-mediated variability of biologically harmful ultraviolet radiation in Antarctica. Photochem. Photobiol. 76: 274–280.

Gordon J.E. and Timmis R.J. 1992. Glacier fluctuations on South Georgia during the 1970s and early 1980s. Antarct. Sci. 4: 215–226.

Grobe C.W., Ruhland C.T. and Day T.A. 1997. A new population of *Colobanthus quitensis* near Arthur Harbor, Antarctica: correlating recruitment with warmer summer temperatures. Arct. Alpine Res. 29: 217–221.

Huiskes A., Lud D., Moerdijk-Poortvliet T. and Rozema J. 1999. Impact of UV-B radiation on Antarctic terrestrial vegetation. In: Rozema J. (ed.), Stratospheric Ozone Depletion: the Effects of Enhanced UV-B Radiation on Terrestrial Ecosystems., Backuys, Leiden.

Johnson D., Campbell C.D., Lee J.A., Callaghan T.V. and Gwynne-Jones D. 2002. Arctic microorganisms respond more to elevated UV-B radiation than CO_2. Nature 416: 82–83.

Kennedy A.D. 1993. Water as a limiting factor in the Antarctic terrestrial environment: a biogeographical synthesis. Arct. Alpine Res. 25: 308–315.

Kennedy A.D. 1994. Simulated climate change: a field manipulation study of polar microarthropod community response to global warming. Ecography 17: 131–140.

Kennedy A.D. 1995a. Antarctic terrestrial ecosystem responses to global environmental change. Ann. Rev. Ecol. Syst. 26: 683–704.

Kennedy A.D. 1995b. Temperature effects of passive greenhouse apparatus in high-latitude climate change experiments. Funct. Ecol. 9: 340–350.

Kennedy A.D. 1995c. Simulated climate change: are passive greenhouses a valid microcosm for testing the biological effects of environmental perturbations? Global Change Biol. 1: 29–42.

King J.C. and Harangozo S.A. 1998. Climate change in the western Antarctic Peninsula since 1945: observations and possible causes. Ann. Glaciol. 27: 571–575.

King J.C., Turner J., Marshall G.J., Connally W.M. and Lachlan-Cope T.A. 2003. Antarctic Peninsula climate variability and its causes as revealed by analysis of instrumental records. In: Domack E., BurnettA. Leventer A., Convey P., Kirby M. and Bindschadler R. (eds), Antarctic Peninsula Climate Variability: Historical and Palaeoenvironmental Perspectives., Antarctic Research Series vol .79. American Geophysical Union, Washington D.C, pp. 17–30.

Lud D., Huiskes A.H.L., Moerdijk T.C.W. and Rozema J. 2001. The effects of altered levels of UV-B radiation on an Antarctic grass and lichen. Plant Ecol. 154: 87–99.

Lud D., Moerdijk T.C.W., van de Poll W.H., Buma A.G.J. and Huiskes A.H.L. 2002. DNA damage and photosynthesis in Antarctic and Arctic *Sanionia uncinata* (Hedw.) Loeske under ambient and enhanced levels of UV-B radiation. Plant Cell Environ. 25: 1579–1589.

Lud D., Schlensog M., Schroeter B. and Huiskes A.H.L. 2003. The influence of UV-B radiation on light-dependent photosynthetic performance in *Sanionia uncinata* (Hedw.) Loeske in Antarctica. Polar Biol. 26: 225–232.

Marshall W.A. 1996. Biological particles over Antarctica. Nature 383: 680.

Marshall W.A. and Convey P. 1997. Dispersal of moss propagules in the maritime Antarctic. Polar Biol. 18: 376–383.

McGraw J.B. and Day T.A. 1997. Size and characteristics of a natural seed bank in Antarctica. Arct. Alpine Res. 29: 213–216.

Montiel P., Smith A. and Keiller D. 1999. Photosynthetic responses of selected Antarctic plants to solar radiation in the southern maritime Antarctic. Polar Res. 18: 229–235.

Newsham K.K. 2003. UV-B radiation arising from stratospheric ozone depletion influences the pigmentation of the Antarctic moss *Andreaea regularis*. Oecologia 135: 327–331.

Newsham K.K., Hodgson D.A., Murray A.W.A., Peat H.J. and Smith R.I.L. 2002. Response of two Antarctic bryophytes to stratospheric ozone depletion. Global Change Biol. 8: 972–983.

Noon P.E., Birks H.J.B., Jones V.J. and Ellis-Evans J.C. 2001. Quantitative models for reconstructing catchment ice-extent using physical-chemical characteristics of lake sediments. J. Paleolimnol. 25: 375–392.

Paul N. 2001. Plant responses to UV-B: time to look beyond stratospheric ozone depletion?. New Phytol. 150: 5–8.

Pugh P.J.A. and Davenport J. 1997. Colonisation vs. disturbance: the effects of sustained ice-scouring on intertidal communities J. Exp. Mar. Biol. Ecol. 210: 1–21.

Quayle W.C., Peck L.S., Peat H., Ellis-Evans J.C. and Harrigan P.R. 2002. Extreme responses to climate change in Antarctic lakes. Science 295: 645.

Quayle W.C, Convey P., Peck L.S., Ellis-Evans J.C., Butler H.G. and Peat H.J. 2003. Ecological responses of maritime Antarctic lakes to regional climate change. In: Domack E., Burnett A., Leventer A., Convey P., Kirby M. and Bindschadler R. (eds), Antarctic Peninsula Climate Variability: Historical and Palaeoenvironmental Perspectives., Antarctic Research Series vol. 79. American Geophysical Union, Washington D.C, pp. 159–170.

Quesada A., Mouget J.L. and Vincent W.F. 1995. Growth of Antarctic cyanobacteria under ultraviolet radiation – UVA counteracts UVB inhibition. J. Phycol. 31: 242–248.

Quesada A., Goff L. and Karentz D. 1998. Effects of natural UV radiation on Antarctic cyanobacterial mats. Proc. NIPR Symp. Polar Biol. 11: 98–111.

Roberts L. 1989. Does the ozone hole threaten antarctic life?. Science 244: 288–289.

Rousseaux M.C., Ballaré C.L., Giordano C.V., Scopel A.L., Zima A.M., Szwarcberg-Bracchitta M., Searles P.S., Caldwell M.M. and Diaz S.B. 1999. Ozone depletion and UV-B radiation: impact on plant DNA damage in southern South America. Proc. Natl. Acad. Sci. U.S.A. 96: 15310–15315.

Rozema J. (ed.), 1999. Stratospheric ozone depletion, the effects of enhanced UV-B radiation on terrestrial ecosystems. Backhuys, Leiden.

Ruhland C.T. and Day T.A. 2000. Effects of ultraviolet-B radiation on leaf elongation, production and phenylpropanoid concentrations of *Deschampsia antarctica* and *Colobanthus quitensis* in Antarctica. Physiol. Plant. 109: 244–251.

Searles P.S., Kropp B.R., Flint S.D. and Caldwell M.M. 2001. Influence of solar UV-B radiation on peatland microbial communities of southern Argentina. New Phytol. 152: 213–221.

Skvarca P., Rack W., Rott H. and Ibarzábaly Donángelo T. 1998. Evidence of recent climatic warming on the eastern Antarctic Peninsula. Ann. Glaciol. 27: 628–632.

Smith R.I.L. 1972. Vegetation of the South Orkney Islands with particular reference to Signy Island. Br. Antarct. Surv. Sci. Reps. 68: 1–124.

Smith R.I.L. 1984. Terrestrial Plant Biology of the Sub-Antarctic and Antarctic. In: Laws R.M. (ed.), Antarctic Ecology., Academic Press, London.

Smith R.I.L. 1987. The bryophyte propagule bank of Antarctic fellfield soils. Symp. Biol. Hung. 35: 233–245.

Smith R.I.L. 1990. Signy Island as a paradigm of biological and environmental change in Antarctic terrestrial ecosystems. In: Kerry K.R., Hempel G. (eds) Antarctic Ecosystems, Ecological Change and Conservation., Springer-Verlag, Berlin, pp. 30–48.

Smith R.I.L. 1991. Exotic sporomorpha as inticators of potential immigrant colonists in Antarctica. Grana 30: 313–324.

Smith R.I.L. 1993. The role of bryophyte propagule banks in primary succession: case study of an Antarctric fellfield soil. In: Miles J. and Walton D.W.H. (eds), Primary Succession on Land, Blackwell, Oxford, pp. 55–78.

Smith R.I.L. 1994. Vascular plants as indicators of regional warming in Antarctica. Oecologia 99: 322–328.

Smith R.I.L. 1996. Introduced plants in Antarctica: potential impacts and conservation issues. Biol. Conserv. 76: 135–146.

Smith R.I.L. 1999. Biological and environmental characteristics of three cosmopolitan mosses dominant in continental Antarctica. J. Veg. Sci. 10: 231–242.

Smith R.I.L. 2000. Plant colonisation on a 45-year sequence of annual micromoraines on a South Georgia glacier foreland. In: Davison W., Howard-williams C. and Hawes I. (eds), Antarctic Ecosystems: Models for Wider Ecological Understanding., Caxton Press, Christchurch, pp. 225–232.

Smith R.I.L. 2001. Plant colonization response to climate change in the Antarctic. Folia Fac. Sci. Nat. Univ. *Masarykianae brunensis*. Geographia 25: 19–33.

Smith R.I.L. 2003. The enigma of *Colobanthus quitensis* and *Deschampsia antarctica* in Antarctica. In: Huiskes A.H.L., Gieskes W.W.C., Rozema J., Schorno R.M.L., van der Vies S.M. and Wolff W.J. (eds), Antarctic Biology in a Global Context., Backhuys, Leiden, pp. 234–239.

Smith R.I.L. and Coupar A.M. 1987. The colonization potential of bryophyte propagules in Antarctic fellfield sites. Comité National Français des Recherches Antarctiques 58: 189–204.

Smith V.R. and Steenkamp M. 1990. Climatic change and its ecological implications at a sub-Antarctic island. Oecologia 85: 14–24.

Sømme L. 1995. Invertebrates in Hot and Cold Arid Environments. Springer-Verlag, Berlin.

Sullivan J. and Rozema J. 1999. UV-B effects on terrestrial plant growth and photosynthesis. In: Rozema J. (ed.), Stratospheric ozone depletion, the effects of enhanced UV-B radiation on terrestrial ecosystems., Backhuys, Leiden, pp. 39–57.

Turner J., Colwell S.R. and Harangozo S. 1997. Variability of precipitation over the coastal western Antarctic Peninsula from synoptic observations. J. Geophys. Res. 102: 13999–14007.

Vaughan D.G., Marshall G.J., Connolley W.C., King J.C. and Mulvaney R. 2001. Devil in the detail. Science 293: 1777–1779.

Vincent W.F. and Quesada A. 1994. Ultraviolet effects on cyanobacteria: implications for Antarctic microbial ecosystems. Antarct. Res. Ser. 62: 111–124.

Voytek M.A. 1990. Addressing the biological effects of decreased ozone on the Antarctic environment. Ambio 19: 52–61.

Walther G.-R., Post E., Convey P., Parmesan C., Menzel M., Beebee T.J.C., Fromentin J.-M., Hoegh-Guldberg O. and Bairlein F. 2002. Ecological responses to recent climate change. Nature 416: 389–395.

Walton D.W.H. 1984. terrestrial environment. In: Laws R.M. (ed.), Antarctic Ecology vol. 1., Academic Press, London, pp. 1–60.

Walton D.W.H., Vincent W.F., Timperley M.H., Hawes I. and Howard-Williams C. 1997. Synthesis: polar deserts as indicators of change. In: Lyons W.B., Howard-Williams C. and Hawes I. (eds), Ecosystem Processes in Antarctic Ice-Free Landscapes., Balkema, Rotterdam.

Wynn-Williams D.D. 1993. Microbial processes and the initial stabilisation of fellfield soil. In: Miles J. and Walton D.W.H. (eds), Primary Succession on Land, Blackwell, Oxford.

Wynn-Williams D.D. 1994. Potential effects of ultraviolet radiation on Antarctic primary terrestrial colonizers: cyanobacteria, algae, and cryptogams. Antarct. Res. Ser. 62: 243–257.

Wynn-Williams D.D. 1996. Response of pioneer soil microalgal colonists to environmental change in Antarctica. Microb. Ecol. 31: 177–188.

Xiong F.S. and Day T.A. 2001. Effect of solar ultraviolet-B radiation during springtime ozone depletion on photosynthesis and biomass production of Antarctic vascular plants. Plant Physiol. 125: 738–751.

Photo. Point-intercept abundance measurements of vascular plant species in the Guisveld lowland *Sphagnum*-*Phragmites* reedland.

Plant Ecology (2006) 182:13–24
DOI 10.1007/s11258-005-9028-9

Vascular plant responses to elevated CO_2 in a temperate lowland *Sphagnum* peatland

Rubén Milla[1,2], Johannes H.C. Cornelissen[1,*], Richard S.P. van Logtestijn[1], Sylvia Toet[1,3] and Rien Aerts[1]
[1]*Department of Systems Ecology, Institute of Ecological Sciences Faculty of Earth and Life Sciences, Vrije Universiteit, De Boelelaan 1085, Amsterdam, 1081, HV, The Netherlands;* [2]*Instituto Pirenaico de Ecología (CSIC), P.O. Box 202, 50080, Zaragoza, Spain;* [3]*Environment Department, University of York Heslington, York, YO10 5DD, United Kingdom; *Author for correspondence: (e-mail: hans.cornelissen@ecology.falw.vu.nl; fax: +31-(0)20-4447123)*

Received 1 September 2004; accepted in revised form 15 December 2004

Key words: CO_2-enrichment, FACE, Litter respiration, Nutrient resorption, Species abundance, *Sphagnum*, The Netherlands

Abstract

Vascular plant responses to experimental enrichment with atmospheric carbon dioxide (CO_2), using MINIFACE technology, were studied in a Dutch lowland peatland dominated by *Sphagnum* and *Phragmites* for 3 years. We hypothesized that vascular plant carbon would accumulate in this peatland in response to CO_2 enrichment owing to increased productivity of the predominant species and poorer quality (higher C/N ratios) and consequently lower decomposability of the leaf litter of these species. Carbon isotope signatures demonstrated that the extra 180 ppmv CO_2 in enriched plots had been incorporated into vegetation biomass accordingly. However, on the CO_2 sequestration side of the ecosystem carbon budget, there were neither any significant responses of total aboveground abundance of vascular plants, nor of any of the individual species. On the CO_2 release side of the carbon budget (decomposition pathway), litter quantity did not differ between ambient and CO_2 treatments, while the changes in litter quality (N and P concentration, C/N and C/P ratio) were marginal and inconsistent. It appeared therefore that the afterlife effects of significant CO_2-induced changes in green-leaf chemistry (lower N and P concentrations, higher C/N and C/P) were partly offset by greater resorption of mobile carbohydrates from green leaves during senescence in CO_2-enriched plants. The decomposability of leaf litters of three predominant species from ambient and CO_2-enriched plots, as measured in a laboratory litter respiration assay, showed no differences. The relatively short time period, environmental spatial heterogeneity and small plot sizes might explain part of the lack of CO_2 response. When our results are combined with those from other *Sphagnum* peatland studies, the common pattern emerges that the vascular vegetation in these ecosystems is genuinely resistant to CO_2-induced change. On decadal time-scales, water management and its effects on peatland hydrology, N deposition from anthropogenic sources and land management regimes that arrest the early successional phase (mowing, tree and shrub removal), may have a greater impact on the vascular plant species composition, carbon balance and functioning of lowland *Sphagnum–Phragmites* reedlands than increasing CO_2 concentrations in the atmosphere.

Introduction

Global atmospheric CO_2 concentrations have steadily risen from 280 ppm before the Industrial Revolution to 370 ppm currently and 560 ppm will be reached by the end of the 21st century according to most predictions (IPCC 2001). Peatlands store a substantial proportion of the global organic carbon pool (Gorham 1991), which is a consequence of a long-term greater productivity compared to decomposition rates. Higher atmospheric CO_2 concentrations could potentially change the balance between productivity and decomposition of peatlands, which could have major repercussions for large-scale carbon budgets. In this paper, we investigate how elevated CO_2 affects the vascular plant contributions to this balance (see Toet et al. In press, for moss contributions to this balance). So far responses in terms of productivity or plant growth have been found to be very limited in realistic *in situ* experiments with CO_2 enrichment in temperate northern ombrotrophic peatlands, both for the dominant peatland moss *Sphagnum* and vascular plants (Berendse et al. 2001; Hoosbeek et al. 2001; Heijmans et al. 2002). However, to our knowledge there is no information on productivity related responses of vascular plants in partly minerotrophic temperate lowland *Sphagnum–Phragmites* reedlands. In such peatlands, a relatively thin mostly rain-fed and nutrient-poor *Sphagnum* layer sits on top of a muddy, more nutrient-rich layer fed mostly by the groundwater, which is in contact with surrounding canals and ditches. While lowland peatlands in general are widespread throughout the temperate northern hemisphere, *Sphagnum–Phragmites* reedlands, which were once more widespread in the western Netherlands and other coastal parts of NW Europe, are now restricted to some nature reserves, mainly in The Netherlands. These peatlands have great conservation value as rare ecosystems with unique compositions of species, several of which are themselves rare or under threat internationally. In these ecosystems, the roots of the predominant vascular plants, including *Phragmites australis* reed, penetrate into this deeper, richer soil horizon. In ombrotrophic peatlands, the lack of CO_2 growth responses may be explained partly by the strong constraint imposed by low nutrient availability, as has been reported from various plants and ecosystems (cf. Curtis and Wang 1998; Stitt and Krapp 1999; Poorter and Pérez-Soba 2001; Hoosbeek et al. 2002). In contrast with these findings, Hoorens et al. (2003a) found significantly increased growth in response to CO_2 enrichment in two graminoids from mesotrophic peatland when grown at corresponding ('mesotrophic') nutrient availability. Therefore, our first hypothesis is that the predominant vascular species in lowland *Sphagnum–Phragmites* reedlands, which can penetrate into deeper, more nutrient-rich soil layers, will show a significant increase in productivity (as represented by abundance) in response to CO_2 enrichment. Increasing dominance of such plants could decrease the conservation value of these rare ecosystems, if they were to outcompete other vascular plants rooting in the *Sphagnum* peat layer (e.g. orchid spp., *Drosera rotundifolia*).

On the other side of the carbon balance, i.e. the organic matter breakdown side, CO_2-induced increases in productivity might result in greater litter amounts entering the soil surface, which may be an important contributor to changing soil carbon dynamics (Norby and Cotrufo 1998). CO_2 enrichment may have indirect effects on the abiotics of peatlands (and thereby on the soil decomposer communities), for instance CO_2 induced plant species replacements or increased water efficiency of extant species could change the hydrology of the peatland (e.g. Heijmans et al. 2001). In lowland *Sphagnum–Phragmites* reedlands, of which the hydrology is tightly controlled by human management, we expect that litter decomposition responses to CO_2 enrichment, if any, would be related mostly to the quantity and quality of the litter produced by plants growing at ambient vs. elevated CO_2 concentrations. Firstly, CO_2 enrichment may change litter quality via changes in species abundances, given the knowledge that vascular peatland species may vary greatly in litter quality and decomposability (Hoorens et al. 2003b; Quested et al. 2003). Second, leaf litter decomposability of a given species at elevated CO_2 may differ from that at ambient CO_2 mostly because of: (1) dilution of nutrient concentrations of green leaves due to increased storage of (mostly mobile, non-structural) organic carbon (Poorter et al. 1997; Curtis and Wang 1998; Saxe et al. 1998; Cornelissen et al. 1999); (2) a different pattern of resorption of mineral nutrients (N, P) or carbon compounds from senescing leaves. In a large meta-analysis of CO_2 responses in terms of nutrient resorption efficiency and litter quality and decomposability no consistent overall response patterns emerged, although a slight decline in litter N concentrations was seen in the less realistic experimental set-ups (Norby et al. 2001a, see also van Heerwaarden 2004). However, peatland species were hardly represented in these datasets. Hoorens et al. (2003a) did find significantly reduced litter N concentrations and litter respiration in the peatland sedge *Carex rostrata* grown at elevated CO_2 (but not in two other vascular peatland species), while Robinson et al. (1997) found either faster or slower decomposition of shoot litter from the subarctic peatland grass *Festuca vivipara* grown at elevated CO_2, depending on the incubation environment. Here, in addition to our first hypothesis outlined above, we predict that the predominant vascular plants in lowland *Sphagnum* reedland respond to CO_2 enrichment by (a) higher C/N and C/P ratios of green leaves; (b) similar N and P resorption efficiencies resulting in higher leaf litter C/N and C/P ratios and correspondingly lower leaf litter decomposability.

We tested our hypotheses in a Dutch lowland *Sphagnum–Phragmites* peatland using a relatively non-intrusive *in situ* MINIFACE (Free air CO_2 enrichment) system (Miglietta et al. 2001; Norby et al. 2001b). In terms of uspscaling of our study to CO_2 responses of peatlands, these two-teer ecosystems can provide insights into the general responses of both oligotrophic (upland) and minerotrophic (lowland) peatlands.

Methods

Study area

We conducted our experiment in a lowland *Sphagnum–Phragmites* reedland in the nature reserve Het Guisveld, Westzaan, The Netherlands (52°29′ N, 4°47′ E. at sea level). These ecosystems used to cover large areas in the northwestern Netherlands, but only small pockets have remained to date. The climate in this region is temperate-maritime, with annual precipitation at 780 mm distributed over all seasons. Mean temperature is 17 °C in the warmest and 3 °C in the coldest month (1971–2000, KNMI weather station at nearby Schiphol airport). The upper soil profile of our experimental site hosts an approx. 50 cm thick *Sphagnum* peat layer, of which the live part consists mostly of *Sphagnum palustre* L., *S. recurvum* var. *mucronatum* (Russ.) Warnst. (= *S. phallax* Klingr.) and *Polytrichum commune* Hedw. (the latter species expanding in recent decades and during the course of our experiment). This layer is probably largely ombrotrophic. The predominant vascular plant is reed *Phragmites australis*, which has its roots and rhizomes mostly in the deeper layer below the peat layer, which is muddy, more nutrient-rich and fed at least partly by groundwater that is in contact with canals and ditches draining the area. Below the 50–120 cm tall reed canopy other predominant vascular species include the woody species *Rubus* cf. *fruticosus* and *Lonicera periclymenum*; the grasses *Calamagrostis canescens* and *Anthoxanthum odoratum*; the forb *Hydrocotyle vulgaris*; and the ferns *Dryopteris carthusiana* and *D. cristata* (for nomenclature of vascular plants see van der Meijden 1996). Other common vascular species include *Cirsium palustre*, *Angelica sylvestris*, *Scirpus lacustris* ssp. *tabernaemontani*, *Dactylorrhiza praetermissa* and *Thelypteris palustris*, with occasional *Drosera rotundifolia*, *Platanthera macrantha* and *Osmunda regalis*. The vegetation is mown once a year in winter at 15 cm above the *Sphagnum* surface, both in common regional reedland management and in our experiment. This management also partly halts strong potential encroachment by shrubs and trees, although they do persist in the system (e.g. *Salix* spp., *Aronia* × *prunifolia*, *Betula pubescens*, *Sorbus aucuparia*). The water table is on average at 20 cm below the *Sphagnum* surface (23 cm during autumn–winter, 18 cm during spring–summer) but there is strong spatial variation within the site. There are visible gradients of productivity (judging from plant heights and densities) from the (more productive) northern edge of the site near the main drainage canal to the (lower productive) more central parts. Slightly elevated, drier parts dominated by *Empetrum nigrum* were excluded from the experiment.

The MINIFACE experiment

Details of the experimental technology, design, and management are in a companion paper focusing on moss responses (Toet et al. In press) and here we only give a summary description. Our CO_2 enrichment system (MINIFACE) was modified from Miglietta et al. (2001). Following a randomised spatial design, there were six control plots and six elevated CO_2 plots, at minimum distances of 7 m to avoid cross-contamination. They were accessed via a board-walk system. It was not possible to find 12 very similar plots initially, but both treatments appeared to have a similar range of variation in productivity among plots. In enrichment plots CO_2 was injected into ambient air that was vented out of two rows of 280 (1 mm diameter) holes each in a 1 m diameter ring and, through a feedback system, maintained at 560 ppmv during daytime. Concentrations stayed within 20% of this target for more than 95% of the operational time and concentrations at 25 cm from the inner edge of the rings deviated on average less than 10% from those in the centre of the plots. CO_2 concentrations in the ambient plots, which had the same rings, but venting ambient air only, ranged generally between 360 and 400 ppmv in daytime. Fumigation started on 26 April 2001 and continued throughout the study period until January 2004, except for a break between 18 December 2001 and 22 March 2002 (before the system was made frost-proof).

Plant abundance measurements

The abundance of each of the vascular plant species as well as litter in each plot was monitored using the point-intercept method (modified from Jonasson 1988), which tends to be an adequate correlate of biomass (Jonasson 1988; Hobbie 1999). We lowered a 5 mm diameter steel pin rod through each experimental plot at 121 points in a grid with 5 cm distances (Figure 1). Thus, we sampled within a 0.5 by 0.5 m square in the central part of each plot. For each species at each point we

Figure 1. Point-intercept abundance measurements of vascular plant species in the Guisveld lowland *Sphagnum–Phragmites* reedland.

recorded the total number of hits of living leaves or stems by the rod, and subsequently calculated the total number of hits per plot. For litter we only recorded the first hit, which we assumed to scale with total amount at the plot level. The initial recording was between 6 and 17 July 2000, i.e. in the summer preceding the start of the CO_2 fumigation treatment, and the second recording between 24 and 31 July 2003.

Leaf and litter chemistry and carbon isotope signatures

Green leaf samples of three abundant species (*Phragmites*, *Rubus*, *Dryopteris carthusiana*) were collected from the plots in mid August 2002 and litter samples of the same species between late October and mid-December 2002. For some ambient CO_2 plots where target species were absent, complementary leaf and litter samples were collected from plants within 1 m distance from the plots. Leaf and litter samples were finely ground and dried at 70 °C for 48 h prior to chemical analyses. For total C and N concentrations of the three most abundant species, these samples were subjected to dry combustion on a Perkin Elmer 2400 CHNS analyzer. Leaf P concentration was measured colorimetrically (Murphy and Riley 1962) after digesting ground material in a 1:4 mixture of 37% (v/v) HCl and 65% HNO_3 (as in Sneller et al. 1999). C isotope compositions of green leaf samples (of the same three species plus *Hydrocotyle* and *Lonicera*) were determined using an elemental analyser (Carlo Erba EA1110) coupled to an isotope ratio mass spectrometer (Thermo Finnigan Delta-Plus). Stable isotope compositions are reported in the δ notation: $\delta = (R_{sample}/R_{standard} - 1) \times 1000‰$ where R represents $^{13}C/^{12}C$. Isotopic results are reported relatively to VPDB. $\delta^{13}C$ of the enriched CO_2 at the source was determined on 27 May, 11 July, 28 November and 4 December 2003. We expected isotope compositions in enriched plants to change in the direction of the composition at the source.

Nutrient resorption efficiency

Nutrient resorption efficiency (RE) was defined as 100%* $([nutrient]_{green\ leaf} - [nutrient]_{litter})/[nutrient]_{green\ leaf}$ In this formula the nutrient pool ([nutrient]) is commonly expressed on a leaf mass basis, but this may produce significant deviations from real resorption efficiency due to simultaneous mass resorption during senescence (van Heerwaarden et al. 2003). We therefore also calculated nutrient resorption efficiency with the nutrient pool expressed on a leaf area basis (Delta-T area meter, Cambridge, UK), for *Phragmites* and *Rubus*, which retained relatively stable leaf area. For *Dryopteris carthusiana*, we expressed the nutrient resorption efficiency on a (presumably stable) lignin basis because senescing leaf fronds tend to shrivel up. See Rowland (1994) for lignin analysis.

Litter respiration assay

Four to six air-dried litter samples per species (*Phragmites*, *Rubus*, *Dryopteris carthusiana*) and treatment (each coming from a different MINI-FACE ring) were used to assess litter decomposability. We followed the procedure described by Aerts and De Caluwe (1997) and Hoorens et al. (2002), which estimates litter decomposability by measuring microbial respiration rates during initial decomposition under standardized, optimal laboratory conditions. The samples were remoistened for 24 h in a filtrate of a mixture of soil and litter from the study site to promote the local microbial community, and to fully hydrate the litter. Each remoistened sample was placed in a 100 ml glass jar. In order to keep jar air humidity as high as possible, 10 ml of a potassium sulphate buffer solution was added to each jar. Some glass marbles were subsequently placed in the buffer so that the top marbles emerged from the solution. A mesh (to host the litter samples) was placed on the top of the marbles, to avoid direct contact between the buffer solution and the samples. The top of the jars was left open to permit free air circulation between the jar and the incubation environment.

The jars were randomly arranged in laboratory trays, and placed in a climate room at 20 °C in the dark, and relative humidity at 95%. Five additional jars without litter were also included in the trays. When necessary (any signs of the samples drying out), we remoistened the samples adding distilled water with a syringe directly to the litter. The litter was incubated for 66 days. During this period, we measured net CO_2 production rate every 7–12 days as follows. The jars were sealed with a lid carrying a silicon septum, and one gas sample of 25 μl was taken from the jar atmosphere with a syringe penetrating the septum. In the gas sample, CO_2 concentration was measured with a Hewlett Packard 5890 gas chromatograph equipped with a thermal conductivity detector. After 4 h of CO_2 build-up in the air-tight jars, CO_2 concentration was measured again. The change in CO_2 concentration during that time period was assumed to be due to microbial respiration. CO_2 concentration was corrected for the CO_2 dissolved in the buffer solution (Stumm and Morgan 1981), for the air volume extracted with the syringe (50 μl), and for the residual CO_2 production measured in the five jars without litter. Litter respiration rates were expressed as mg CO_2 g l^{-1} h^{-1}. Total estimated CO_2 production per gram of litter in each jar throughout the 66 days period of the experiment (mg CO_2 g l^{-1}) was calculated by Newton integration, after the average CO_2 production respiration rate for each time interval between two measuring dates had been computed.

Statistical analyses

Point-intercept abundance data by species were $\log(x + 1)$ transformed before analyses in order to account for zero values and to improve homogeneity of variances. We subjected these to a three-way repeated measures analysis of variance (ANOVA) with species (the five with occurrence in a sufficient number of plots: *Phragmites*, *Calamagrostis*, *Anthoxanthum*, *Rubus*, *Hydrocotyle*) and CO_2 treatment as between-subject factors and year (2000 vs. 2003) as the within-subject factor. A combination of a CO_2 effect and a CO_2 *year interaction would be interpreted as a significant overall CO_2 response, while the combination of a CO_2 effect and a species *CO_2 *year interaction would be interpreted as a possible CO_2 response of one or more species. With a similar rationale, log-transformed data for the total number of live vascular plant hits or litter hits per plot were subjected to a two-way repeated measures ANOVA, with CO_2 as between-subjects and year as within-subjects factor.

To test for treatment effects on $\delta^{13}C$ signatures, on the chemistry of green leaves and litter, on nutrient resorption efficiency, and on litter respiration rates, two-way ANOVAs with treatment and species as fixed factors were performed for each variable. We explored the relationship between litter chemistry and litter decomposability by simple linear regressions between C:N or C:P ratios and total CO_2 production. Prior to the analyses, normality and homoscedasticity were checked. $\delta^{13}C$ signatures had to be $\log(-x)$ transformed and percentage data were arcsine [square-root($X/100$)] transformed where necessary to improve variance homogeneity. All statistical analyses were carried out using SPSS 11.0.

Results

Plant abundance varied significantly among species (Table 1, $F = 21.7$, $p < 0.001$). However, the three-way repeated measures ANOVA revealed no significant effect of CO_2 ($F = 0.036$, $p = 0.85$), CO_2 *year ($F = 0.562$, $p = 0.69$) or species *CO_2 *year interaction ($F = 0.405$, $p = 0.80$). The two-way repeated measures ANOVAs for total vascular plant abundance (CO_2: $F = 0.462$, $p = 0.51$, CO_2 *year: $F = 0.482$, $p = 0.48$) or for litter abundance (CO_2: $F = 0.027$, $p = 0.87$, CO_2 *year: $F = 1.55$, $p = 0.24$) did not reveal any significant CO_2 effects on abundance either. The apparently greater total vascular plant abundance after CO_2 treatment (Table 1) could partly be attributed to the expansion of patches of *Rubus*, *Lonicera* or *Dryopteris carthusiana* in some of the plots (authors' unpublished data) and was not necessarily related to CO_2 enrichment. Thus, no obvious CO_2 enrichment effects on vascular plant or litter abundance were detected at all. Correspondingly, there were no CO_2 enrichment effects on vascular plant species richness (initial in July 2000: ambient treatment 8.3 ± 0.3, CO_2 enrichment 9.0 ± 0.5 species per plot; July 2003: ambient 8.3 ± 0.6, CO_2 8.5 ± 0.5 species per plot).

Foliar $\delta^{13}C$ values were consistently lower in CO_2 enriched plots than in ambient plots (Figure 2). While all individual species showed this pattern, the significant Species *CO_2 interaction supports the observation that one species, i.e. *Phragmites australis*, had a smaller difference in $\delta^{13}C$ values between treatments than others. This may be attributed to the elevated position of *Phragmites* leaves, which probably experienced lower CO_2 concentrations than the 560 ppmv

Table 1. Point intercept abundance (number of hits) of the main vascular plant species and total vascular plant litter in Summer 2003 and the change in abundance (difference in number of hits) between both recordings. SE, standard error of the mean; *N*, number of plots. Plots in which the species was absent were included in Summer 2000 (data not shown) and Summer 2003 (zero values), but plots in which a species was absent at both recordings were not used to calculate change in abundance. In each plot a 0.25 m^2 square was sampled.

Summer 2003	Ambient			Elevated CO_2		
	Mean	SE	*N*	Mean	SE	*N*
Phragmites australis	123	38	6	103	28	6
Anthoxanthum odoratum	4	1	6	3	2	6
Calamagrostis canescens	3	2	6	1	1	6
Hydrocotyle vulgaris	40	27	6	18	15	6
Rubus cf. *fruticosus*	8	5	6	43	28	6
Total hits vascular plants	199	48	6	297	97	6
Litter from vascular plants	46	6	6	66	13	6
Summer 2003 – Summer 2000 Change						
Phragmites australis	49	36	6	49	28	6
Anthoxanthum odoratum	−8	4	6	−9	8	5
Calamagrostis canescens	−16	30	2	3	2	2
Hydrocotyle vulgaris	13	35	4	1	15	5
Rubus cf. *fruticosus*	4	5	4	37	31	4
Total hits vascular plants	64	61	6	139	79	6
Litter from vascular plants	7	8	6	31	15	6

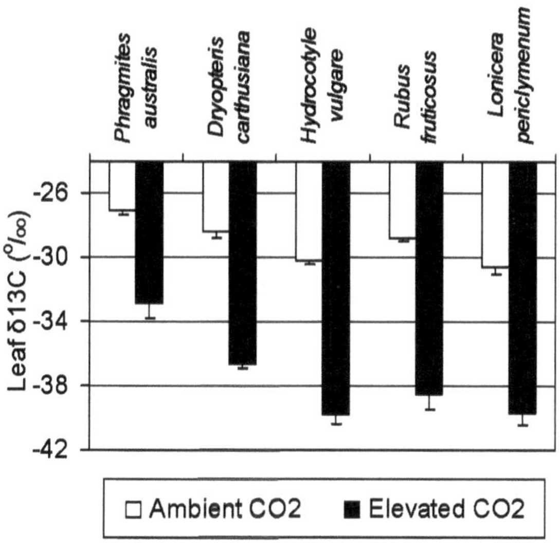

Figure 2. Response of $\delta^{13}C$ of green leaves of five vascular plant species to CO_2 enrichment. Standard errors of the means are shown one-sided only. Results of two-way ANOVA: Species: $F = 35.5$, $p < 0.001$; CO_2: $F = 539.7$, $p < 0.001$; Species * CO_2: $F = 3.28$, $p = 0.023$.

maintained lower down. For the other four species, the differences between treatments deviated only little from the calculated expected difference of 7.7‰ based on the contribution of enriched CO_2 to the total CO_2 supply in enriched plots ((560−380)/560), where ambient CO_2 had an approximate $\delta^{13}C$ value of −8‰ and enriched CO_2 of −31.9 ± 2.1‰ (see also Toet et al. In press).

In green leaves of the three focal species (*Phragmites*, *Rubus*, *Dryopteris carthusiana*) [N] and [P] were generally lower and C/N ratios and C/P ratios generally higher in CO_2 enrichment plots than in ambient plots (Table 2), although the significant CO_2 * Species interaction for C/P could be attributed to *Phragmites* not showing a CO_2 response (data not shown). Two further species with poorer replication (*Hydrocotyle*, *Lonicera*) showed similarly reduced [N] and [P] and higher C/N and C/P ratios in green leaves of CO_2 enriched plants (data not shown). Such a chemical CO_2 response was no longer detectable in leaf litter of the same species from the same plots, except for litter C/N ratio which was still somewhat higher in elevated CO_2 plots (Table 2), mainly owing to the contribution of *Dryopteris carthusiana*. The difference in response for green leaves and litter translated into lower mass-based resorption efficiency in response to CO_2 enrichment, both for N and P, but there was no significant CO_2 effect (only a trend) on area-based N or P resorption efficiency (Table 2, Figure 3). Area-based resorption efficiencies were on average 15% higher than mass-based ones in both treatments.

Table 2. Results of two-way ANOVAs on variables relation to leaf and litter chemistry, nutrient resorption efficiency and respiration, with fixed factors CO_2 treatment and species (*Phragmites australis, Rubus* cf. *fruticosus, Dryopteris carthusiana*). * $p<0.05$; ** $p<0.01$; *** $p<0.001$; ns, not significant. Codes are as follows:

	CO_2 Effect Sign	CO_2	Species	CO_2*Sp.	Error df
N_g	–	***	***	ns	17
P_g	–	**	**	*	16
N_l		ns	ns	ns	18
P_l		ns	*	ns	18
C/N_g	+	***	***	ns	17
C/P_g	+	*	**	ns	16
C/N_l	+	*	ns	ns	18
$C/P_l^{(a)}$		ns	**	ns	18
$N\ r_m$	–	*	**	ns	16
$P\ r_m$	–	**	***	ns	16
$N\ r_A^{(b)}$		ns	***	ns	15
$P\ r_A^{(a)\ (b)}$		ns	**	ns	15
$L_{respiration}$		ns	ns	ns	18

N_g, N% in green leaves; N_l, N% in litter; P_g, P% in green leaves; P_l, P% in litter; C/N_g, C/N ratio in green leaves; C/P_g, C/P ratio in green leaves; C/N_l, C/N ratio in litter; C/P_l, C/P ratio in litter; Nr_m, N resorption efficiency (mass basis); Pr_m, P resorption efficiency (mass basis); Nr_a, N resorption efficiency (area basis); Pr_a, P resorption efficiency (area basis); $L_{respiration}$, Cumulative CO_2 production over 66 days (mg CO_2/g litter).

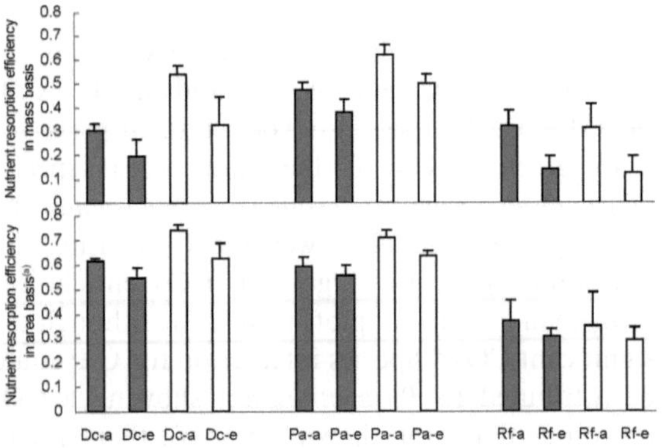

Figure 3. Mass and area based nutrient resorption efficiency. Dark bars: N resorption efficiency, white bars: P resorption efficiency. Horizontal axis: Dc, *Dryopteris carthusiana*; Pa, *Phragmites australis*; Rf, *Rubus* cf. *fruticosus*; a – ambient CO_2; e – elevated CO_2 concentration. For the bottom graph, resorption in Dc was calculated on a lignin basis, in Pa and RF on an area basis. Standard errors are shown one-sided. See Table 2 for statistical analyses.

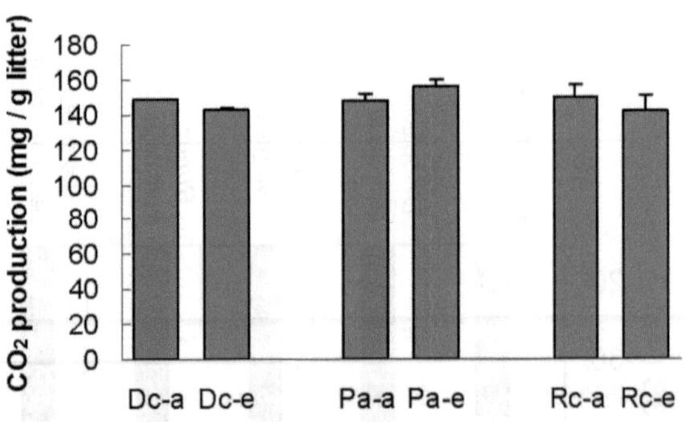

Figure 4. Cumulative CO_2 production due to initial respiration (66 days) of litter collected from ambient and elevated CO_2 plots during incubation under laboratory conditions. Horizontal axis: Dc, *Dryopteris carthusiana*; Pa, *Phragmites australis*; Rf, *Rubus* cf. *fruticosus*; a – ambient CO_2; e – elevated CO_2 concentration. Standard errors are shown one-sided. See Table 2 for statistical analyses.

There was no significant CO_2 effect on litter respiration for any of the three species investigated (Table 2), neither for patterns over time (data not shown) nor for cumulative CO_2 production (Figure 4). There was no relationship between initial litter C/N ratio and cumulative CO_2 production (negative slope, $R^2=0.11$, $p=0.11$) or between initial litter C/P ratio and cumulative CO_2 production ($R^2<0.01$, $p=0.88$), irrespective of the CO_2 treatment.

Discussion

The most striking finding from this 3-year experimental study in a Dutch lowland *Sphagnum–Phragmites* peatland was the lack of CO_2 response of the vascular vegetation component, both on the production (gains) and on the 'destruction' (losses) side of the organic carbon balance. This is particularly striking in view of the clear empirical evidence that the vegetation had taken up and processed the extra 180 ppmv of CO_2 supplied by our MINIFACE enrichment. This evidence was provided by the consistently lower $\delta^{13}C$ signatures and higher C/N and C/P ratios of green leaves of some of the predominant vascular plant species in CO_2 enriched plots. Below we shall discuss some of the factors that may together explain the lack of vegetation response to CO_2 in this study.

Contrary to our first hypothesis, we did not detect plant abundance responses and presumably, therefore, biomass responses (see Methods). Hoosbeek et al. (2002) found the same pattern analyzing vascular plants from several ombrotrophic bogs in NW Europe, and argued that low availability of K or P could limit the potentially fertilizing effect of elevated CO_2. However, in our minerotrophic system, nutrient availability should not have posed such a limiting effect, and the lack of response may be attributed to several other factors acting together. Firstly, the higher C/N and C/P ratios of green leaves in CO_2 enriched plots, combined with the apparently mobile nature of the additional carbon as deduced from increased mass resorption (*sensu* van Heerwaarden et al. 2003) during senescence at high CO_2 (affirmative data not shown), together point towards predominant storage of the surplus carbon as non-structural carbohydrates (starch). This has also been found in many previous studies (for reviews see Poorter et al. 1997; Curtis and Wang 1998, Körner 2000). Carbon that is translocated from the leaves to other plant parts may be processed for growth in subsequent years, for instance in periods of assimilate limitation due to shading, but it will contribute little or not at all to further plant growth in the shorter term. Second, annual winter mowing at 15 cm above the *Sphagnum* carpet removed not only litter, but also some living perennial shoots of woody and semi-woody plants such as *Rubus* and *Lonicera*. This could have reduced the overall biomass accumulation of such species. At the same time, this mowing regime does represent the typical management for this ecosystem, also in other areas. Third, we can not rule out experimental artifacts that could have obscured CO_2 effects on vegetation productivity. Our site was spatially most heterogeneous, with clear internal gradients in water tables (Toet et al. In press), productivity (visual observations) and species composition. Combined with logistic constraints such as (a) minimum distances between plots to prevent CO_2 contamination of control plots and (b) small plot size relative to the large size, clonal habit and patchiness of the predominant plants, this resulted inevitably in substantial initial variability among plots. We can not exclude the possibility that this heterogeneity has obscured an underlying trend of greater productivity under CO_2 enrichment. At the same time, the spatial heterogeneity is a real characteristic of this ecosystem type. Besides employing more and larger plots, longer time periods for CO_2 response of the vegetation would be recommendable too, in view of the apparent substantial belowground resource storage of the clonal plants (pre-treatment 'memory') and possible feedbacks provided by long-term CO_2-induced soil responses.

On the 'destruction' side of the ecosystem carbon balance, viz. the decomposition pathway, the lack of clear CO_2 responses was similarly obvious. Firstly, litter quantity (i.e. litter abundance as point-intercept hits), which largely determines how much organic matter could potentially be decomposed, did not differ significantly between ambient and elevated CO_2 treatments. Second, litter quality, which determines the rate at which this organic matter can be decomposed, showed little significant CO_2 response other than a somewhat higher C/N ratio in *Dryopteris carthusiana* (and *Hydrocotyle vulgare*; data not shown) in enriched plots, similar to the findings of Hoosbeek et al. (2002) in ombrotrophic bogs. Although the N and P concentrations of green leaves were lower and C/N and C/P ratios higher in CO_2 plots, in support of our second hypothesis, this chemical response was partly cancelled out by resorption of presumably mobile carbohydrates during leaf senescence. As a consequence, leaf litter quality was somewhat convergent between both treatments, which is at odds with our third hypothesis. Our findings seem to correspond with the current body of literature, where green leaf N concentrations are generally lower and C/N ratios higher in response to CO_2 enrichment (Poorter et al. 1997; Cotrufo et al. 1998; Curtis and Wang 1998; Cornelissen et al. 1999; Hoorens et al. 2003a), while the same responses (lower [N], higher C/N) are much less consistent and less strong in leaf litter (Norby et al. 2001a). It seems therefore that differences in resorption pattern are an important moderating factor in the translation from green plant responses to CO_2 enrichment to responses of leaf litter and their decomposition. If 'surplus' mobile carbon is diverted towards perennial structures (e.g rhizhomes, wood) of long-lived plants, adding to their year-to-year biomass, their changing potential for C sequestration would be partly a function of carbon and nutrient resorption from senescing leaves.

Indeed, a large meta-analysis of CO_2 enrichment studies covering various biomes (Norby et al. 2001a) revealed that there was an overall average reduction of litter N concentration of 7.1%, but this was not seen in the subset of experiments where litter was collected from *in situ* CO_2 exposure such as ours. Hoorens et al. (2003a) found significantly lower litter N concentrations and higher C/N ratios in CO_2-enriched plants in three out of five vascular plant species (and one species with a trend in the same direction of response), but the plants from which their leaf litter was derived had been grown and enriched in greenhouses.

The total lack of CO_2 response of litter respiration was at odds with our third hypothesis (see Introduction). However, this can not automatically be interpreted as the consequence of the small and inconsistent response of litter quality, because litter C/N or C/P ratios were no correlates of litter respiration across all three species and treatments. This is surprising, provided that, in our type of litter respiration set-up, no other factors than litter chemistry are supposed to influence decomposition. It is likely that the allocation of C to secondary compounds vs. carbohydrates, and consequently parameters such as Lignin content or Lignin/N, are the litter chemistry factors that are really controlling litter decomposition rates (Aerts 1997). Similarly, Norby et al. (2001a) in their meta-analysis found no CO_2 effect on litter decomposability whether measured as mass loss or as respiration, in spite of the on average lower N concentrations in CO_2-enriched litters. In contrast, Hoorens et al. (2003a) in a smaller meta-analysis did find that a CO_2 induced reduction of litter N correlated with a reduction in litter decomposition (measured as mass loss or respiration). However, in the latter, a substantial proportion of the studies involved had derived litter from plants grown and enriched in less natural settings.

Conclusions

When combining our findings with those from other *in situ* MINIFACE experiments with CO_2 enrichment in temperate or boreal *Sphagnum*-dominated peatlands (Berendse et al. 2001; Hoosbeek et al. 2001; Heijmans et al. 2002), our preliminary conclusion is that vascular vegetation in these ecosystems is not very responsive to CO_2. Indirect longer term responses via changes in the *Sphagnum* turf, e.g. *Polytrichum* moss outcompeting *Sphagnum* (Toet et al. In press) should however not be ruled out as yet. At least in the shorter to medium term, any possible CO_2 effects will probably be very small compared to other anthropogenic environmental changes. The composition and functioning of *Sphagnum* peatlands are very strongly dependent on local and regional hydrology and water quality (e.g. Glaser et al. 1990; Moore et al. 2002), both of which are under strong control of human management, at least in many temperate regions and especially in The Netherlands. Also, high N deposition is known to drastically alter such peatlands, as evidenced for instance by vascular plant increases in a fertilized Dutch peat bog (Heijmans et al. 2002). Moreover, in the real (future) world the 'greenhouse effect' comprises both higher atmospheric CO_2 concentrations and higher temperatures simultaneously and the interaction of the two factors (in combination with variation in hydrology and N deposition) may perhaps produce different vegetation responses if studied experimentally *in situ* (see Norby and Luo 2004).

The composition, productivity and functioning of most of the lowland *Sphagnum–Phragmites* reedlands, like the one under study here, also depend strongly on the timing, frequency and intensity of reed mowing, on whether this is done manually or by heavy vehicles compacting the *Sphagnum* foundation, and on whether and how frequently trees and shrubs are pulled out of the soil. All these measures are aimed at arresting the relatively early successional phase of these peatlands in order to preserve biodiversity. It is likely that a cessation of such management, resulting in a development towards woodland within several decades, would alter carbon sequestration and processing more than the extra atmospheric carbon availability on a similar time scale.

Acknowledgements

We thank the Dutch Forestry Authority (Staatsbosbeheer, Alkmaar office) for allowing us to use their land and facilities, and particularly the Guisveld wardens, Erik Gerrevink and Wouter Maatje, for all their help, hospitality and anecdotes. Rob Stoevelaar, Cor Stoof, Martin van-Vilsteren and co-workers helped with the development and maintenance of the FACE-related equipment. Ellen Dorrepaal and Bart Hoorens helped with the litter respiration work and Miranda de Beus, Adrie van Beem, Martin Stroetenga and Nancy de Bakker with some of the fieldwork. Thanks to all.

References

Aerts R. 1997. Climate, leaf litter chemistry and leaf litter decomposition in terrestrial ecosystems: a triangular relationship. Oikos 79: 439–449.

Aerts R. and de Caluwe H. 1997. Initial litter respiration as indicator for long-term leaf litter decomposition of Carex species. Oikos 80: 353–361.

Berendse F., van Breemen N., Rydin H., Buttler A., Heijmans M., Hoosbeek M.R., Lee J.A., Mitchell E., Saarinen T., Vasander H. and Wallén B. 2001. Raised atmospheric CO_2 levels and increased N deposition cause shifts in plant species composition and production in *Sphagnum* bogs. New Phytol. 7: 591–598.

Cornelissen J.H.C., Carnelli A.L and Callaghan T.V. 1999. Generalities in the growth, allocation and leaf quality responses to elevated CO_2 in eight woody species. New Phytol. 141: 401–409.

Cotrufo M.F., Ineson P. and Scott A. 1998. Elevated CO_2 reduces the nitrogen concentration of plant tissues. Glob. Change Biol. 4: 43–54.

Curtis P.S. and Wang X. 1998. A meta-analysis of elevated CO2 effects on woody plant mass, form, and physiology. Oecologia 113: 299–313.

Glaser P.H., Janssens J.A. and Siegel D.I. 1990. The response of vegetation to chemical and hydrological gradients in the Lost River Peatland, northern Minnesota. J. Ecol. 78: 1021–1048.

Gorham E. 1991. Northern peatlands: role in the carbon cycle and probable responses to climatic warming. Ecol. Appl. 1: 182–195.

Heijmans M.M.P.D., Arp W.J. and Berendse F. 2001. Effects of elevated CO_2 and vascular plants on evapotranspiration in bog vegetation. Glob. Change Biol. 7: 817–827.

Heijmans M.M.P.D., Klees H., De Visser W. and Berendse F. 2002. Response of a Sphagnum bog community to elevated CO2 and N supply. Plant Ecol. 162: 123–134.

Hobbie S.E., Shevtsova A. and Chapin F.S. III 1999. Plant responses to species removal and experimental warming in Alaskan tussock tundra. Oikos 84: 417–434.

Hoorens B., Aerts R. and Stroetenga M. 2002. Litter quality and interactive effects in litter mixtures: more negative interactions under elevated CO_2?. J. Ecol. 90: 1009–1016.

Hoorens B., Aerts R. and Stroetenga M. 2003a. Is there a trade-off between the plant's growth response to elevated CO_2 and subsequent litter decomposability?. Oikos 103: 17–30.

Hoorens B., Aerts R. and Stroetenga M. 2003b. Does initial litter chemistry explain litter mixture effects on decomposition?. Oecologia 137: 578–586.

Hoosbeek M.R., van Breemen N., Berendse F., Grosvenier P., Vasander H. and Wallén B. 2001. Limited effect of increased atmospheric CO_2 concentration on ombrotrophic bog vegetation. New Phytol. 150: 459–463.

Hoosbeek M.R., van Breemen N., Vasander H., Buttler A. and Berendse F. 2002. Potassium limits potential growth of bog vegetation under elevated atmospheric CO_2 and N deposition. Glob. Change Biol. 8: 1130–1138.

IPCC 2001. Climate Change 2001 – Synthesis Report Third Assessment Report of the Intergovernmental Panel on Climate Change (IPCC). Cambridge University Press, Cambridge.

Jonasson S. 1988. Evaluation of the point intercept method for the evaluation of plant biomass. Oikos 52: 101–106.

Körner C. 2000. Biosphere responses to CO_2 enrichment. Ecol. Appl. 10: 1590–1619.

Miglietta F., Hoosbeek M.R., Foot J., Gigon F., Hassinen A., Heijmans M., Peresotti A., Saarinen T., van Breemen N. and Wallén B. 2001. Spatial and temporal performance of the Miniface (Free air CO_2 enrichment) system on bog ecosystems in northern and central Europe. Environ. Monit. Assess. 66: 107–127.

Moore T.R., Bubier J.L., Frolking S.E., Lafleur P.M. and Roulet N.T. 2002. Plant biomass and production and CO_2 exchange in an ombrotrophic bog. J. Ecol. 90: 25–36.

Murphy J. and Riley J.P. 1962. A modified single solution method for the determination of phosphate in natural waters. Analytica Chimica Acta 27: 31–36.

Norby R.J. and Cotrufo M.F. 1998. A question of litter quality. Nature 396: 17–18.

Norby R.J., Cotrufo M.F, Ineson P., O'Neill E.G. and Canadell J.G. 2001a. Elevated CO_2, litter chemistry, and decomposition: a synthesis. Oecologia 127: 153–165.

Norby R.J., Kobayashi K. and Kimball B.A. 2001b. Rising CO_2 – future ecosystems. New Phytol. 150: 215–221.

Norby R.J. and Luo Y. 2004. Evaluating ecosystem responses to rising atmospheric CO_2 and global warming in a multi-factor world. New Phytol. 162: 281–293.

Poorter H., van Berkel Y., Baxter R., Den Hertog J., Dijkstra P., Gifford R.M., Griffin K.L., Roumet C., Roy J. and Wong S.C. 1997. The effect of elevated CO_2 on the chemical composition and construction cost of leaves of 27 C_3 species. Plant Cell Environ. 20: 472–482.

Poorter H. and Pérez-Soba M. 2001. The growth response of plants to elevated CO_2 under non-optimal environmental conditions. Oecologia 129: 1–20.

Quested H.M., Cornelissen J.H.C., Press M.C., Callaghan T.V., Aerts R., Trosien F., Riemann P., Gwynn-Jones D., Kondratchuk A. and Jonasson S. 2003. Litter decomposition of sub-arctic plants with different nitrogen economies: a functional role for hemiparasites?. Ecology 84: 3209–3221.

Robinson C.C., Michelsen A., Lee J.A., Whitehead S.J., Callaghan T.V., Press M.C. and Jonasson S. 1997. Elevated atmospheric CO_2 affects decomposition of *Festuca vivipara* (L) Sm Litter and roots in experiments simulating environmental change in two contrasting ecosystems. Glob. Change Biol. 3: 37–49.

Rowland A.P. 1994. Lignin and cellulose fractionation in decomposition studies using acid-detergent fibre methods. Comm. Soil Sci. Plant Anal. 25: 269–278.

Saxe H., Ellsworth D.S. and Heath J. 1998. Tree and forest functioning in an enriched CO_2 atmosphere. New Phytol. 139: 395–436 Tansley Review No. 98.

Sneller F.E.C., van Heerwaarden L.M., Kraaijeveld-Smit F.J.L., Ten Bookum W.M., Koevoets P.L.M., Schat H and Verkleij J.A.C. 1999. Toxicity of arsenate in *Silene vulgaris*, accumulation and degradation of arsenate-induced phytochelatins. New Phytol. 144: 223–232.

Stitt M. and Krapp A. 1999. The interaction between elevated carbon dioxide and nitrogen nutrition: the physiological and molecular background. Plant Cell Environ. 22: 583–621.

Stumm W. and Morgan J.J. 1981. Aquatic Chemistry. An Introduction Emphasizing Chemical Equilibria in Natural Waters. Wiley, New York.

Toet S., Cornelissen J.H.C., Aerts R., van Logtestijn R.S.P., de Beus M.A.H. and Stoevelaar R. 2005. Moss responses to elevated CO_2 and hydrology in a temperate peatland. Plant Ecology, DOI: 10.1007/s11258-005-9029-8.

van der Meijden R. 1996. Heukels' Flora van Nederland. Wolters-Noordhoff, Groningen 22nd Ed.

van Heerwaarden L.M. 2004. From Leaf to Litter: Nutrient Resorption in a Changing Environment. Vrije Universiteit, Amsterdam The Netherlands PhD Thesis. ISBN 90-901-8060-5.

van Heerwaarden L.M., Toet S. and Aerts R. 2003. Current measures of nutrient resorption efficiency lead to a substantial underestimation of real resorption efficiency: facts and solutions. Oikos 101: 664–669.

Photo. The *Sphagnum* lowland peatland Het Guisveld, The Netherlands, with MiniFACE rings to enhance atmospheric CO_2 levels.

Plant Ecology (2006) 182:27–40
DOI 10.1007/s11258-005-9029-8

Moss responses to elevated CO_2 and variation in hydrology in a temperate lowland peatland

Sylvia Toet[1,2,*], Johannes H.C. Cornelissen[1], Rien Aerts[1], Richard S.P. van Logtestijn[1], Miranda de Beus[1] and Rob Stoevelaar[1]

[1]*Department of Systems Ecology, Institute of Ecological Sciences, Faculty of Earth and Life Sciences, Vrije Universiteit Amsterdam, De Boelelaan 1085, 1081, HV, Amsterdam, The Netherlands;* [2]*Environment Department, University of York, Heslington, York, YO10 5DD, United Kingdom;* **Author for correspondence (e-mail: st501@york.ac.uk; fax: +44-0-1904-432998)*

Received 1 September 2004; accepted in revised form 15 December 2004

Key words: Abundance, C/N ratio, *Polytrichum commune*, *Sphagnum palustre*, *Sphagnum recurvum*, Stable C isotope composition

Abstract

We studied the effects of elevated CO_2 (180–200 ppmv above ambient) on growth and chemistry of three moss species (*Sphagnum palustre*, *S. recurvum* and *Polytrichum commune*) in a lowland peatland in the Netherlands. Thereto, we conducted both a greenhouse experiment with both *Sphagnum* species and a field experiment with all three species using MiniFACE (Free Air CO_2 Enrichment) technology during 3 years. The greenhouse experiment showed that *Sphagnum* growth was stimulated by elevated CO_2 in the short term, but that in the longer term (≥1 year) growth was probably inhibited by low water tables and/or down-regulation of photosynthesis. In the field experiment, we did not find significant changes in moss abundance in response to elevated CO_2, although CO_2 enrichment appeared to reduce *S. recurvum* abundance. Both *Sphagnum* species showed stronger responses to spatial variation in hydrology than to increased atmospheric CO_2 concentrations. *Polytrichum* was insensitive to changes in hydrology. Apart from the confounding effects of hydrology, the relative lack of growth response of the moss species may also have been due to the relatively small increase in assimilated CO_2 as achieved by the experimentally added CO_2. We calculated that the added CO_2 contributed at most 32% to the carbon assimilation of the mosses, while our estimates based on stable C isotope data even suggest lower contributions for *Sphagnum* (24–27%). Chemical analyses of the mosses showed only small elevated CO_2 effects on living tissue N concentration and C/N ratio of the mosses, but the C/N ratio of *Polytrichum* was substantially lower than those of the *Sphagnum* species. Continuing expansion of *Polytrichum* at the expense of *Sphagnum* could reduce the C sink function of this lowland *Sphagnum* peatland, and similar ones elsewhere, as litter decomposition rates would probably be enhanced. Such a reduction in sink function would be driven mostly by increased atmospheric N deposition, water table regulation for agricultural purposes and land management to preserve the early successional stage (mowing, tree and shrub removal), since these anthropogenic factors will probably exert a greater control on competition between *Polytrichum* and *Sphagnum* than increased atmospheric CO_2 concentrations.

Introduction

Peatlands are usually net carbon sinks because of very low decomposition rates that are surpassed by rates of productivity (Coulson and Butterfield 1978; Clymo 1984; Brock and Bregman 1989). *Sphagnum* mosses are often dominant in ombrotrophic and oligotrophic peatlands and play a major role in carbon sequestration as they create conditions for low decomposition rates. Litter of *Sphagnum* is very resistant to decomposition, while *Sphagnum* also generates acidic, anoxic, heat-insulating and nutrient-poor conditions, and exudes secondary metabolites reducing microbial decay of plant litter (Coulson and Butterfield 1978; Clymo and Hayward 1982; Van Breemen 1995; Verhoeven and Toth 1995; Verhoeven and Liefveld 1997). The long-term C accumulation in peat soils has resulted in a large C pool and, though considerable uncertainty of its size still prevails (Maltby and Immirzi 1993), the organic C pool in peatlands has been estimated to amount to 455 Pg, which is 20–30% of the global soil pool (Bolin 1986; Gorham 1991). Changes in the C storage capacity of these peatlands due to disturbances such as drainage and increases in atmospheric CO_2 concentration and nitrogen deposition may therefore have a substantial impact on the global C cycle.

Atmospheric CO_2 concentration has increased since pre-industrial time and is expected to have at least doubled by the end of the century. As a consequence, global surface temperature is predicted to rise by 1.4–5.8 °C, while annual precipitation patterns will presumably change (Houghton et al. 2001). Elevated CO_2 may influence the functioning of plants through these climatic changes but also directly. Many studies have shown that increased CO_2 concentrations can affect photosynthesis, transpiration, growth, morphology, allocation and chemical composition of vascular plants (see Makino and Mae 1999; Mooney et al. 1999; Stitt and Krapp 1999; Körner 2000; Poorter and Navas 2003). Much less knowledge exists on the responses of mosses to CO_2 enrichment. The available studies, mostly on mosses from peatlands, show various reactions to elevated CO_2. In all moss studies, a short-term positive effect of raised CO_2 concentrations on photosynthetic rates was observed. Length growth and biomass production of peatland mosses were generally either enhanced or not influenced by elevated CO_2. In addition, soluble sugar concentration of *Sphagnum* mosses increased in response to CO_2 enrichment, whereas N concentration was reduced or did not respond in peatland mosses (Silvola 1985; Jauhiainen et al. 1994, 1997, 1998; Jauhiainen and Silvola 1999; Tuba et al. 1999; Van der Heijden et al. 2000a, b; Berendse et al. 2001; Heijmans et al. 2002).

The inconsistent moss responses to raised CO_2 levels are possibly related to differences in experimental design including monocultures vs. mixed stands (with vascular plants), species specific properties and/or different environmental conditions such as water availability, nutrient status and temperature. Water table is often a strong regulator of biomass production and abundance of mosses, particularly for mosses without specialised internal water-conducting tissues such as *Sphagnum* (Clymo and Hayward 1982; Rydin 1993; Jauhiainen et al. 1997; Weltzin et al. 2000, 2001). Hydrology may therefore confound elevated CO_2 responses of moss growth and abundance, although hydrology may also be influenced by elevated CO_2 effects itself through enhanced water use efficiency or cover of vascular plants (Field et al. 1995; Heijmans et al. 2001). Besides, moss growth in wet conditions may not only depend on atmospheric CO_2 but also on CO_2 originating from the substrate, since dissolved CO_2 concentrations in the upper peat layers are usually higher than atmospheric concentrations, reducing potential elevated atmospheric CO_2 responses of mosses (Lansdown et al. 1992; Lamers et al. 1999; Hornibrook et al. 2000; Smolders et al. 2001).

Species-specific responses of moss species to elevated CO_2 might eventually lead to changes in species abundance and composition in peatlands, possibly altering overall biomass and litter production and litter quality, and accordingly changing the C sink capacity of these systems. However, the influence of elevated CO_2 on mixtures of moss species in peatlands has only been studied in a few cases (Van der Heijden et al. 2000b; Mitchell et al. 2002; Saarnio et al. 2003).

In this study, we investigated the response of peatland mosses to elevated CO_2 and to variation in hydrology both in the field and in the greenhouse. In the field experiment, we studied the influence of elevated CO_2 (180 ppmv above ambient) on the abundance and chemical composition of the three dominant moss species, i.e. *Sphagnum palustre*, *S. recurvum* and *Polytrichum commune*, during 3 years of fumigation. We hypothesised (1) that changes in moss species abundance in response to elevated CO_2 at field conditions were minor during this period, although *Polytrichum* was expected to be more responsive and competitive than *Sphagnum* under prevailing conditions of increased N availability, (2) that chemical responses to CO_2 enrichment were stronger resulting in higher tissue C/N ratios, and (3) that responses of moss abundance were more determined by variation in hydrology than by elevated CO_2.

In order to provide a base-line to interpret realised response in the field, a parallel greenhouse experiment was conducted to study growth and chemical responses of the two *Sphagnum* species to an almost similar increase in atmospheric CO_2 concentration (200 ppmv above ambient) but at probably more favourable conditions than in the field including the lack of competition of other mosses and vascular plants, a higher water table and higher winter air temperature.

Methods

Study site

The lowland *Sphagnum* peatland Het Guisveld in polder Westzaan is situated in the north-western part of the Netherlands (52° 29′ N, 4° 47′ E, at sea level). This 2.06 km^2 peatland nature reserve forms part of a larger area of remaining lowland peatland, with extensive management including reed (*Phragmites australis*) cutting. This ecosystem used to cover large parts of the province of Noord-Holland until much of it was drained and turned into more intensively managed agricultural or urban land. The Guisveld peatland has a surface *Sphagnum* peat layer of approximately 50 cm with a muddy layer underneath. The peat layer probably has only minor contact with the surrounding surface water (canals and ditches) and is therefore largely ombrotrophic (pH of the top 10 cm layer was around 4.5). The water table was on average 23 and 18 cm below the *Sphagnum* surface during the growing season and autumn–winter period, respectively. The water table has probably been lowered during the last decade (by decreasing water levels in surrounding ditches for agricultural purposes). Mean annual precipitation is 780 mm and mean temperature of the warmest and coldest month 17 and 3 °C, respectively (1971–2000, KNMI weather station Schiphol airport, situated 22 km from the study site).

The vegetation consists of an almost closed moss cover dominated by *Sphagnum palustre* L., *S. recurvum* var. *mucronatum* (Russ.) Warnst. (or *S. fallax* (Klingr.) Klingr.) and *Polytrichum commune* Hedw., with some *Aulacomnium palustre* (Hedw.) Schwaegr.. The dominant vascular plant in the upper canopy (50–120 cm) is *Phragmites australis*. Predominant species of the lower canopy are *Rubus fruticosus s.l.*, *Dryopteris carthusiana*, *D. cristata*, *Hydrocotyle vulgaris* and *Lonicera periclymenum*, with occasional *Calamagrostis canescens*, *Anthoxanthum odoratum*, *Scirpus tabernaemontani*, *Drosera rotundifolia*, *Dactylorrhiza majalis* ssp. *praetermissa* and saplings of woody invaders (*e.g. Salix* spp. and *Aronia x prunifolia*; for nomenclature of vascular plants see Van der Meijden 1996). Drier parts dominated by *Empetrum nigrum* were not included in the plots. The vegetation is mown 15 cm above the *Sphagnum* surface in winter each year and the mown biomass is removed, which corresponds both to the traditional reed culture and current nature management.

Field experiment

Experimental design

Two atmospheric CO_2 levels were established in twelve plots using MiniFACE rings with a diameter of 1 m, shown in Figure 1 (modified from Miglietta et al. 2001). These plots were positioned within parts of the peatland where *Polytrichum* was not (yet) dominant. There was substantial initial heterogeneity in vegetation composition among the plots, since no 12 very similar plots were available within the site. The spatial variability is, however, a feature of this type of ecosystem and could only be overcome by more replicate plots or larger plots but this was not financially feasible. The plots were accessed by wooden boardwalks so as not to tred and compact

Figure 1. The *Sphagnum* lowland peatland Het Guisveld, The Netherlands, with MiniFACE rings to enhance atmospheric CO_2 levels.

the direct surroundings. In six randomly selected plots, the atmospheric CO_2 concentration was kept at ambient levels (on average 360–400 ppmv in the daytime), while in the other plots a mean elevated CO_2 concentration of 560 ppmv was maintained in the daytime, i.e. from sunrise to sunset with day length programmed to follow the Julian calender. The plots were at least 7 m apart to avoid CO_2 enrichment in the ambient plots. No data were collected in the outer 10 cm of the plots to minimise edge effects.

The hollow, corrugated polyethylene MiniFACE rings with an internal diameter of 50 mm were positioned 2 cm above the moss surface. Ambient air was supplied to the rings by blowers (EBM G2E 140 A1-40-01) with a 385 m^3 h^{-1} load, which were connected to the rings by a 10 cm diameter pipe. The blowers were located at about 1.5 m distance from the rings. Air was vented out of each ring through two rows of 280 holes (diameter of 1 mm), one row facing upwards, and one diagonally outwards. Pure CO_2 gas was injected into the connecting pipe 20 cm ahead of the rings to create elevated CO_2 conditions.

CO_2 concentration was measured continuously in the centre of the elevated CO_2 rings at 3 and 10 cm above the moss surface in winter and summer, respectively (the difference being related to the vascular plant canopy in summer), with two infrared gas analysers (IRGA, PP systems SBA-1) coupled to a control unit. A proportional differential integral control algorithm (PID) in the control unit calculated the CO_2 supply needed to reach the target CO_2 concentration of 560 ppmv. The PID also made use of wind speed data recorded with an anemometer (Campbell Scientific A100R) at 2.5 m height located in the vicinity of the rings. CO_2 injection into the connecting pipe was controlled by converting the output voltage (0–24 V) of the control unit to a microprocessor connected to 24 V solenoid on/off valves (Fluid Automation Systems, 2/2/NC-NF-NC), resulting in intervals of 6 s that the valves were open (at 24 V) or closed (at 0 V). Each ring was monitored for 4 min out of 12. CO_2 concentration, output voltage and wind speed were stored as 1 min averages.

CO_2 concentrations in these plots remained within 20% of the pre-set target concentration for more than 95% of operational time. CO_2 measurements in the centre and at the upwind and downwind side of one elevated CO_2 plot during a 2-week period, and at two individual days in all elevated CO_2 plots showed that CO_2 concentrations at 25 cm from the edge remained within 10% of those in the centre of the plots. CO_2 concentration was always higher at the upwind side and lower at the downwind side of the plots (details can be requested from the authors).

The fumigation started on 26 April 2001. *Polytrichum* expanded fast during the first fumigation year, covering large parts of the plots in early spring 2002. It became apparent that it was impossible to study the effect of elevated CO_2 on the competition between *Polytrichum* and *Sphagnum* during the following few years. To set this interspecific moss competition back in time, all visible *Polytrichum* shoots were manually pulled out of the soil in March 2002 both in the plots and in a strip of 0.3 m around the plots. Disturbance to the *Sphagnum* carpet was kept to a minimum by pulling *Polytrichum* shoots out individually. The removed shoots usually had a length of 20–30 cm and often included rhizoids, but residual *Polytrichum* biomass was still present below the green *Sphagnum* cover, presumably providing some storage and buds for regrowth.

Fumigation was interrupted between 18 December 2001 and 22 March 2002 to prevent frost damage to the MiniFACE equipment. Next, the equipment was made frost proof, so that fumigation was continued throughout successive winters as moss growth was also expected to occur during milder winter periods.

Moss species abundance

Before the *Polytrichum* removal, the abundance of the three dominant moss species *Polytrichum commune*, *Sphagnum palustre* and *S. recurvum* was determined in August 2000, i.e. 8 months before fumigation was started, and in March 2002 after 7.5 months of fumigation. After allowing the vegetation to recover from possible disturbance due to *Polytrichum* removal, abundance measurements of the mosses were also performed one (March 2003) and almost 2 years (January 2004) after *Polytrichum* removal. A 60× 60 cm permanent quadrat was selected in each plot. A point-intercept frame was positioned over this quadrat for each plant abundance recording (modified from Jonasson 1998). The frame allowed a 5 mm diameter pin to pass vertically through two lined-up holes. There were 11× 11 points in the frame at 5 cm distances between neighbour points. At each point, the pin was lowered through the vegetation and the first contact with the green parts of individual moss species (and in August 2000 also vascular plants, Milla et al. 2005) was noted. The numbers of *Polytrichum* shoots hit at each point were also registered in 2003 and 2004 to obtain a more accurate measure of abundance.

Dry weight of the harvested *Polytrichum* shoots removed from the permanent quadrats in March 2002 was measured after drying at 70 °C for 2 days. As a result, the correspondence of *Polytrichum* point-intercept abundances and biomass data among plots could be tested (Jonasson 1998).

Moss C and N analyses

The upper 2 cm of 25 *Polytrichum* shoots, and the upper 1 cm of 10 *S. palustre* and 20 *S. recurvum* shoots (i.e. mainly the capitulum) were randomly collected from each plot in March 2002. These plant parts represented the plant tissues largely grown under elevated CO_2 conditions between April 2001 and March 2002 (personal communications of J. Rozema and R. Aerts). Samples were finely ground and dried for 2 days at 70 °C prior to successive analysis for C and N concentration, and stable C isotope composition on an elemental analyser (Carlo Erba EA1110) coupled to an isotope ratio mass spectrometer (Thermo Finnigan Delta-Plus). Separate runs were carried out for the C and N analyses with sample weights of 1 and 3 mg, respectively. Stable isotope compositions are reported in the δ notation: $\delta = (R_{sample}/R_{standard} - 1) \times 1000$, where R represents $^{13}C/^{12}C$. Isotopic results are reported relatively to VPDB and precision was better than $\pm 0.15‰$.

CO_2 exposure in elevated CO_2 plots

The degree to which the moss species had processed the added CO_2 in the elevated CO_2 plots was roughly estimated from the stable C isotope composition of the mosses, the commercial CO_2 added to accomplish the target CO_2 concentration in the elevated CO_2 plots and ambient CO_2 in air (Pepin and Körner 2002; Pataki et al. 2003). Only an indication of the actual CO_2 exposure in the elevated CO_2 plots could be obtained, because the moss species are C_3 plants. The initial carbon-fixing enzyme used by plants with this photosynthetic pathway (ribulose-1,5-biphosphate carboxylase-oxygenase or Rubisco) has a relatively high fractionation factor compared to that used by C_4 plants. As a consequence, stable C isotope composition of the mosses were relatively sensitive to changes in the ratio of chloroplastic and ambient CO_2 concentration (c_c/c_a) caused by varying environmental factors such as humidity, light and temperature (Farquhar et al. 1989). Differences in atmospheric CO_2 concentration probably also influenced $\delta^{13}C$ of the mosses, because these non-vascular plants cannot regulate c_c/c_a as they lack stomata (White et al. 1994).

The stable carbon isotope composition of the commercial CO_2 was determined four times during the CO_2 enrichment experiment by manually injecting small volumes with an equivalent of 6 nmol CO_2 on a PreCon/Gasbench II (preconcentration, Thermo Finnigan) coupled to an isotope ratio mass spectrometer (Thermo Finnigan Delta-Plus). Precision was better than $\pm 0.2‰$.

Water table

A dipwell piezometer with a length of 1 m was installed 10–20 cm from each plot. Water tables relative to the median *Sphagnum* surface were recorded weekly from 28 August 2002 on.

Greenhouse experiment

Experimental design

Mesocosms of 25× 25 cm were cut in peatland Het Guisveld close to the field experiment and put in plastic containers in June 2001. Total height of these mesocosms was 12 cm, including approximately 5 cm of peat soil. Vascular plants and *Polytrichum* were removed at the start and upon emergence during the experiment. In two separate compartments of a greenhouse, mesocosms were exposed to either ambient atmospheric CO_2 concentration (on average 400 ppmv in the daytime due to the location in the city) or to a 200 ppmv increased CO_2 concentration in the daytime, almost similar to the field experiment. Each of the two CO_2 treatments included eight mesocosms.

Temperature in the greenhouse reflected general daily and seasonal variation in outdoor temperature between spring and autumn (minimum and maximum temperature ranges of 10–15 °C and 15–22 °C). Greenhouse temperatures in winter (minimum temperature ranges of 10–12 °C and maximum temperature of 15 °C) frequently surpassed temperatures outside, since lower temperatures could not be realised in the greenhouse. No artificial illumination was applied. The mesocosms were regularly sprayed with demineralised water to maintain the water table 4 cm below the initial moss surface. Nitrogen (NH_4Cl dissolved in the spraying water) was supplied weekly, so that the annual supply simulated the regional estimate for atmospheric N deposition of 4.7 g N m^{-2} yr^{-1} (Heij and Schneider 1991).

Sphagnum growth and abundance

Length increment of the *Sphagnum* cover and the relative abundance of the two dominant species *S. palustre* and *S. recurvum* in the mesocosms were determined at the start of the experiment on 19 July 2001, and after 5 months (12 December 2001), 1 year (5 August 2002) and 2 years (15 August 2003) of fumigation. *Sphagnum* length increment was measured at six randomly selected, permanent locations in each mesocosm using stainless steel crank wires (Clymo 1970). The separate covers of *S. palustre* and *S. recurvum* were estimated visually as the mean of cover percentages in each of the 12 rectangular sections comprising the entire mesocosm.

Sphagnum C and N analyses

The top part of 12 *S. palustre* and 24 *S. recurvum* shoots grown during the first year of fumigation (largely the capitulum) were collected in each mesocosm in August 2002. Equal amounts of shoots were sampled near the six crank wires of each mesocosm. Sampled shoot length corresponded to that determined at the individual crank wires. C and N concentrations of the *Sphagnum* samples were determined as described for the field experiment.

Statistical analysis

All statistical analyses were performed using SPSS 10.1 for Windows (SPSS, Chicago, Illinois, USA). If variances were unequal, data were log-transformed.

In the field experiment, the effects of elevated CO_2, moss species and time on moss abundance were analysed with repeated measures three-way analysis of variance (ANOVA). The interaction between species and time was significant and, therefore, repeated measures two-way ANOVAs were performed for each separate moss species. Similar two-way ANOVAs were used for *Polytrichum* abundance in the field corrected for the number of shoots at each hit, and for relative abundances of the two *Sphagnum* species and *Sphagnum* length increment in the mesocosms. Repeated contrasts were used to compare subsequent dates. Independent samples *t*-tests were carried out to test elevated CO_2 effects on moss variables for individual dates and species.

Water table values were averaged for each plot between August 2002 and mid March 2003, and between mid March 2003 and January 2004. The water table between ambient and elevated CO_2 plots, and between periods were compared in a repeated measures two-way ANOVA. To relate the abundance of the three moss species to the water table (of the preceding period), regression analyses were carried out. For abundance data during periods without preceding water table measurements (July 2000 and March 2002), the average water table values of the entire period of August 2002 to January 2004 were used, since water table differences among plots remained quite constant in time.

The effects of elevated CO_2 and moss species on C/N ratio and stable C isotope composition (only field experiment) of moss samples were tested with two-way ANOVAs, followed by a posteriori Tukey tests.

Results

Field experiment

Responses of moss abundance to elevated CO_2
Abundances of *Polytrichum*, *Sphagnum palustre* and *S. recurvum* were not significantly affected by elevated CO_2 during the first 3 years of the field experiment (Figures 2a–d). *Polytrichum* abundance already seemed to be higher in the CO_2 enriched than in the ambient plots before commencing the CO_2 fumigation, though the difference was not significant. A significant positive relation of *Polytrichum* abundance in July 2000 and March 2002 was observed ($r^2=0.36$, $n=12$, $p=0.04$). The apparent but non-significant greater abundance of *Polytrichum* in 2003 and 2004 in the elevated CO_2 plots (Figure 2a, b) is difficult to interpret, since we can not distinguish between inherited differences from before removal, due to regrowth from remaining plant parts lower down in the peat layer, and possible genuine CO_2 responses. Total *Polytrichum* biomass collected in the permanent quadrats in March 2002 was not significantly different between the two CO_2 treatments either (176 ± 39 g and 218 ± 26 g (mean $\pm$ se)) in ambient and elevated CO_2 plots, respectively). *Polytrichum* biomass could be estimated rather well by the *Polytrichum* point-intercept data in March 2002 ($r^2=0.65$, $p=0.001$).

Sphagnum recurvum seemed to show a progressively greater decrease in abundance in the CO_2 enriched plots during the experiment, but the difference between the average values for the two CO_2 levels, i.e. a 32% lower abundance in the elevated CO_2 plots, was still not significant in January 2004 ($p=0.37$, Figure 2d).

The abundance of the moss species also changed in time irrespective of the CO_2 treatments, but in different ways (Figure 2a–d). *Polytrichum* abundance based on the point-intercept data did not change significantly between July 2000 and March 2002, though substantial increases in shoot density were observed visually. The point-intercept measurements solely based on the number of hits within each quadrat did not reflect these substantial changes in abundance, because the increases in density largely occurred in areas already covered by *Polytrichum* shoots in July 2000. After *Polytrichum* removal, the abundance of this species was significantly lower in 2003 ($p<0.001$), and increased further the next year ($p<0.04$; Figure 2a, b). *S. palustre* abundance was significantly enhanced the year after *Polytrichum* removal ($p=0.001$), whereas *S. recurvum* abundance decreased significantly during the last year ($p=0.01$; Figure 2c, d).

Responses of moss abundance to variation in hydrology
The water table relative to the *Sphagnum* surface was on average significantly higher between August 2002 and March 2003 (-17.5 ± 2.5 cm and -20.8 ± 1.7 cm for ambient and elevated CO_2 plots, respectively (mean $\pm$ se)) than between March 2003 and January 2004 (-20.2 ± 3.3 cm and -23.2 ± 1.5 cm, respectively). This was probably mainly due to a record warm summer in 2003 in which evapotranspiration was very high. The

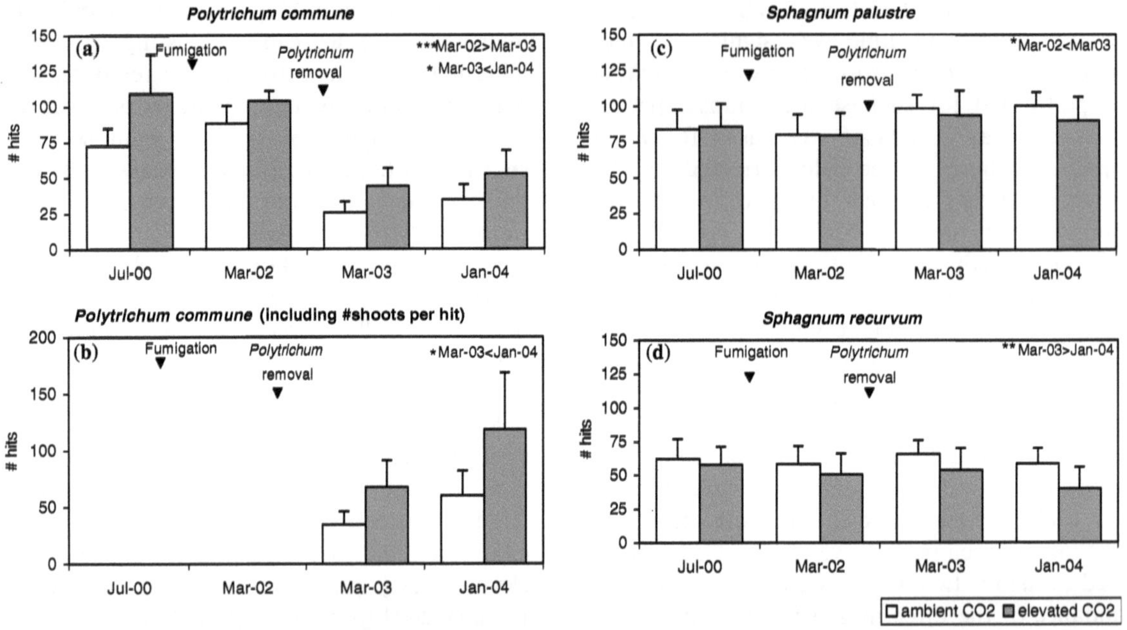

Figure 2. The abundance (# hits) of *Polytrichum commune* without (a) and with correction for the number of shoots per hit (b), and the abundance of *Sphagnum palustre* (c) and *S. recurvum* (d) at ambient (white bars) and elevated CO_2 conditions (180 ppmv above ambient, grey bars) in the field between July 2000, before fumigation had started, and January 2004. Significant effects in time are included: *$p<0.05$, **$p<0.01$, ***$p<0.001$. Data are means ± SE ($n=5$–6).

water table was not influenced by elevated CO_2 ($p=0.288$ and 0.432 for the first and second period, respectively).

Polytrichum abundance was not related to water table ($0.01 \leq r^2 \leq 0.03$ for all six recordings, with and without numbers of shoots per hit). On the other hand, *S. palustre* abundance was positively related to water table and *S. recurvum* negatively ($p<0.05$), except for *S. palustre* in 2004 and *S. recurvum* in 2003 (Figure 3a, b). This means that *S. recurvum* was in general more abundant at higher water tables, while the opposite held for *S. palustre*.

Responses of moss N concentrations and C/N ratios to elevated CO_2

Elevated CO_2 brought about a 10% overall increase in tissue C/N ratio of the mosses ($p=0.034$, Figure 4), which was due to the lower N concentration ($p=0.030$). Such a CO_2 effect on moss C/N ratio and N concentration was not significant for each species separately. Tissue C/N ratio was highest for *S. recurvum*, intermediate for *S. palustre* and lowest for *Polytrichum* shoots (Figure 4). N concentration was significantly higher for *Polytrichum* than for the *Sphagnum* species. C/N ratio and N concentrations of the three moss species were not related to water table.

CO_2 exposure in elevated CO_2 plots

The stable C isotope composition of the mosses was significantly lower (more negative) at elevated than at ambient CO_2 conditions ($p<0.001$, Table 1). The $\delta^{13}C$ of the three moss species showed a similar response to elevated CO_2 (interaction CO_2 treatment × species: $p=0.145$), but it was significantly higher for *Polytrichum* than for *Sphagnum* ($p<0.001$). The contribution of added CO_2 to moss C estimated from the stable C isotope data was rather close to the 32% contribution quantified from the continuous CO_2 concentration measurements in the CO_2 enriched plots for *Polytrichum*, but differences were larger for the *Sphagnum* species (Table 1).

Greenhouse experiment

Responses of Sphagnum growth and abundance to elevated CO_2

After 5 months of exposure, *Sphagnum* length increment in the mesocosms was greater at

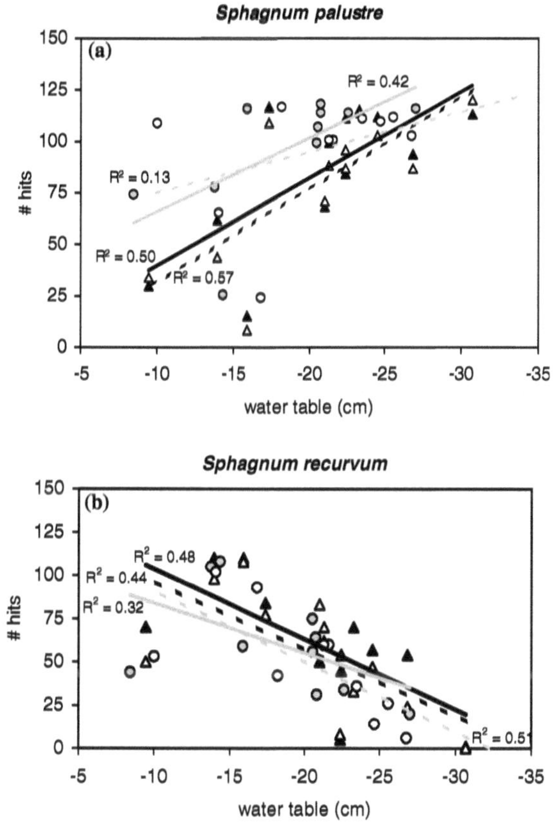

Figure 3. Relations between the abundance (# hits) of *Sphagnum palustre* (a) or *S. recurvum* (b) and water table in the field on July 2000 (black lines with filled triangles), March 2002 (black dotted line with open triangles), March 2003 (grey lines with filled circles) and January 2004 (grey dotted line with open circles).

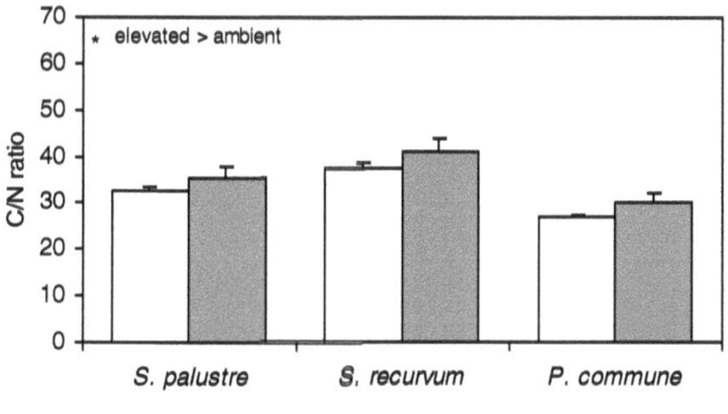

Figure 4. Living tissue C:N ratio of *Sphagnum palustre*, *S. recurvum* and *Polytrichum commune* in the field at ambient (white bars) and elevated CO_2 conditions (180 ppmv above ambient, grey bars) collected in March 2002 after the first year of CO_2 fumigation (April–December 2001) . Significant elevated CO_2 effects (overall effect in left top corner) are included: $^*p < 0.05$, $^{**}p < 0.01$, $^{***}p < 0.001$. Data are means ± SE ($n = 5$–6).

elevated than at ambient CO_2 (day 0–146: $p = 0.001$, Figure 5). This difference was not present anymore after 1 and 2 years of CO_2 fumigation (day 0–402 and 0–777), because *Sphagnum* length increased relatively faster in the ambient plots during the two time periods following the first 5 months (interaction CO_2 treatment × time: $p = 0.003$). Length growth rate was high during the first year of the experiment, while in the second year it was 3–4 times lower (Figure 5).

The relative abundances of *S. palustre* and *S. recurvum* were not affected by CO_2 enrichment (Figure 6). *S. palustre* significantly increased in abundance in the course of the experiment, whereas *S. recurvum* abundance showed the opposite pattern (repeated contrasts: $p < 0.05$, Figure 6).

Responses of Sphagnum C and N concentrations, and C/N ratio to elevated CO_2

Tissue C/N ratio of both *Sphagnum* species was significantly higher at elevated than at ambient CO_2 (*S. palustre*: 47.8 ± 4.2 and 32.5 ± 1.3 ($p < 0.01$) and *S. recurvum*: 60.5 ± 3.1 and 32.5 ± 1.3 ($p < 0.001$) (mean ± se)). The lower tissue N concentration in CO_2 enriched conditions ($p < 0.001$) mainly caused this response, while C concentration was significantly yet only slightly enhanced at elevated CO_2 ($p = 0.003$). *S. recurvum* shoots had a significantly higher tissue C/N ratio than *S. palustre* shoots ($p < 0.001$) resulting from a significantly lower tissue N concentration ($p < 0.001$).

Discussion

Moss abundance responses to elevated CO_2

In agreement with our first hypothesis, the abundance of the three dominant moss species in this Dutch lowland *Sphagnum* peatland was not significantly influenced by 3 years of CO_2 enrichment, even though *S. recurvum* appeared to show a negative response. These findings are in line with other peatland moss studies showing often only small or no effects of elevated CO_2 on length growth and biomass production (Jauhiainen et al. 1994, 1997, 1998; Van der Heijden et al. 2000a, b; Berendse et al. 2001; Heijmans et al. 2002). The small or absent abundance responses to raised CO_2 may be attributed to the relatively small share

Table 1. Stable C isotopic composition of the moss species *Polytrichum commune*, *Sphagnum palustre* and *S. recurvum* (moss $\delta^{13}C$) at ambient (on average 380 ppmv during daytime) and elevated atmospheric CO_2 concentration (on average 560 ppmv during daytime for 1 year), and the estimated share of plant C derived from the added CO_2 in the elevated CO_2 plots based on the $\delta^{13}C$ of the mosses, and the $\delta^{13}C$ of ambient (−8‰) and added CO_2 (−31.91 ± 2.08‰, mean ± se, $n=4$). The moss $\delta^{13}C$ data are means ± SE ($n=6$).

Moss species	Moss $\delta^{13}C$ (‰)		Estimated share of moss C derived from added CO_2 (%)
	Ambient CO_2	Elevated CO_2	
P. commune	−27.71 ± 0.14	−34.56 ± 0.18	29
S. palustre	−29.56 ± 0.33	−35.38 ± 0.24	24
S. recurvum	−29.45 ± 0.28	−35.81 ± 0.34	27

of the added CO_2 in overall assimilated CO_2 of the mosses, and to other growth constraints such as water, nutrient and/or light availability. The added CO_2 in the elevated CO_2 plots could at the most contribute 32% of the C in the mosses. This estimate is based on the mean atmospheric CO_2 concentration of the ambient and elevated CO_2 treatment of 380 and 560 ppmv in the daytime, respectively. The estimated contributions of added CO_2 to moss C calculated from the stable C isotope composition of the *Sphagnum* mosses, and the added and ambient atmospheric CO_2 were even substantially lower (24–27%), suggesting that *Sphagnum* also assimilated CO_2 from the usually CO_2-rich upper peat soil (see Introduction). Thus, large abundance and growth effects on mosses may not be expected in studies that simulate future atmospheric CO_2.

The $\delta^{13}C$ approach, however, only estimated the shares of moss C derived from the added CO_2 roughly. Interactions between water availability and atmospheric CO_2 concentration may have led to differences in moss $\delta^{13}C$ whilst such elevated CO_2 effects were not taken into account in the calculations. Elevated CO_2 can influence water availability in peatlands through higher water use efficiency and cover of vascular plants altering both the degree of evapotranspiration and wind speed in the vegetation (Field et al. 1995; Heijmans et al. 2001). However, such an effect was probably negligible in our study because CO_2 enrichment did not significantly affect water table or vascular plant cover (Milla et al. 2005). Differences in discrimination against fixation of $^{13}CO_2$ (depending on the relative importance of diffusive CO_2 limitation and fractionation by Rubisco, Farquhar et al. 1989) or in CO_2 uptake from the substrate (soil respiration) between mosses in ambient vs. CO_2-enriched plots could have been additional sources of variability in moss C stable isotope signatures.

Moss abundance as dependent on hydrology

Spatial variation in water table has probably obscured and lowered elevated CO_2 responses (if any) of the *Sphagnum* mosses, because water table had a strong control over abundances of both species. Water table had opposite effects on the abundance of the *Sphagnum* species corresponding to earlier observations indicating that of the two

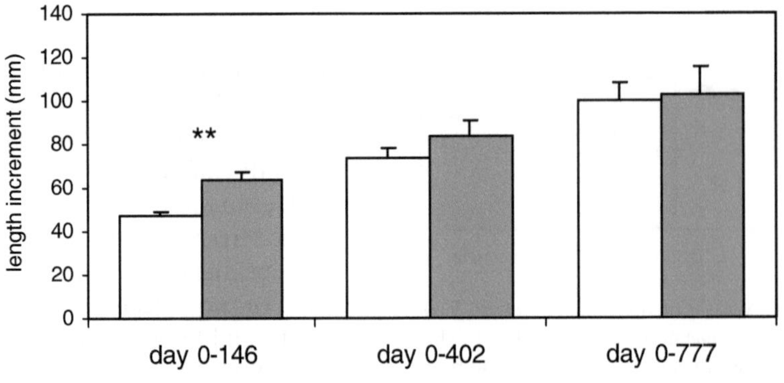

Figure 5. Length increment of *Sphagnum* spp. at ambient (white bars) and elevated CO_2 conditions (200 ppmv above ambient, grey bars) in the greenhouse after 146, 402 and 777 days of CO_2 fumigation. Significant elevated CO_2 effects are included: **$p < 0.01$. Data are means ± SE ($n=8$).

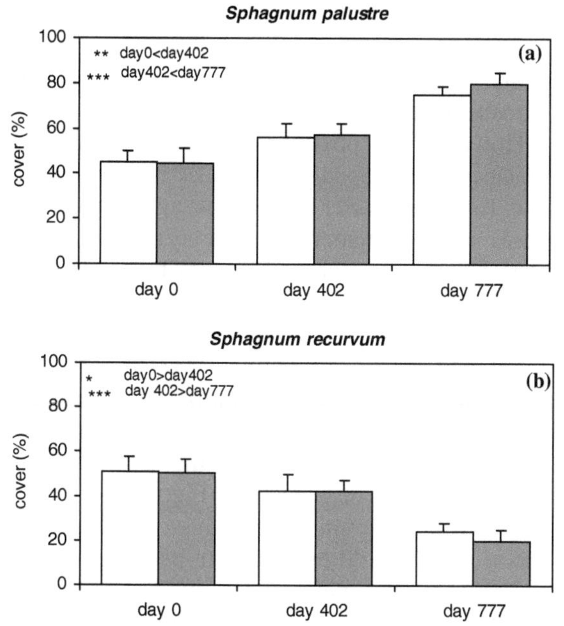

Figure 6. The relative abundance (cover in %) of *Sphagnum palustre* (a) and *S. recurvum* (b) at ambient (white bars) and elevated CO_2 conditions (200 ppmv above ambient, grey bars) in the greenhouse after 146, 402 and 777 days of CO_2 fumigation. Significant effects in time are included: $*p<0.05$, $**p<0.01$, $***p<0.001$. Data are means ± SE ($n=8$).

species of relatively wet peatland habitats *S. palustre* was better adapted to lower water tables (in Clymo and Hayward 1982). Changes in abundance of the two species in the mesocosms during incubation in the greenhouse also suggested the same opposite preference for water availability. The increase in relative abundance of *S. palustre* at the expense of the abundance of *S. recurvum* coincided with a continual increase of the distance between *Sphagnum* surface and water table due to length growth of the plant shoots. In the mesocosms, the *Sphagnum* species did, however, respond to elevated CO_2 during the first summer-autumn period by enhanced length growth at more favourable conditions than in the field including high water levels and lack of competing *Polytrichum* and vascular plants. The CO_2 response was only short-term, already fading during the second part of the first year, which may have been the result of down-regulation of photosynthesis by non-structural carbohydrate accumulation or other limitations suggested by the significantly lower tissue N concentrations and higher C concentrations at elevated CO_2 (Tuba et al. 1999; Van der Heijden et al. 2000a, b). Exposure to drier, growth-restricting conditions in the CO_2 enriched mesocosms may also have played a role, since the taller *Sphagnum* cover in the elevated CO_2 mesocosms started to approach the upper side of the container (i.e., more distant from the water table) during the second year, while at the same time the *Sphagnum* surface in the ambient mesocosms were still experiencing more humid conditions. During winter, the frequently higher air temperatures and relatively high water table in the greenhouse compared to those in the field in connection with low light availability, may have led to differences in form of *Sphagnum* growth such as number of capitula, length growth and biomass per unit length growth in the greenhouse, though length growth in the greenhouse was much lower during winter than during the remaining part of the year.

Contrary to *Sphagnum*, *Polytrichum* abundance was not measurably regulated by the water levels occurring within the peatland. The water table was probably crucial for competition between *Sphagnum* and *Polytrichum*, because in other parts of the Guisveld area with very high water tables *Sphagnum* dominated over *Polytrichum* (personal observation). In other peatlands, *Polytrichum* was also more common at less wet stands, while habitats with higher water saturation were more dominated by *Sphagnum* species (Walbridge 1994; Jalink 1996). The recent strong expansion of *Polytrichum* in the studied peatland seems to confirm unquantified observations that the water table has been lowered (by decreasing water levels in surrounding ditches for agricultural purposes) favouring the growth of *Polytrichum* directly or through the enhancement of mineralisation rates or increased CO_2 release from the soil.

Moss abundance as dependent on N availability and spatial heterogeneity

The increased atmospheric N deposition in the Netherlands during the last few decades could also have contributed to the recent competitive advantage of *Polytrichum* over *Sphagnum*, because of its higher N demand (Beltman and Van den Broek 1993; Van Breemen 1995; Bowden 1991; Mitchell et al. 2002). However, N mineralisation rates in Dutch lowland peatlands can be considerably higher than N deposition inputs, thus N mineralisation may have been a more important N source for *Polytrichum* (Koerselman and Verhoeven 1992). The rate of atmospheric N deposition was unlikely detrimental to the *Sphagnum* mosses, since tissue N concentrations (0.8–1.4% and 1.1–1.4% in greenhouse and field, respectively) remained below the proposed threshold of 1.5% to detect N pollution stress (Van der Heijden et al. 2000a). Moreover, *Sphagnum* in the greenhouse with similar N supply as in the field flourished and showed high growth rates, and *Sphagnum* in the field did not show signs of necrosis due to fungal attack either (Limpens and Berendse 2003; Limpens et al. 2003). On the other hand, the high N deposition in Europe seems to have led to a change in growth limitation of mosses and vascular plants by N to limitation by P, K or to co-limitation of K with N or P, which may in turn attenuate growth responses to elevated CO_2 (Aerts et al. 1992; Hoosbeek et al. 2002; Bragazza et al. 2004).

The great spatial heterogeneity among plots probably also has reduced moss responsiveness to elevated CO_2. Not only variability in moss abundance was considerable, but also the large heterogeneity in vascular plant abundance influencing light, water and nutrient conditions in the plots may have played a role, though vascular plant abundance was not influenced by elevated CO_2 either (Milla et al. 2005). Nevertheless, *S. recurvum* appeared to show early, albeit weak signs of a response to CO_2 enrichment. The 3-year time span of the elevated CO_2 treatment was probably still rather short for these slow-growing plant species, and clearer shifts in moss species abundance in response to enhanced CO_2 can not yet be ruled out in the medium to longer term.

Consequences of elevated CO_2 responses of mosses to C sink function of peatland

The merely small changes in tissue N concentration and C/N ratio of the mosses in response to elevated CO_2 in the field will probably not result in significant changes of these parameters in litter. Thus, we do not expect substantial effects on C and N cycling in this peatland. Other studies under natural field conditions usually also showed only subtle or no elevated CO_2 effects on N concentration of aboveground litter and subsequent litter decomposition both for mosses and many vascular plant species (Norby et al. 2001; Hoosbeek et al. 2002; Milla et al. 2005).

However, the possible medium to longer term replacement of *Sphagnum* species by *Polytrichum*, due to lower water tables for agricultural purposes and/or greater nutrient availability partly caused by atmospheric N deposition, may have consequences for the C balance. The living tissue N concentration of *Polytrichum* (1.6% at elevated CO_2) was markedly higher and the C/N ratio substantially lower compared to those of the *Sphagnum* species (N concentration of 1.1 and 1.3% for *S. recurvum* and *S. palustre* at elevated CO_2, respectively; Figure 4). On the speculative assumption that these relative differences in C/N ratios between *Polytrichum* and *Sphagnum* are maintained during senescence, resulting in lower litter C/N ratio and greater litter decomposability of the overall litter input, a future scenario of *Polytrichum* expansion could increase the C turnover in these peatlands. Annual biomass and litter input to the peatland would increase as well in such a scenario, since *Polytrichum* can be more productive than *Sphagnum* (Mitchell et al. 2002). However, the turnover rate of *Polytrichum* biomass would probably still be higher than that of *Sphagnum*, due to the relatively high C/N ratio and high content of secondary compounds of *Sphagnum* biomass and litter (Coulson and Butterfield 1978; Clymo and Hayward 1982; Verhoeven and Toth 1995; Verhoeven and Liefveld 1997). Together, these responses would mean that the C sink of this peatland was to be reduced eventually if *Polytrichum* gained considerable competitive advantage over *Sphagnum*.

Polytrichum abundance is already increasing fast in this peatland at the moment. Patches with only *Polytrichum* and no *Sphagnum* underneath, probably as a consequence of strong shading by *Polytrichum*, occur. The fast expansion of *Polytrichum* is most likely caused by the anthropogenic factors mentioned above, combined with land management as an artificial control on vascular plants and thereby of succession (Milla et al. 2006). These factors will at least in the short term have a larger impact on the competitive balance between *Polytrichum* and *Sphagnum* than elevated CO_2.

Acknowledgements

The authors thank the Dutch Forestry Authority (Staatsbosbeheer, Alkmaar office) for allowing the use of their land and facilities. We gratefully acknowledge the Guisveld wardens Erik Gerrevink and Wouter Maatje for all their support and hospitality, and Cor Stoof, Martien van Vilsteren and co-workers for assistance in development and maintenance of the MiniFACE equipment. The study was supported by USF grant 98.24 of the Vrije Universiteit Amsterdam to RA.

References

Aerts R., Wallén B. and Malmer N. 1992. Growth-limiting nutrients in Shagnum-dominated bogs subject to low and high atmopsheric nitrogen supply. J. Ecol. 80: 131–140.

Beltman B. and Van den Broek T. 1993. Verzuring van kalkrijke venen – een studie naar effectgerichte maatregelen. Landschap 10: 17–32 (in Dutch).

Berendse F., Van Breeemen N., Rydin H., Buttler A., Heijmans M., Hoosbeek M.R., Lee J.A., Mitchell E., Saarinen T., Vasander H. and Wallén B. 2001. Raised atmospheric CO_2 levels and increased N deposition cause shifts in plant species composition and production in *Sphagnum* bogs. Global Change Biol. 7: 591–598.

Bolin B. 1986. How much CO_2 will remain in the atmosphere?. In: Bolin B.R., Döös B., Jäger J. and Warrick R.A. (eds.), The Greenhouse Effect, Climate Change, and Ecosystems, John Wiley & Sons, Chichester, pp. 93–155.

Bowden R.D. 1991. Inputs, outputs, and accumulation of nitrogen in an early successional moss (*Polytrichum*) ecosystem. Ecol. Monogr. 61: 207–223.

Bragazza L., Thavanainen T., Kutnar L., Rydin H., Limpens J., Hájek M., Grosvernier P., Hájek T., Hajkova P., Hansen I., Iacumin P. and Gerdol R. 2004. Nutritional constraints in ombotrophic *Sphagnum* plants under increasing atmospheric nitrogen deposition in Europe. New Phytol. 163: 609–616.

Brock T.C.M. and Bregman R. 1989. Periodicity in growth, productivity, nutrient content and decomposition of *Sphagnum recurvum* var. *mucronatum* in a fen woodland. Oecologia 80: 44–52.

Clymo R.S. 1970. The growth of *Sphagnum*: methods of measurement. J. Ecol. 58: 13–49.

Clymo R.S. 1984. The limits to peat bog growth. Philos. Trans. Roy. Soc. Lond. B303: 605–654.

Clymo R.S. and Hayward P.M. 1982. The ecology of *Sphagnum*. In: Smith A.J.E. (ed.), Bryophyte Ecology., Chapman and Hall, New York, pp. 229–289.

Coulson J.C. and Butterfield J. 1978. An investigation of the biotic factors determining the rates of plant decomposition on blanket bog. J. Ecol. 66: 631–650.

Farquhar G., Ehleringer J.R. and Hubick K.T. 1989. Carbon isotope discrimination and photosynthesis. Annu. Rev. Plant Physiol. Plant Mol. Biol. 40: 503–537.

Field C.B., Jackson R.B. and Mooney H.A. 1995. Stomatal responses to increased CO_2: implications from the plant to the global scale. Plant Cell Environ. 18: 1214–1225.

Gorham E. 1991. Northern peatlands: role in the carbon cycle and probable responses to climatic warming. Ecol. Appl. 1: 182–195.

Heij G.J. and Schneider T. (eds), 1991. Acidification Research in the Netherlands, Studies in Environmental Science 46. Elsevier, Amsterdam.

Heijmans M.M.P.D., Arp W.J. and Berendse F. 2001. Effects of elevated CO_2 and vascular plants on evapotranspiration in bog vegetation. Global Change Biol. 7: 817–827.

Heijmans M.M.P.D., Klees H. and Berendse F. 2002. Competition between *Sphagnum magellanicum* and *Eriophorum angustifolium* as affected by raised CO_2 and increased N deposition. Oikos 97: 415–425.

Hoosbeek M.R., Van Breemen N., Vasander H., Buttler A. and Berendse F. 2002. Potassium limits potential growth of bog vegetation under elevated CO_2 and N deposition. Global Change Biol. 8: 1130–1138.

Hornibrook E.R.C., Longstaffe F.J. and Fyfe W.S. 2000. Factors influencing stable isotope ratios in CH_4 and CO_2 within subenvironments of freshwater wetlands: implications for δ-signatures of emissions. Isotopes Environ. Health Stud. 36: 151–176.

Houghton J.T., Ding Y., Griggs D.J., Noguer M., Van der Linden P.J., Dai X., Maskell K. and Johnson C.A. 2001. In: Climate Change 2001: The Scientific Basis – Contribution of Working Group I to the Third Assessment Report of the Intergovernmental Panel on Climate Change. Cambridge University Press, Cambridge.

Jalink M.H. 1996. Indicatorsoorten voor Verdroging, Verzuring en Eutrofiëring in Laagveenmoerassen. Staatsbosbeheer Driebergen, Driebergen(in Dutch).

Jauhiainen J. and Silvola J. 1999. Photosynthesis of *Sphagnum fuscum* at long-term raised CO_2 concentrations. Ann. Bot. Fenn. 36: 11–19.

Jauhiainen J., Vasander H. and Silvola J. 1994. Response of *Sphagnum fuscum* to N deposition and increased CO_2. J. Bryol. 18: 83–95.

Jauhiainen J., Vasander H. and Silvola J. 1998. Nutrient concentration in *Sphagna* at increased N-deposition rates and raised atmospheric CO_2-concentrations. Plant Ecol. 138: 149–160.

Jauhiainen J., Silvola J., Tolonen K. and Vasander H. 1997. Response of *Sphagnum fuscum* to water levels and CO_2 concentration. J. Bryol. 19: 391–400.

Jonasson S. 1988. Evaluation of the point intercept method for the estimation of plant biomass. Oikos 52: 101–106.

Koerselman W. and Verhoeven J.T.A. 1992. Nutrient dynamics in mires of various trophic status: nutrient inputs and outputs and the internal nutrient cycle. In: Verhoeven J.T.A. (ed.), Fens and Bogs in the Netherlands: Vegetation, History, Nutrient Dynamics and Conservation, Kluwer, Dordrecht, pp. 397–432.

Körner C. 2000. Biosphere responses to CO_2 enrichment. Ecol. Appl. 10: 1590–1619.

Lamers L.P.M., Farhoush C., Van Groenendael J.M. and Roelofs J.G.M. 1999. Calcareous groundwater raises bogs; the concept of ombotrophy revisited. J. Ecol. 87: 639–648.

Lansdown J.M., Quay P.D. and King S.L. 1992. CH_4 production via CO_2 reduction in a temperate bog: A source of ^{13}C-depleted CH_4. Geochim. Cosmochim. Acta 56: 3493–3503.

Limpens J. and Berendse F. 2003. Growth reduction of *Sphagnum magellanicum* subjected to high nitrogen deposition: the role of amino acid nitrogen concentration. Oecologia 135: 339–345.

Limpens J., Raymakers T.A.G., Baar J., Berendse F. and Zijlstra J.D. 2003. The interaction between epiphytic algae, a parasitic fungus and *Sphagnum* as affected by N and P. Oikos 103: 59–68.

Makino A. and Mae T. 1999. Photosynthesis and plant growth at elevated levels of CO_2. Plant Cell Physiol. 40: 999–1006.

Maltby E. and Immirzi P. 1993. Carbon dynamics in peatlands and other wetland soils: regional and global prespectives. Chemosphere 27: 999–1023.

Miglietta F., Hoosbeek M.R., Foot J., Gigon F., Hassinen A., Heijmans M., Peresotti A., Saarinen T., Van Breemen N. and Wallén B. 2001. Spatial and temporal performance of the MiniFACE (free air CO_2 enrichment) system on bog ecosystems in northern and central Europe. Environ. Monitor. Assess. 66: 107–127.

Milla R., Cornelissen J.H.C., Van Logtestijn R.S.P., Toet S. and Aerts R. 2006. Vascular plant responses to elevated CO_2 in a temperate lowland *Sphagnum* peatland. Plant Ecol. DOI: 10.1007/s11258-005-9028-9.

Mitchell E.A.D., Buttler A., Grosvernier P., Rydin H., Siegenthaler A. and Gobat J.-M. 2002. Contrasted effects of increased N and CO_2 supply on two keystone species in peatland restoration and implications for global change. J. Ecol. 90: 529–533.

Mooney H.A., Canadell J., Chapin F.S. III, Ehleringer J.R., Körner Ch., McMurtrie R.E., Parton W.J., Pitelka L.F. and Schulze E.-D. 1999. Ecosystem physiology responses to global change. In: Walker B., Steffen W., Canadell J. and Ingram J. (eds.), The Terrestrial Biosphere and Global Change – Implications for Natural and Managed Ecosystems, Cambridge University Press, Cambridge, pp. 141–189.

Norby R.J., Cotrufo M.F., Ineson P., O'Neill E.G. and Canadell J.G. 2001. Elevated CO_2, litter chemistry, and decomposition: a synthesis. Oecologia 127: 153–165.

Pataki D.E., Ellsworth D.S., Evans R.D., Gonzalez-Meler M., King J., Leavitt S.W., Lin G., Matamala R., Pendall E., Siegwolf R., Van Kessel C. and Ehleringer J.R. 2003. Tracing changes in ecosystem function under elevated carbon dioxide conditions. BioScience 53: 805–818.

Pepin S. and Körner C. 2002. Web-FACE: a new canopy free-air CO_2 enrichment system for tall trees in mature forests. Oecologia 133: 1–9.

Poorter H. and Navas M.-L. 2003. Plant growth and competition at elevated CO_2: on winners, losers and functional groups. New Phytol. 157: 175–198.

Rydin H. 1993. Mechanisms of interactions among *Sphagnum* species along water-level gradients. Adv. Bryol. 5: 153–185.

Saarnio S., Järviö S., Saarinen T., Vasander H. and Silvola J. 2003. Minor changes in vegetation and carbon gas balance in a boreal mire under a raised CO_2 or NH_4NO_3 supply. Ecosystems 6: 46–60.

Silvola J. 1985. CO_2 dependence of photosynthesis in certain forest and peat mosses simulated photosynthesis at various actual and hypothetical CO_2 concentrations. Lindbergia 11: 86–93.

Smolders A.J.P., Tomassen H.B.M., Pijnappel H.W., Lamers L.P.M. and Roelofs J.G.M. 2001. Substrate-derived CO_2 is important in the development of *Sphagnum* spp. New Phytol. 152: 325–332.

Stitt M. and Krapp A. 1999. The interaction between elevated carbon dioxide and nitrogen nutrition: the physiological and molecular background. Plant Cell Environ. 22: 583–621.

Tuba Z., Proctor M.C.F. and Takács Z. 1999. Dessiccation-tolerant plants under elevated air CO_2: a review. Zeitschrift für Naturforschung C 54: 788–796.

Van Breemen N. 1995. How*Sphagnum* bogs down other plants. Trends Ecol. Evol. 10: 270–275.

Van der Heijden E., Verbeek S.K. and Kuiper P.J.C. 2000a. Elevated atmospheric CO_2 and increased nitrogen deposition: effects on C and N metabolism and growth of the peat moss *Sphagnum recurvum* P. Beauv. var. *mucronatum* (Russ.) Warnst. Global Change Biol. 6: 200–212.

Van der Heijden E., Jauhiainen J., Silvola J., Vasander H. and Kuiper P.J.C. 2000b. Effects of elevated atmospheric CO_2 concentration and increased nitrogen deposition on growth and chemical composition of ombotrophic *Sphagnum balticum* and oligo-mesotrophic *Sphagnum papillosum*. J. Bryol. 22: 175–182.

Van der Meijden R. 1996. Heukels' Flora van Nederland 22th ed. Wolters Noordhoff, Groningen (in Dutch).

Verhoeven J.T.A. and Toth E. 1995. Decomposition of *Carex* and *Sphagnum* litter in fens: effect of litter quality and inhibition by living tissue homogenates. Soil Biol. Biochem. 27: 271–275.

Verhoeven J.T.A. and Liefveld W.M. 1997. The ecological significance of organochemical compounds in *Sphagnum*. Acta Bot. Neerland. 46: 117–130.

Walbridge M.R. 1994. Plant community composition and surface water chemistry of fen peatlands in West Virginia's Appalachian plateau. Water Air Soil Pollut. 77: 247–269.

Weltzin J.F., Harth C., Bridgham S.D., Pastor J. and Vonderharr M. 2001. Production and microtopography of bog bryophytes: response to warming and water-table manipulations. Oecologia 128: 557–565.

Weltzin J.F., Pastor J., Harth C., Bridgham S.D., Updegraff K. and Chapin C.T. 2000. Response of bog and fen plant communities to warming and water-table manipulations. Ecology 81: 3464–3478.

White J.W.C., Ciais P., Figge R.A., Kenny R. and Markgraf V. 1994. A high-resolution record of atmospheric CO_2 content from carbon isotopes in peat. Nature 367: 153–156.

Photo. Rocky Mountain Biological Laboratory Meadow Warning Experiment (Photo courtesy: John Harte).

Plant Ecology (2006) 182:43–62
DOI 10.1007/s11258-005-9030-2

From transient to steady-state response of ecosystems to atmospheric CO_2-enrichment and global climate change: conceptual challenges and need for an integrated approach

Lindsey E. Rustad*
*USDA Forest Service, 271 Mast Road, Durham, NH, UK; *Author for correspondence (e-mail: rustad@maine.edu; phone: +1-207-829-6817; fax: +1-207-829-6817)*

Received 1 September 2004; accepted in revised form 15 December 2004

Key words: Climate change, CO_2 enrichment, Ecosystem modeling, Global change, Temporal scaling

Abstract

Evidence continues to accumulate that humans are significantly increasing atmospheric CO_2 concentrations, resulting in unprecedented changes in the global climate system. Experimental manipulations of terrestrial ecosystems and their components have greatly increased our understanding of short-term responses to these global perturbations and have provided valuable input to ecosystem, dynamic vegetation, and global scale models. However, concerns exist that these initial experimental responses may be transitory, thereby limiting our ability to extrapolate short-term experimental responses to infer longer-term effects. To do these extrapolations, it will be necessary to understand changes in response patterns over time, including alterations in the magnitude, direction, and rate of change of the responses. These issues represent one of our largest challenges in accurately predicting longer-term changes in ecosystems and associated feedbacks to the climate system. Key issues that need to be considered when designing future experiments or refining models include: linear vs. non-linear responses, direct vs. indirect effects, lags in response, acclimation, resource limitation, homeostasis, buffers, thresholds, ecosystem stoichiometry, turnover rates and times, and alterations in species composition. Although experimental and landscape evidence for these response patterns exist, extrapolating longer-term response patterns from short-term experiments will ultimately require a unified multidisciplinary approach, including better communication and collaboration between theoreticists, experimentalists and modelers.

Introduction

An overwhelming consensus exists that 20th century human activities have induced dramatic and unprecedented changes in the global chemical and physical environment, including a ~33% increase in atmospheric CO_2 concentrations, a ~0.6 °C increase in mean annual temperature and changes in both the magnitude and degree of variability of precipitation (IPCC 2001). Current predictions indicate that, unless greenhouse gas emissions are significantly curtailed, atmospheric CO_2 concentrations will double in the next century, inducing an additional 1.4–5.8 °C increase in mean global temperature, and further alterations in the amount, timing, and intensity of regional and global patterns of precipitation (IPCC 2001).

The response of terrestrial ecosystems to these predicted alterations in atmospheric CO_2 and climate has been the subject of intense scientific scrutiny over the past several decades, and the focus of a growing number of single and multi-factor ecosystem-scale manipulation experiments. The accumulating evidence from these experiments has greatly increased our understanding of short-term responses of terrestrial ecosystems and their components to elevated atmospheric CO_2, warming and changes in water availability (for syntheses, see Curtis and Wang 1998; Peterson et al. 1999; Medlyn et al. 1999, 2001; Norby et al. 2001a,b; Rustad et al. 2001; Zak et al. 2003; Badeck et al. 2004; Pendall et al. 2004; Nowalk et al. 2004; Ainsworth and Long 2005), and has provided valuable input for dozens of ecosystem and global scale models that are allowing us to better understand and predict future response patterns (e.g., Potter et al. 1993; Running and Hunt 1993; Aber and Driscoll 1997; Tian et al. 1999). Concern exists, however, that these initial responses may be transitory, and caution should be used when attempting to extrapolate short-term experimental responses from a limited number of experiments to infer longer-term effects (Norby and Luo 2004). To do these extrapolations and to better construct conceptual and empirical models of ecosystem response to global change, it will be necessary to improve our understanding of the *change in response patterns over time*, including alterations in the magnitude, direction, and rate of change of the response. These issues represent one of the biggest challenges in accurately predicting long-term changes in ecosystems and associated feedbacks to the atmospheric and climate system.

Previous papers have explored conceptual challenges in evaluating ecosystem response to global warming (Shaver et al. 2000) and global warming in combination with elevated atmospheric CO_2 (Norby and Luo 2004). This paper (a) focuses on current and emergent issues that are important to guide our conceptual understanding of the *temporal patterns of response* of ecosystems to elevated CO_2 and global climate change, and (b) presents an integrated approach to future work in this field.

Definitions

The terms 'steady state', 'transient state', and 'transient response' have been loosely defined in the ecological literature. Here the term 'steady state' is used to describe a system where the sum of all fluxes of material and energy going into any individual component of the system equals the sum of all fluxes of material and energy going out of that same component of the system or $a+b=c+d$ (Figure 1). A 'static steady state' describes a system where there is no flux of material and energy going into or out of any of the components of the system, or $a=b=c=d=0$. Biological systems, by definition, are never in a static steady state as life is defined by the flow of materials and energy. A 'dynamic steady state' describes a system where the fluxes of material and energy going into each component of the system equal the fluxes going out of each component of the system, or $a+b \neq 0$, $c+d \neq 0$, $a+b=c+d$. For example, soil carbon is considered to be in a dynamic steady state when biotic and abiotic carbon inputs into the soil system equal biotic and abiotic carbon exports from the soil system. A 'cyclic steady state' describes a system where the cumulative flux of material and energy going into each individual component of the system over the period of the cycle equals the cumulative flux of material and energy going out of each component of the system over the same period. For example, annual cycles of leaf area index (LAI) in mature hardwood forest ecosystems and annual hydrologic cycles can be considered in cyclic steady states if they return to the same value or stage year after year. A cyclic steady state can either be externally driven (e.g., diel, monthly, or annual temperature or radiation cycles), or can arise from

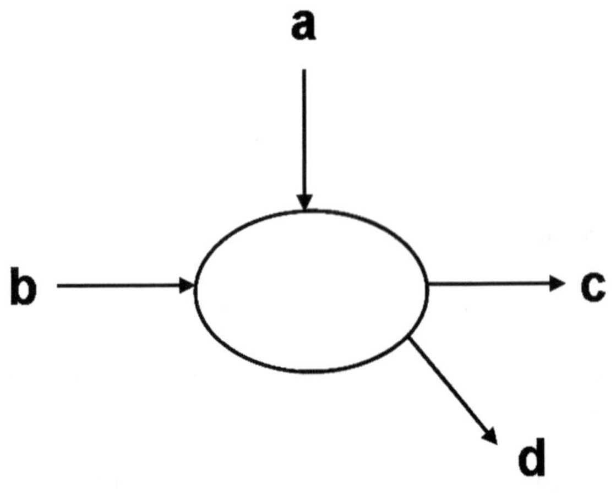

Figure 1. System states. Arrows represent flux of material and energy going into and out of a system.

interactions of the internal components of the system (e.g., lynx-hare cycle). The failure to recognize these cycles could lead to erroneous conclusions about the trajectory of the system.

A 'transient state' refers to a system where the sum of all the fluxes of material and energy going into all components of a system do not equal the sum of all fluxes of material and energy going out of all components of a system, or $a+b \neq 0$, $c+d \neq 0$, $a+b \neq c+d$. Although theoretically the distinction between a transient state and a steady state is clear, in practice it can be difficult to distinguish whether a system is in a transient state, a part of a cyclical steady state, or truly in a steady state.

A 'transient response' describes the dynamics of a system as it approaches a steady state following a perturbation. Examples include increased N mineralization following the physical disturbance of a soil system (Lamontagne 1998; Kristensen et al. 2000; Jefts et al. 2004) or the initial rapid increase in soil respiration following a step increase in soil warming (Peterjohn et al. 1994; Rustad and Fernandez 1998). The magnitude of both of these responses typically decreases over time as the systems come into a new equilibrium (Kristensen et al. 2000; Melillo et al. 2002). Questions have been raised as to whether transient responses such as these are important, and this term has increasingly been used to describe an undesired artifact at the beginning of an experiment (Lukewille and Wright 1997). These initial 'transient' responses are now often disregarded, with greater emphasis placed on understanding the longer-term responses at time scales of years or decades. By definition, however, even these 'longer-term' responses remain transitory until a new steady state is reached, and short-term transient responses can be important in determining the trajectory of the longer-term response. This would occur if, for example, the transient response resulted in changes in key nutrients or resources, or if the transient response displaced the system onto a trajectory leading to an alternate steady state.

Temporal response patterns: considerations and controls

Understanding temporal response patterns and the underlying mechanisms that control them will be fundamental to making longer-term predictions of ecosystem response to a changing environment. As illustrated in Figure 2, a perturbation, such as elevated CO_2 or a change in climate can move a system from one state (A) to another state (B), but the trajectory of the response may vary.

If the response is assumed to be approximately linear (i.e., a change in the perturbation results in a uni-directional change in the response over time; hypothetical line 1 in Figure 2), when, in fact, the response is non-linear or cyclic (hypothetical lines 2–6 in Figure 2), then extrapolations from measurements of the initial response may lead to false conclusions concerning the future state of the system. The following sections describe issues that should be considered when evaluating and modeling temporal patterns of ecosystem response to global change.

Direct vs. indirect effects

The direct effects of CO_2 enrichment, warming, and changes in moisture on ecosystem processes are relatively well understood. Indirect effects, which will likely regulate long-term changes in ecosystem response, are more complex and will require considerably more effort to accurately predict and model because they can involve a complex web of interactions (Shaver et al. 2000). An example of direct vs. indirect effects is the influence of warming on soil respiration. In general, and up to a temperature optima, warming directly increases both autotrophic and heterotrophic soil respiration (Rustad and Norby 2002). However, if higher temperatures increase evapotranspiration and thereby reduce soil moisture, then warming can indirectly result in a decrease in soil respiration. This was demonstrated by Rustad et al. (2001) who showed that experimental soil warming (either with electrical heating cables, infra-red heaters, or glasshouses) at 16 different research sites generally increased soil respiration (Figure 3). However, at one site (The Rocky Mountain Biological Laboratory, or 'RMBL' in Figure 3), a warming-induced decline in soil moisture resulted in lower rates of soil respiration in the heated plots compared to the controls (Rustad et al. 2001; Figure 3).

Interestingly, soil carbon also declined over time in the heated plots compared to the controls at the

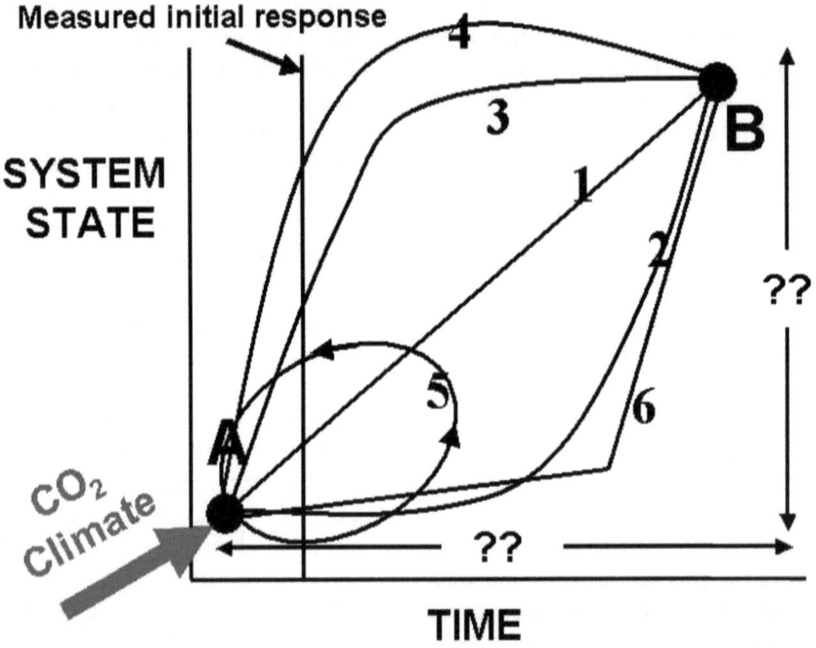

Figure 2. Hypothetical trajectories as a system moves from state A to state B. The lines represent the following hypothetical responses: 1 = a linear response, 2 = lag, 3 = acclimation, 4 = resource limitation, 5 = homeostasis, and 6 = threshold. (Diagram courtesy of Gus Shaver, TERACC Workshop, 2002.)

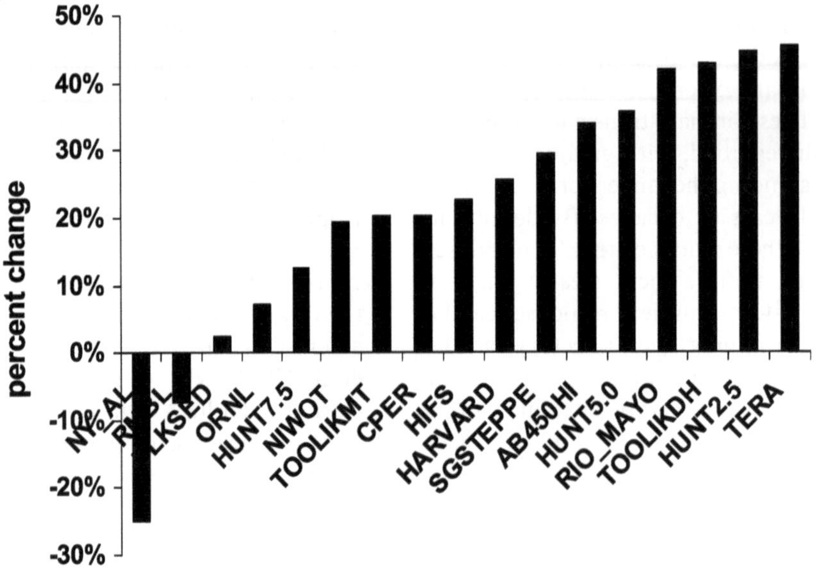

Figure 3. Percent change in soil respiration at 16 ecosystem warming experiments (from Rustad et al. 2001).

RMBL (Saleska et al. 2002b, Figure 4). This decline in soil carbon was not due to an increased loss of carbon through soil respiration (since soil respiration had declined) but rather was due indirectly to a change in plant community dynamics, with a shift from forbs (characterized by high productivity) to shrubs (characterized by low productivity), and consequent declines in above and belowground plant detrital quantity and quality (deValpine and Harte 2001). Results from observations across an associated climate gradient, however, suggest that the temporal response patterns for this study will be even more complicated, and that the observed decline in soil carbon is a transient response that will eventually be reversed as lower quality detrital inputs from the increasingly dominant shrub species reduce soil respiration losses (Saleska et al. 2002a).

At a larger scale, direct effects of warming on snow and/or ice cover, LAI, and/or changes in disturbance frequency such as fire may alter local, regional, or even global albedo, or the fraction of incoming solar radiation that is reflected back to the atmosphere (Ingram et al. 1989; Betts et al. 1997; Lynch and Wu 1999; Betts 2000). Decreases in albedo will increase the radiation absorbed by the region which will amplify warming whereas increases in albedo will cause a greater amount of radiation to be reflected back to the atmosphere, and will thus have a negative feedback to warming.

'Lags' in response

Lags in response occur when some responses take longer to come to a new equilibrium with the environment than others because of either internal (e.g., life span, seed dispersal, vegetative propagation, etc) or external (e.g., fire, pathogens, etc.) factors (hypothetical line 2 in Figure 2). For example, Vetaas (2002) showed that although mature individuals of long-lived Rhododendron species were not able to survive outside the cold limit of their realized niche, they could survive and continue to reproduce vegetatively when planted outside their natural high temperature range. The future distribution of this species may thus show a lag in response to gradual increases in mean annual temperature. 'Lags' due to limitations to seed dispersal and changes in disturbance regimes have been suggested by Chapin and Starfield (1997) who modeled a lag of 150–250 years in forestation of an arctic tundra following climatic warming due to (1) slow tree establishment and growth under slow climatic warming and (2) higher frequencies of fire and insect attack under more rapid climatic warming.

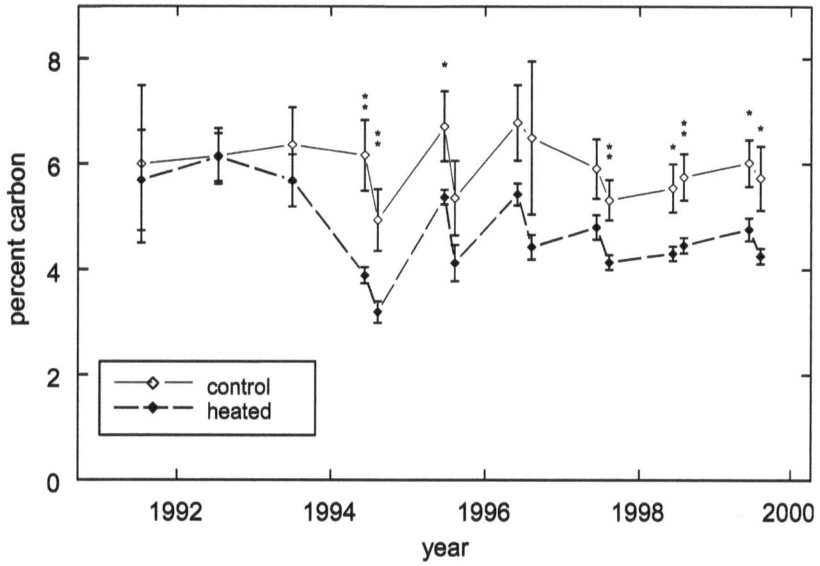

Figure 4. Percent carbon in heated and control plot soils at the Rocky Mountain Biological laboratory (RMBL) (from Saleska et al. 2002b; reproduced by permission of the American Geophysical Union).

Acclimation

Acclimation is the often misused term that refers to a non-heritable, reversible change in the physiology or morphology of an organism in response to changing environmental conditions (Ricklefs 1990; hypothetical line 4 in Figure 2). Plants, for example, can acclimate to changing conditions by various mechanisms including changing enzyme concentrations (e.g., Maroco et al. 1999; Watling et al. 2000; Gesch et al. 2002), altering shoot:root ratios (e.g., Equiza et al. 2001; Kozlowski and Pallardy 2002; Horacio 2003; Matsuki et al. 2003), or changes in phenology (e.g., Campbell and Sorensen 1973; Adam et al. 2001). Evidence is accumulating that many ecosystem processes acclimate to elevated CO_2 and warming at the physiological level, thereby reducing their sensitivity to these perturbations, and invalidating many predictions of future responses. Considerable effort must be made to (a) understand the mechanisms underlying physiological acclimation at the organism level and (b) incorporate acclimation into existing ecosystem models.

Three examples of physiological acclimation that have received considerable attention in recent years are the acclimation, or down regulation, of photosynthesis in response to elevated CO_2, the acclimation of photosynthesis to elevated temperature, and the acclimation of autotrophic respiration to elevated temperature. Photosynthetic down regulation in response to elevated CO_2 was initially reported in dozens of CO_2 enrichment studies (e.g., Gunderson and Wullscheleger 1994; Luo et al. 1994; Drake et al. 1997; Egli et al. 1998; Rey and Jarvis 1998; Ziska 1998; Medlyn et al. 1999; Sims et al. 1999; Hymus et al. 2002b; Rogers and Ellsworth 2002) and was generally attributed to decreases in leaf nitrogen and ribulose 1,5-biphosphate carboxylase/oxygenase (Rubisco) which lead to declines in photosynthesis (Rogers and Humphries 2000). More recently, however, the role of photosynthetic down regulation has been questioned, and its prevalence, particularly in earlier pot or chamber studies has been attributed, in part, to root restriction within experimental pots (e.g., Stitt 1991; Farage et al. 1998), inadequate N supply (e.g. Webber et al. 1994; Drake et al. 1997; Kubiske et al. 2002; Ainsworth et al. 2003), or the age class of needles in conifers (Medlyn et al. 1999).

Photosynthetic acclimation to increased temperature including both shifts in temperature optima and uniform shifts across all temperatures, has been long recognized (e.g., Barry and Bjorkman 1980; Ferrar et al. 1989; Read 1990; Gunderson et al. 2000), and has been attributed to various factors including different thermal properties of key photosynthetic enzymes, different temperatures at which membranes are damaged, and differential thermal stability of photochemical reactions (Nilsen and Orcutt 1996). The acclimation of autotrophic respiration to elevated temperature has also been demonstrated (e.g., Kirshbaum and Farquhar 1984; Tjoelker et al. 1999, 2001; Atkin et al. 2000a,b; Will 2000; Griffin et al. 2002; Bolstad et al. 2003), and has been attributed variously to decreased number of mitochondria (Miroslavov and Kravkina 1991), decreased respiratory capacity per mitochondria (Klikoff 1966), limitations in substrate supply (Lambers et al. 1996), changes in the concentration of plant soluble sugars (Atkin et al. 2000a), changes in demand for respiratory energy (Atkin and Lambers 1998), and/or changes in enzymatic capacity (Atkin et al. 2002).

Although the acclimation of both photosynthesis and autotrophic respiration to warming is well established, the potential acclimation of 'soil respiration' (i.e., the combined respiration of roots and soil micro- and macro-biota) to warming is more controversial. Historically, dozens of studies have demonstrated strong positive relationships between soil respiration and temperature (for syntheses see Raich and Nadelhoffer 1989; Raich and Schlesinger 1992; Raich and Potter 1995; Kirschbaum 1996; Rustad et al. 2001), and soil respiration is typically and effectively modeled with an exponential or Arrehenius function (Rustad et al. 2000). Recently, however, the temperature dependence of soil respiration has been challenged by Luo et al. (2001) who suggested that soil respiration 'acclimates' to elevated temperature. They conducted a warming × grazing experiment in a tall grass prairie in Oklahoma, USA using overhead infra-red lamps and clipping, and reported a decline in the respiration quotient Q_{10} from 2.70 in the unheated, unclipped plots to 2.43 in the heated, unclipped plots, and from 2.25 in the unheated, clipped plots to 2.10 in the heated, clipped plots. However, a physiological mechanism for the acclimation of soil respiration to temperature has yet to be elucidated. This is in part because, unlike photosynthesis, soil respiration is not a single process but is instead the sum of the combined respiration of plants and the complex community of micro- and macro-heterotrophic soil organisms, and includes several alternate chemical pathways. In addition, the direct effect of warming on soil respiration is complicated by a host of indirect effects, including warming-induced changes in above and belowground biomass, soil moisture, N mineralization, substrate quality and/or quantity, and microbial community activity, biomass, and composition.

Given that gross primary productivity (GPP), aboveground respiration, and soil respiration represent three of the largest fluxes in the terrestrial global carbon cycle (estimated at ~120, 60, and 60 Pg C yr^{-1}, respectively; Schlesinger 1997), it is imperative to understand if and to what degree these processes will acclimate to changing environmental conditions such as CO_2 enrichment and global warming. Even slight changes in the direction and or magnitude of these fluxes could equal or exceed the annual input of CO_2 to the atmosphere via combined fossil fuel combustion and land-use changes (estimated at ~6 Pg C yr^{-1}), and could therefore significantly accelerate – or decelerate – the rate of atmospheric build-up of CO_2, with consequent feedbacks to climate change.

Resource limitation/initial conditions

The sustainability of the magnitude and even direction of a response may be governed by the availability of resources which will be governed in part by initial conditions. For example, ecosystems with large stocks of relatively labile C may show a larger and more sustained increase in soil respiration in response to warming than an ecosystem with low initial labile C stocks, or an N-rich ecosystem may show a more sustained increase in photosynthesis and NPP under CO_2-enrichment then a N-poor ecosystem. In either case, if the systems receive no new inputs of labile C or atmospheric or fertilizer N, the magnitude of the response will decline over time as either labile C or N are depleted (hypothetical line 2 in Figure 2). For example, at the Harvard Forest soil warming experiment, Peterjohn et al. (1994) initially reported an approximately 40% increase in soil respiration during the first six months of the experiment. However, the magnitude of this increase diminished over time such that after 10 years of warming soils at 5 °C above ambient, soil respiration rates in the heated plots were not significantly different than rates in the control plots (Figure 5; Melillo et al. 2002). One explanation is that labile C supplies were depleted during the course of the experiment suggesting resource limitation. The lack of a treatment effects

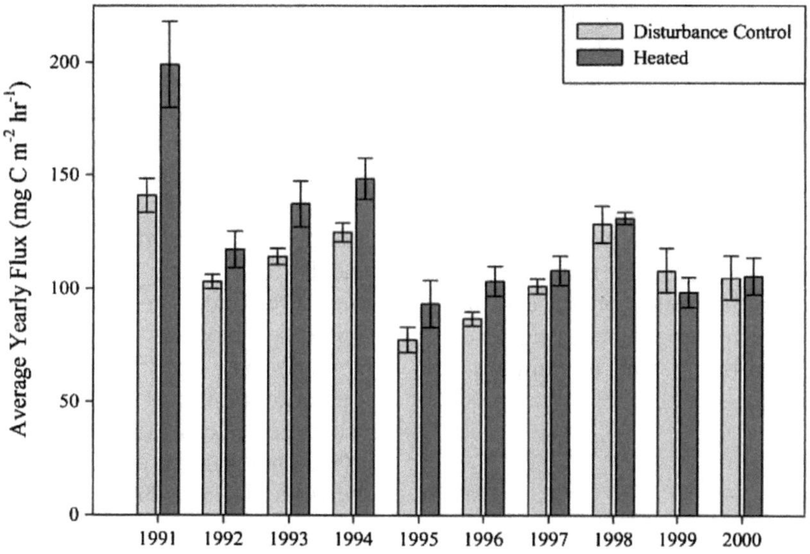

Figure 5. Annual soil respiration at the Harvard Forest soil warming experiment (reprint with permission from Melillo et al. SCIENCE 298: 2173–2176 (2002)).

in the latter years of the experiment also lend support to the hypothesis that different carbon fractions have different temperature sensitivities, with labile carbon fractions (consistently predominantly of simple sugars and amino acids) being highly temperature sensitive but recalcitrant carbon fractions (consisting of more complex aromatic compounds) being relatively temperature insensitive (Liski et al. 1999; Giardina and Ryan 2000; Melillo et al. 2002; Gu et al. 2004).

Homeostasis

Homeostasis is the 'maintenance of' or 'return to' constant internal conditions in the face of a varying external environment (Ricklefs 1990). Classic examples include (a) the thermal regulation of homeotherms despite external fluctuations in temperature, (b) the ability of organisms to maintain their internal chemical composition despite fluctuations in the chemical content of their environment or food source, and (c) predator–prey cycles where as the population of prey increases so does the population of predators, thereby decreasing the population of prey and consequently the population of predators (hypothetical line 5 in Figure 2). Local, regional, and even global ecological systems also exhibit homeostatic behavior. An example is elevated atmospheric CO_2 and the global carbon cycle. Within limits, as atmospheric CO_2 increases, leaf level photosynthesis and NPP should increase, thereby removing CO_2 from the atmosphere and stabilizing atmospheric CO_2 concentrations. Concerns exist, however, that the current anthropogenic input of carbon to the atmosphere from fossil fuel combustion and land-use changes, particularly in combination with possible positive (rather than negative) feedbacks from warming-induced increases in the release of soil carbon to the atmosphere or decreases in albedo, may exceed the capacity of the earth's systems to maintain this homeostatic balance.

Buffers

Buffers are mechanisms or attributes that allow systems to resist change in response to external perturbation or impact. In chemistry, solutions that contain a weak acid and its salt or a weak base and its salt, and which thereby can resist changes in pH, are called buffers. Similarly, ecological systems have certain attributes that allow them to resist moderate changes in environmental variables. Examples of ecosystem properties that may provide 'buffers' against impacts of CO_2 fertilization and climate change may include soil C quality and quantity [i.e., systems with more protected, chemically stable C will be less vulnerable to soil C loss than systems with less stable C e.g. (Collins et al. 1997; Paustian et al. 1997, 2000; Six et al. 2000)], soil depth and water holding capacity (i.e., ecosystems with deeper soils with better water holding capacity will be less sensitive to fluctuations in precipitation than those with shallower soils with limited water holding capacity), albedo [i.e., ecosystems with higher albedo will reflect more solar radiation back to the atmosphere and will thus be less sensitive to warming than systems with lower albedo (e.g., Betts et al. 1997; Betts 2000; IPCC 2001; Berbet and Costa 2002)] and biodiversity [i.e., ecosystems with greater species or functional group diversity may be more resistant and resilient to environmental change than those with lower diversity (e.g., Naeem and Li 1997; Walker et al. 1999; Chapin et al. 2000; Ives and Cardinale 2004)].

Thresholds

A process is said to have a threshold if below that threshold there is either no change or proportionate change in the response of the process to a perturbation and above that threshold there is a dramatic, non-proportional response (hypothetical line 6 in Figure 2). Arnold et al. (1999) provide an example of the former, where, using laboratory incubations under controlled temperature and moisture conditions, they showed no difference in microbial biomass at gravimetric soil moisture contents between 120 and 320%, but a dramatic reduction of almost 95% of the microbial biomass when gravimetric soil moisture was decreased to 20%. They suggest a soil moisture threshold exists between 20 and 120% for their soils above which moisture is not limiting and temperature largely controls microbial biomass dynamics, and below which moisture is too low to sustain viable microbial biomass, regardless of temperature.

At a larger scale, thresholds also appear to exist in the climate system. Reconstruction of past climates, for example, show gradual changes in climate over geologic time scales, punctuated by dramatic changes in temperature and precipitation on time scales as small as decades (IPCC 2001). Examples include a 5–10 °C increase in temperature and a doubling of snowfall that occurred in Greenland over a period of 40 years following the last glaciation and the rapid transition from shrubland to desert that occurred in the Sahara approximately 5500 years ago (Rahmstorf 2002). The causes of these rapid changes are uncertain but may be associated with thresholds in ocean circulation and sea ice dynamics, or vegetation-induced changes in albedo (Rahmstorf 2002). Concerns exist, including those expressed by the National Academy of Science Committee on Abrupt Climate Change (2001) and by Gregory et al. (2004), that similar mechanisms will come into play such that CO_2-induced global warming will lead to increased precipitation in high northern latitudes, which, combined with melting of the polar ice sheets, will increase freshwater input to the North Atlantic Ocean, leading to a precipitous reduction in the global ocean's thermohaline circulation, thereby shutting down the Gulf Stream, and resulting in decreases in temperature, particularly over much of Europe.

Ecosystem stoichiometry

Ecosystem stoichiometry is based on principals of (1) the conservation of matter, (2) the stoichiometry of chemical reactions, and (3) the observation that plants, animals and even ecosystems are constructed of multiple elements in relatively fixed forms (Sterner and Elser 2002). Ratios between elements are therefore also relatively fixed, which puts constraints on element distribution and cycling, and implies that a change or disruption in the ecosystem- or global-scale cycle of one element, such as C, N, or P, will necessarily impact the cycling of other elements. For example a CO_2 enrichment-induced increase in photosynthesis and NPP will require an increase in belowground nitrogen acquisition in order to maintain leaf C:N ratios within a relatively fixed range, and will thereby impose a change in the nitrogen cycle. Or, as pointed out by Nadelhoffer et al. (1999) and Hungate et al. (2003), the amount of C that can be sequestered by an ecosystem with increasing N deposition will depend largely on whether the N is immobilized in bacteria (C:N ratios typically between 5 and 15) or soil organic matter (C:N ratios typically between 10 and 50), or whether the added N is taken up and stored in foliage (C:N ratios typically 30–100) or wood (C:N ratios typically >300). Results from the decadal-scale N fertilization experiment at the Bear Brook Watershed in Maine and the decadal-scale soil warming experiment at the Harvard Forest in Massachusetts both show that most of the added or warming-induced mineralized N is stored in soil organic matter with relatively low C/N ratios, thus limiting the potential for these systems to sequester large amounts of additional carbon (Nadelhoffer et al. 1999; Melillo et al. 2002).

Turnover rates and times

The concepts of turnover rates and turnover times are fundamental to understanding and modeling ecosystem response to global change. Assuming a steady state, turnover 'rate' is defined as the net mass of a material entering or leaving a system or reservoir in a given time period (i.e., flux) divided by the total mass of the material present in that system or reservoir (i.e., pool; units are percent/time period). Turnover 'time' is the inverse, or the total mass of a material in a system or reservoir (i.e., pool) divided by the net mass of the material going into or out of that system or reservoir over a given time (i.e., flux; units are time). Turnover times can also be interpreted as the mean life span of a system or component of a system (e.g. mean tree or root lifespans) or the mean residence time of material in a system or component of a system (e.g., mean residence times for greenhouse gases in the atmosphere and for the amount of carbon in a particular soil carbon pool).

Within the global change literature, concepts of turnover rates and times have been most frequently applied to greenhouse gas concentrations, above- and below-ground biomass pools, and carbon and nutrient cycles, and questions have arisen as to whether global change will alter the *fluxes* of material into or out of atmospheric, biomass or nutrient pools, or the *pool sizes* themselves. For example, over the long term, the amount of carbon that can be sequestered by an ecosystem will depend on both the size of the carbon pool in that system and its turnover time. More carbon can be stored in an ecosystem only if either the same amount of carbon is retained for a longer time (longer turnover times) or more carbon is added to the total pool than is lost from the pool (larger pool size). Elevated CO_2 and warming will generally increase photosynthesis and will thus increase the flux of carbon going into an ecosystem. However, if this carbon is stored in labile carbon pools with fast turnover times, and if elevated CO_2 and temperature directly or indirectly increase the turnover time of these labile carbon pools, then little or no carbon will be sequestered. Mitigation efforts to reduce the rise in atmospheric CO_2 must therefore be focused not just on stabilizing or increasing terrestrial or oceanic carbon pool sizes but also either decreasing or slowing turnover rates of existing pools (for example, by increasing the chemical and physical protection of soil carbon through better soil management practices) or transferring carbon from pools with short turnover times to pools with longer turnover times (for example, converting pasture land to forest and forest to wood products).

Community composition, biodiversity and ecosystem function

It is widely accepted that species composition and community dynamics will be strongly affected by the combined effects of elevated CO_2, warming, and changes in precipitation, and that these community changes will, in turn, have significant feedbacks on ecosystem function. However, despite this consensus, the underlying mechanisms driving plant community responses to global change are not well understood, and it has been difficult to accurately predict both community response to global change and the ecosystem consequences of these responses. This is due, in part, to the variable influence in time and space of global change on individual species, functional groups, and/or entire communities.

The responses of plant communities to simulated global change can be strongly influenced by individual plant species. A few CO_2 enrichment experiments have shown that even a single species can dominate responses of an entire plant community. For instance, Grünzweig and Körner (2001) reported significant ecosystem-scale changes in aboveground biomass, reproduction, and plant nitrogen content in response to CO_2 enrichment in semi-arid grassland assemblages from Israel. Surprisingly, these ecosystem-scale responses were attributable to CO_2-induced changes in just one out of 32 plant species. Morgan et al. (2004a) also reported that CO_2-induced increases in aboveground biomass in native Colorado shortgrass steppe were driven primarily by one of 36 plant species, and that enhanced seedling recruitment appeared to be an important mechanism behind this response. How would these plant communities have responded without the CO_2-responsive species, and what would have been the long-term implications for the ecosystems? These questions are difficult to answer, since species interact complexly in plant communities where microclimatic feedbacks and competition for resources occur. Absence of the CO_2-sensitive plant species would not necessarily result in a non-responsive plant community since more resources would be available to the remaining plants, and the reaction of individual species to CO_2 often interact with resource availability (Smith et al. 2000; Poorter and Perez-Soba 2001; Belote et al. 2003; Zavaleta et al. 2003). While a single species may drive a plant community response, plant community production and related responses to CO_2 are generally enhanced by plant species diversity (Niklaus et al. 2001; Reich et al. 2001b). Species-rich plant communities are thus more likely to exhibit strong reactions to global changes. Greater responsiveness of species-rich over species-poor communities can involve one of several forms of synergy whereby the presence of one species enhances the capability of another species to respond to CO_2 (Morse and Bazzaz 1994; Lüscher et al. 1996; Reich et al. 2001b), or may simply be attributed to the greater likelihood of having global change-sensitive species in a community with more species.

Less work has been done on the role of belowground biological diversity in global change experiments (Pendal et al. 2004). Linkages of aboveground and belowground biota indicate that global change may indirectly affect a number of belowground biological activities that will have powerful potential to feedback on plant communities, invoking both positive and negative responses (Wardle et al. 2004). Belowground biotic diversity will likely be important in determining the long-term reactions of plant communities to global change which are expected to be strongly conditioned by soil nutrient cycling (Zak et al. 2000).

Functional groups may also show differential responses to global change, and may be useful in streamlining approaches to understanding plant community responses to global change. However, contradictory results from field studies show that more work is needed to elucidate these differences (Morgan et al., 2004b; Nowak et al., 2004). For example, it has generally been predicted that C_3 species will show greater photosynthetic response to CO_2 enrichment compared to C_4 species (Strain and Bazaaz 1983). In a higher CO_2 world, an increase in the ecosystem abundance of C_3 relative to C_4 species over time would thus be accompanied by increased ecosystem productivity (Arp et al. 1993). Although numerous studies have demonstrated the greater photosynthetic response to CO_2 enrichment in C_3 compared to C_4 species (e.g., Bazaaz 1990; Bowes 1993; Ehrlinger and Monson 1993; Poorter 1993; Reich 2001a), other studies have shown few differences between species with these very different photosynthetic pathways, particularly under conditions of water or nutrient stress (e.g., Wand et al. 1999; Derner et al. 2003). Failure of the C_3 vs. C_4 functional group paradigm to manifest may be attributed, in part, to the fact that stomates of most herbaceous species close under elevated CO_2, which induces a water relations benefit that minimizes differences among photosynthetic functional groups. This is especially important in dry environments where CO_2-induced water relations responses often drive CO_2 responses (Morgan et al. 2004b). Legumes are another functional group that has been predicted to respond strongly to elevated CO_2, because of their capability to fix atmospheric N. While this has been confirmed in several studies (Hebeisen et al. 1997; Tissue et al. 1997; Lüscher et al. 1998; Grünzweig and Körner 2001), other experiments show little or advantage of N-fixing capability under elevated CO_2 (Niklaus et al. 1998; Stöcklin and Körner 1998; Nowak et al. 2004). In some cases, lack of a legume CO_2 response may be attributable to insufficient soil P levels such that N fixation capacity is impaired (Körner 2000), or to super-optimal N levels (Poorter et al. 1996). However, in many cases, failure of legumes, C_3 plants and other functional groups to respond simply indicates that one response mechanism may be insufficient to account for a species response in a plant community and other factors may need to be considered (e.g. water relations, nutrition, plant morphology, phenology). The temporal and spatial variability of the environment, which can interact with species and plant communities in complex ways, may also be important to determine species responses.

Different plant communities are also expected to show different responses to global change. Updegraff et al. (2001), for example, reported greater seasonal CH_4 emissions, aboveground net primary productivity, and dissolved N retention in bog compared to fen mesocosms under conditions of warming and water table manipulation. All of these examples underscore the linkage between species composition and ecosystem function, and illustrate that temporal patterns of ecosystem response to global change will be determined, in part, by the changing assemblages of species within that ecosystem.

For pragmatic reason, much of the experimental work on the effects of global change on species diversity has been done on species with short life spans such as annuals and short-lived perennials (Wand 1999; Reich et al. 2001a,b; Morgan et al. 2004a,b). Exceptions include the work on tree species response to (a) elevated CO_2 in a coastal scrub-oak community in Florida, USA (Hymus et al. 2002a,b), (b) elevated CO_2 and warming for two species of maple in Tennessee, USA (Norby et al. 1997, 2000), and (c) elevated CO_2 and ozone in a northern forest ecosystem in Wisconsin, USA (Karnosky et al. 2003). The time scale of response will vary directly with the life span of the biota, with changes occurring on the scale of days to months for soil microbes, to years for annual plants, to decades and even centuries for longer-lived perennials and woody species. Although experimental manipulations will continue to be useful to evaluate the effects of changes in species composition on ecosystem function for short-lived species, alternative approaches, using space-for-time substitutions such as gradients and chronosequences, along with ecosystem- and regional-scale models may be necessary to elucidate species change and ecosystem consequences over time for species with longer life spans (e.g. decade or greater).

Observations, experiments, and models: towards an integrated approach

The foregoing discussion underscores some of the complexities involved in understanding short and longer-term responses of ecosystems to global change. A challenge remains to better integrate these concepts into current and future global change research, particularly those efforts aimed at understanding and predicting longer-term responses. As discussed by Rastetter (1996), these efforts include (1) reconstructions of past events, (2) observations across existing elevational and/or latitudinal gradients (space-for-time substitutions), (3) long-term monitoring, (4) experimental manipulations, and (5) modeling. All five approaches have their strengths and weaknesses.

Reconstruction of past events

Over the last 20,000 years, regional and global environments have experienced dramatic shifts in their physical and chemical environments. Examples include the approximate doubling of atmospheric CO_2 concentrations from a low of ~160–200 μmol/mol in the Last Glacial Maximum (about 18,000 years ago) to the current high of ~360 μmol/mo, and the relatively regular oscillations in temperature and moisture between cool/dry and warm/moist periods on an ~1500 yr cycle (IPCC 2001). More subtle changes, including a 0.6 °C increase in mean global temperature and alterations in the timing and magnitude of precipitation, have also been documented during the more recent past (i.e., past 100–150 yrs) (IPCC 2001). Reconstructing temporal response patterns of terrestrial ecosystems to these past changes have provided valuable insights on predicting temporal patterns of response to future changes in these same factors. For example, in the northeastern United States, the mean annual temperature has increased by 1.0 °C and the mean length of the growing season has increased by an average of 8 days since 1899 (Wake and Markham 2005). Recent evidence documents a decrease of 4–8 days in Julian date of bloom for three horticultural woody perennial species in New England during this same time period (Wolfe et al. 2005). Similarly, in the southeastern United States, a dramatic increase in winter precipitation was documented for the period 1977–1992. This increase was accompanied by a 3-fold increase in the density of woody shrubs, and the gain and loss of several species of small mammals (Brown et al. 1997). However, because (a) the rate and magnitude of climatic change predicted for the 21st century is likely to be greater and more rapid than any experienced during the last 20,000 yrs (IPCC 2001; NERA 2001) and (b) atmospheric CO_2 concentrations are likely to be higher in the 21st century than at any time in the past, caution must be exercised when making extrapolations from past ecosystem responses to relatively slow changes in climate in a low CO_2 world to future ecosystem responses to possibly more rapid climate change in a higher CO_2 world.

Space-for-time substitutions

Observations across latitudinal or elevational gradients also allow for the evaluation of ecosystem response to gradual changes in climate, and have added to our understanding of ecosystem response to changes in temperature (Ineson et al. 1998; Saleska et al. 2002a,b), moisture (Davidson et al. 1998), and other global change factors such as atmospheric N deposition (McNulty et al. 1990, 1991; Aber et al. 2003). However, three drawbacks exist for space-for-time substitutions. First, even with the greatest attention to detail in site selection, it is impossible to hold all ecosystem properties constant, and thus differences between ecosystems at different positions along the gradient may not reflect the same changes that might occur in a single ecosystem over time. Second, the characteristics of ecosystems at different positions along the gradient have typically evolved over the millennia time scale providing sufficient time for different short and longer-term processes to operate and for the ecosystems to come into equilibrium with their local climate. The characteristics of ecosystems responding to rapid changes in temperature and precipitation, such as those predicted to occur over the next 50–200 years, may be different. Third, spatial gradients can be identified for temperature, precipitation, and combinations of temperature and precipitation. However, with the exception of a few studies on CO_2 emission from hot springs, no comparable gradients exist for atmospheric CO_2 and thus it is not possible to use space-for-time substitutions for either CO_2 effects alone or the effects of simultaneous changes in CO_2, temperature, and precipitation. Despite these cautions, space-for-time substitutions remain a valuable approach to studying ecosystem response to changing conditions.

Long-term monitoring

Long-term monitoring, or the methodical collection of environmental data at single or multiple sites over time, has provided a wealth of data and invaluable insights on changes in vectors such as atmospheric CO_2 (Keeling and Whorf 2004), climate (IPCC 2001), and atmospheric sulfur and nitrogen deposition (Likens et al. 1972). These measurements also provide insights on ecosystem responses to daily, seasonal, annual, and decadal climatic variability (e.g., NERA 2001; Fitzhugh et al. 2003; Park et al. 2003). Results from long-term monitoring studies will also, eventually, provide the ultimate validation for ecosystem and global scale models, as the results of humankind's global CO_2 enrichment 'experiment' unfold. The drawbacks to monitoring are that historic records rarely go back more then 100 years, and future responses are as yet unknown, making later validations of models of limited use to policy makers and land managers now.

Experimental manipulations

Experimental manipulations of whole ecosystems or ecosystem components are powerful tools that allow for the elucidation of cause-and-effect relationships and provide for a mechanistic understanding of short-term (typically < 20 years) responses of ecosystems to single or multiple vectors of global change. These experiments further provide a much needed means to validate (or not) current ecosystem models of global change, both highlighting processes that are well understood and those that need further study. Experimental manipulations also provide the opportunity for 'surprises' that might not be anticipated based on the current understanding of ecosystem dynamics. These anomalies can point to areas where more work is needed and can lead to new directions and discoveries. Experimental manipulations have several drawbacks. First, experimental manipulations typically involve a step increase in state factors such as CO_2 or temperature. Global change will involve gradual changes in these factors over time, and the response to a step change may be different than the response to a gradual change. Second, because of financial, logistical, and intellectual constraints, few manipulations vary more then two or three factors in any one ecosystem-scale experiment at any one time. An exception is the Jasper Ridge Experiment where four factors (CO_2, temperature, water, nutrients) were varied in a full factorial design (Shaw et al. 2002). Although single factor experiments provide critical information on the response to single vectors of change, and two or three-factor experiments provide some insight on the nature of interactions, it is recognized that these single or few factor experiments can not directly inform us on ecosystem response to simultaneous changes in multiple factors, including atmospheric CO_2, temperature, moisture, N deposition, UVB radiation, ozone, and a host of other factors, some of which may not have been identified yet. Third, even decadal-scale experiments still only generate short-term data. Concern exists that short term data are only useful for testing short-term mechanisms, and that long-term mechanisms will likely dominate the longer-term response. For example, understanding the short-term response of soil respiration to increasing temperature sheds little light on longer-term effects of elevated temperature on the turnover of soil organic matter, which will ultimately be controlled by longer term changes in plant productivity and the quantity and quality of litter inputs, or stochastic events such as fire.

Models

Models are essential tools for conceptually and empirically integrating existing knowledge and for making longer-term predictions of ecosystem response to multiple interacting vectors of global change at multiple spatial scales. Models can also be used to generate testable hypotheses, and because they integrate the current understanding of ecosystem processes, their failure highlights gaps or errors in that knowledge. For example, due to a strong mechanistic understanding of leaf photosynthesis and canopy radiation interception, most models work well at predicting ecosystem carbon uptake. Conversely, due to a lack of a theoretical foundation, they typically do not work well at predicting carbon loss through respiration or plant carbon allocation to leaves, stems, and roots, pointing to the need for more empirical work in these areas (Classen and Langeley 2005). Further, the current generation of models needs to better incorporate (a) spatial heterogeneity within the existing structures, (b) the ecological ramifications of extreme events, and (c) the temporal scaling issues discussed in the preceding section. Finally, as pointed out by Rastetter (1996), a fundamental drawback of using models to make longer-term predictions, is that it is not possible in the short term to validate models of longer-term effects. Despite these drawbacks, the field of ecological modeling has seen major advances during the past several decades and these models remain critical tools for continuing to integrate our understanding of ecosystem response to global change and making projections of how ecosystems will continue to evolve under projected future global change scenarios.

Conclusions

Considerable progress has been made during the past several decades to better understand and model short and longer-term responses of ecosystems to global change. A growing consensus, however, exists, that in order to more rapidly advance this field of inquiry, it will be necessary to better integrate observational, experimental and modeling techniques into a *unified multidisciplinary approach* to evaluate ecosystem response to global change (Norby and Luo 2004; Classen and Langeley 2005). For example, combining experimental studies with gradient studies or superimposing experiments across gradients would provide powerful tools to bracket the decadal to century-scale response (which is of most interest to policy-makers) between the short-term experimental response and the longer-term response that has developed over the millennia across a landscape. Improved communication between experimentalists and modelers and closer data-model integration will also help move the global change research agenda forward more rapidly. A better match, for example, between empirical data from observation, gradient or experimental studies and model requirements could be achieved if empirical scientists and modelers interacted more closely during the design stage of experiments or studies. The empiricists could thus better understand what types of data are needed and at what temporal and spatial scales for models, and modelers could better understand what types of data are available to be used in model construction. Models also can be used more advantageously to (a) help generate testable hypotheses for observational, gradient and experimental studies, (b) scale-up empirical results in time and space, and (c) extrapolate results from single- or few-factor experiments to a better understanding of ecosystem response to multiple interacting vectors of global change. Communication efforts should also focus on identifying processes that are poorly represented in models, such as respiration and carbon allocation, and designing empirical studies to help develop a better mechanistic understanding of these processes that can then be incorporated into models. Finally, if we are truly committed to understanding longer-term responses of terrestrial ecosystems to global change, it is imperative to increase the number of decadal and longer term experiments and to provide more stable funding for long-term monitoring.

All of these considerations, along with more frequent and extensive data-model comparisons and model-model comparisons, will require increased communication and information exchange amongst scientists from the theoretical, experimental, and modeling communities. Research coordination networks, such as the NSF funded TERACC (Terrestrial Ecosystem Response to Atmospheric and Climatic Change) network of global change scientists, are an effective mechanism to bring multidisciplinary communities of scientists together. It is this larger community of scientists that will ultimately move our understanding of, and ability to effectively model, transient and steady state responses of ecosystems to CO_2 enrichment and global climate change.

Acknowledgements

This paper summarizes themes leading to and expressed at the workshop on 'From Transient to Steady State Response of Ecosystems to

Atmospheric CO_2 Enrichment and Global Climate Change, Durham, New Hampshire, April 28 to May 1, 2002. Special recognition goes to Gus Shaver, Ed Rastetter, and Jack Morgan for contributions and critical review of the manuscript. This workshop was sponsored by the Terrestrial Ecosystem Response to Atmospheric and Climatic Change (TERACC), a research coordination network supported by the National Science Foundation. This paper was prepared with support from the USDA Forest Service, and is based upon work supported by the National Science Foundation under Grant No. 0090238.

References

Aber J.D. and Driscoll C.T. 1997. Effects of land use, climate variation and N deposition on N cycling and C storage in northern hardwood forests. Global Biogeochem. Cycles 11: 639–648.

Aber J.D., Goodale C.L., Ollinger S.V., Smith M.-L., Magill A.H., Martin M.E. and Stoddard J.L. 2003. Is nitrogen deposition altering the nitrogen status of northeastern forests? BioScience 53: 375–390.

Adam N.R., Wall G.W., Kimball B., Pinter P. Jr., LaMorte R.L., Hunsaker D., Adamsen F.J., Thompson T.L., Matthias A.D., Leavitt S.W. and Webber A. 2001. Acclimation response of the photosynthetic apparatus in a wheat ecosystem under free-air CO_2 enrichment (FACE) and variable soil nitrogen regimes: leaf position and phenology determine acclimation response. Photosynth. Res. 66: 79–95.

Ainsworth E.A. and Long S.P. 2005. What have we learned from 15 years of free-air CO_2 enrichment (FACE)? A meta-review of the responses of photosynthesis, canopy properties and plant production to rising CO_2. New Phytol. 165: 351–372.

Ainsworth E.A., Davey P.A., Hymus G.J., Osborne C.P., Rogers A., Blum H., Nosberger J. and Long S.P. 2003. Is stimulation of leaf photosynthesis by elevated carbon dioxide concentration maintained in the long term? A test with *Lolium perenne* grown for 10 years at two nitrogen fertilization levels under free air CO_2 enrichment (FACE). Plant, Cell Environ. 26: 705.

Arnold S.S., Fernandez I.J., Rustad L.E. and Zibilske L.M. 1999. Microbial response of an acid forest soil to experimental soil warming. Biol. Fertil. Soils 30: 239–244.

Arp W.J., Drake B.G., Pockman W.T., Curtis P.S. and Whigham D.F. 1993. Interactions between C_3 and C_4 salt marsh plant species during four years of exposure to elevated atmospheric CO_2. Vegetatio 104(105): 133–143.

Atkin O.K. and Lambers H. 1998. Slow-growing alpine and fast-growing lowland species: a case study of factors associated with variation in growth rate among herbaceous higher plants under natural and controlled conditions. In: Lambers H., Poorter H. and VanVuuren M.M.I. (eds), Inherent Variation in Plant Growth: Physiological Mechanisms and Ecological Consequences, Backhuys Publishers, Leiden, pp. 259–288.

Atkin O.K., Holly C. and Ball M.C. 2000a. Acclimation of snow gum (*Eucalyptus pauciflora*) leaf respiration to seasonal and diurnal variations in temperature: the importance of changes in the capacity and temperature sensitivity of respiration. Plant Cell Environ. 23: 15–26.

Atkin O.K., Edwards E.J. and Loveye B.R. 2000b. Response of root respiration to changes in temperature and its relevanc to global warming. New Phytol. 147: 141–154.

Badeck F.W., Bondeau K., Bottcher D., Doktor D., Lucht W., Schaber J. and Sitch S. 2004. Responses of spring phenology to climate change. New Phytol. 162: 295–310.

Barry J. and Bjorkman O. 1980. Photosynthetic temperature response and adaptation to temperature in higher plants. Annu. Rev. Plant Physiol. 31: 491–543.

Bazzaz F.A. 1990. The response of natural ecosystems to the rising global CO_2 levels. Annu. Rev. Ecol. Systemat. 21: 167–196.

Belote R.T., Weltzin J.F. and Norby R.J. 2003. Response of an understory plant community to elevated [CO_2] depends on differential responses of dominant invasive species and is mediated by soil water availability. New Phytol. 161: 827–835.

Berbet M.L.C. and Costa M.H. 2002. Climate change after tropical deforestation: seasonal variability of surface albedo and its effects on precipitation change. J. Climate 16: 2099–2104.

Betts R.A. 2000. Offset of the potential carbon sink from boreal forestation by decreases in surface albedo. Nature 408: 187–190.

Betts R.A., Cox P., Lee S.E. and Woodward F.I. 1997. Contrasting physiological and structural vegetation feedbacks in climate change simulations. Nature 387: 796–799.

Bolstad P.V., Reich P. and Lee T. 2003. Rapid temperature acclimation of leaf respiration rates in *Quercus alba* and *Quercus rubra*. Tree Physiol. 23: 969–976.

Bowes G. 1993. Facing the inevitable: plants and increasing atmospheric CO_2. Annu. Rev. Plant Physiol. Plant Mol. Biol. 44: 309–332.

Brown J.H., Valone T.J. and Curtin C.G. 1997. Reorganization of an arid ecosystem in response to recent climate change. Proceed. Natl. Acad. Sci. 94: 9729–9733.

Campbell R.K. and Sorensen F.C. 1973. Cold-acclimation in seedling Douglas-Fir related to phenology and provenance. Ecology 54: 1148–1151.

Chapin F.S. III and Starfield A.M. 1997. Time lags and novel ecosystems in response to transient climate change in arctic Alaska. Climatic Change 35: 449–461.

Chapin F.S. III, Zavaleta E.S., Eviners V.T., Naylor R.L., Vitousek P.M., Reynolds H., Hooper D.U., Lavorel S., Sala O.E., Hobbie S.E., Mack M. and Diaz S. 2000. Consequences of changing biodiversity. Nature 405: 234–242.

Classen A.T. and Langley J.A. 2005. Data-model integration is not magic: Modeling ecosystem responses to global change: techniques and recent advances. New Phytol. 166: 367–370.

Collins H.P., Paul E.A., Paustian K. and Elliott E.T. 1997. Characterization of soil organic matter relative to its stability and turnover. In: Paul E.A., Paustian Elliott E.T. and Cole C.V. (eds), Soil Organic Matter in Temperate

Agroecosystems: Long-term Experiments in North America, CRC Press, Boca Raton, FL, pp. 51–72.

Curtis P.S. and Wang X. 1998. A Meta-analysis of elevated CO_2 on woody plant mass, form, and physiology. Oecologia 113: 299–313.

Davidson E.A., Belk E. and Boone R. 1998. Soil water content and temperature as independent or cofounded factors controlling soil respiration in a temperate mixed hardwood forest. Global Change Biol. 4: 217–227.

Derner J.D., Johnson H.B., Kimball B.A, Pinter P.J. Jr., Polley H.W., Tischler C.T., Boutton T.W., Lamorte R.L., Wall G.W., Adam N.R., Leavitt S.W., Ottman M.J., Matthias A.D. and Brooks T.J. 2003. Above- and below-ground responses of C_3-C_4 species mixtures to elevated CO_2 and soil water availability. Global Change Biol. 9(3): 452–460.

deValpine P. and Harte J. 2001. Plant responses to experimental warming in a montane meadow. Ecology 82: 637–648.

Drake B.G., Gonzalez-Meler M.A. and Long S.P. 1997. More efficient plants: a consequence of rising atmospheric CO_2? Annu. Rev. Plant Physiol. Plant Mol. 48: 609–639.

Egli P., Maurer S., Gunthardt-Goerg M.S. and Korner C. 1998. Effects of elevated CO_2 and soil quality on leaf gas exchange and aboveground growth in beech-spruce model ecosystems. New Phytol. 140: 185–196.

Ehleringer J.R. and Monson R.K. 1993. Evolutionary and ecological aspects of photosynthetic pathway variation. Annu. Rev. Ecol. Systemat. 24: 411–439.

Equiza M., Mirave J. and Tognett J. 2001. Morphological, anatomical, and physiological responses related to differential shoot vs root growth inhibition at low temperature in spring and winter wheat. Ann. Bot. 87: 67–76.

Farage P.K., McKee I.F. and Long S.P. 1998. Does a low nitrogen supply necessarily lead to acclimation of photosynthesis to elevated CO_2? Plant Physiol. 118: 573–580.

Ferrar P.J., Slatyer R.O. and Vranjic J.A. 1989. Photosynthetic temperature acclimation in *Eucalyptus* species from diverse habitat, and comparison with *Nerium oleander*. Aust. J. Plant Physiol. 16: 199–217.

Fitzhugh R.D., Likens G.E., Driscoll C.T., Mitchell M.J., Groffman P.M., Fahey T.J. and Hardy J.P. 2003. The role of soil freezing events in interannual patterns of stream chemistry at the Hubbard Brook Experimental Forest. Environ. Sci. Technol. 37: 1575–1580.

Gesch R.W., Vu J.C.V., Boote K.J., Allen L.H. Jr. and Bowes G. 2002. Sucrose-phosphate synthase activity in mature rice leaves following changes in growth CO_2 is unrelated to sucrose pool size. New Phytol. 154: 77–84.

Giardina C.P. and Ryan M.G. 2000. Evidence that decomposition rates of organic carbon in mineral soil do not very with temperature. Nature 404: 858–861.

Gregory J.M., Huybrechts P. and Raper S.C.B. 2004. Climatology: threatened loss of the Greenland ice-sheet. Nature 428: 616.

Griffin K.L., Turnball M. and Murphy R., , 2002. Leaf respiration is differentially affected by leaf vs stand night-time warming. Global Change Biol. 8: 479–485.

Grünzweig J.M. and Körner C. 2001. Growth, water and nitrogen relations in grassland model ecosystems of the semi-arid Negev of Israel exposed to elevated CO_2. Oecologia 128: 251–262.

Gu L., Post W.M., King A.W. 2004. Fast labile carbon turnover obscures sensitivity of heterotrophic respiration from soil to temperature: a model analysis. Global Biogeochem. Cycles 18, GB1022, doi:10.1029/2003GB002119.

Gunderson C.A. and Wullschleger S.D. 1994. Photosynthetic acclimation in trees to rising atmospheric CO_2: a broader perspective. Photosynth. Res. 39: 369–388.

Gunderson C.A., Norby R.J. and Wullschleger S.D. 2000. Acclimation of photosynthesis and respiration to simulated climatic warming in northern and southern populations of *Acer saccharum*: laboratory and field evidence. Tree Physiol. 20: 87–96.

Hebeisen T., Löscher Zanetti S., Fischer B.U., Hartwig U.A., Frehner M., Hendrey G.R., Blum G. and Nrger J. 1997. Growth response of *Trifolium repens* L. and *Lolium perenne* L. as monocultures and bi-species mixture to free air CO_2 enrichment and management. Global Change Biol. 3: 149–160.

Horacio H. 2003. Root/shoot allocation and root architecture in seedlings: variation among forest sites, microhabitats, and ecological. Biotropica 35: 318–332.

Hungate B.A., Dukes J.S., Shaw M.R., Luo Y. and Field C.B. 2003. Nitrogen and climate change. Science 302: 1512–1513.

Hymus G.J., Pontailler J.Y., Li J., Stiling P., Hinkle C.R., Drake B.G. 2002a. Seasonal variability in the effect of elevated CO_2 on ecosystem leaf area index in a scrub-oak ecosystem. Global Change Biol. 8: 931–940.

Hymus G.J., Snead T.G., Johnson D.P., Hungate B.A. and Drake B.G. 2002. Acclimation of photosynthesis and respiration to elevated atmospheric CO_2 in two scrub oaks. Global Change Biol. 8: 318.

Ineson P, Taylor K., Harrison A.F., Poskitt J., Benham D.G., Tipping E. and Woof C. 1998. Effects of climate change on nitrogen dynamics in upland soils I A transplant approach. Global Change Biol. 4: 143–152.

Ingram W.J., Wilson C.A. and Mitchell J.F.B. 1989. Modeling climate change: an assessment of sea ice and surface albedo feedbacks. J. Geophys. Res. 94: 8609–8622.

Intergovernmental Panel on Climate Change 2001. In: Houghton J.T., Ding Y., Griggs D.J., Noguer M., van der Linden P.J. and Xiaosu D. (eds.), Climate Change 2001: The Scientific Basis. University Press, Cambridge, United Kingdom.

Ives R.A. and Cardinale B.J. 2004. Food-web interactions govern the resistance of communities after non-random extinctions. Nature 429: 174–177.

Jefts S., Fernandez I.J., Rustad L.E. and Dail D.B. 2004. Decadal responses in soil N dynamics at the Bear Brook Watershed in Maine, USA. Forest Ecol. Manage. 189: 189–205.

Karnosky D.F., Zak D.R., Pregitzer K.S., Awmack C.S., Bockheim J.G., Dickson R.E., Hendrey G.R., Host G.E., King J.S., Kopper B.J., Kruger E.L., Kubiske M.E., Lindroth R.L., Mattson W.J., McDonald E.P., Noormets A., Oksanen E., Parsons W.F.J., Percy K.E., Podila G.K., Riemenschneider D.E., Sharma P., Thakur R.C., Sober A., Sober J., Jones W.S., Anttonen S., Vapaavuori E., Mankovska B., Heilman W.E. and Isebrands J.G. 2003. Tropospheric O_3 moderates responses of temperate hardwood forests to elevated CO_2: A synthesis of molecular to ecosystem results from the Aspen FACE project. Funct. Ecol. 17: 289–304.

Keeling C.D., Whorf T.P. 2004. Atmospheric CO_2 records from sites in the SIO air sampling network. In: Trends: A Compendium of Data on Global Change. Carbon Dioxide Information Analysis Center, Oak Ridge National Laboratory, US Department of Energy, Oak Ridge, Tennessee, USA.

Kirscbaum M.U.F. and Farquhar G.D. 1984. Temperature dependence of whole-leaf photosynthesis in *Eucalyptus pauciflora* Sieb ex Spreng. Plant Physiol. 11: 519–538.

Klickoff L.G. 1966. Temperature dependence of the oxidative rates of mitochondria in *Danthonia intermedia, Penstemmon davidsonii* and *Sitanion hystrix*. Nature 212: 529–530.

Körner Ch. 2000. Biosphere responses to CO_2 enrichment. Ecol. Appl. 10: 1590–1619.

Kozlowski T.T. and Pallardy S.G. 2002. Acclimation and adaptive responses of woody plants to environmental stresses. Bot. Rev. 68: 270–334.

Kristensen H.L., McCarty G.W. and Meisinger J.J. 2000. Effects of soil structure disturbance on mineralization of organic soil nitrogen soil. Sci. Soc. Am. J. 64: 371–378.

Kubiske M.E., Zak D.R., Pregitzer K.S. and Takeuchi Y. 2002. Photosynthetic acclimation of overstory *Populus tremuloides* and understory *Acer saccharum* to elevated atmospheric CO_2 concentration: interactions with shade and soil nitrogen. Tree Physiol. 22: 321–329.

Lambers H., Atkin O.K. and Scheurwater I. 1996. Respiratory patterns in roots in relation to their functioning. In: Waisel Y., Eschel A. and Kafkaki U. (eds), Plant Roots: The Hidden Half, Marcel Dekker, New York, pp. 323–362.

Lamontagne S. 1998. Nitrogen mineralization in upland Precambrian Shield catchments: contrasting the role of lichen-covered bedrock and forested areas. Biogeochemistry 441: 53–69.

Likens G.E., Bormann F.H. and Johnson N.M. 1972. Acid rain. Environment 14: 33–40.

Liski J., Ilvesniemi H., Makela A. and Westman C.J. 1999. CO_2 emissions from soil in response to climatic warming are overestimated - the decomposition of old soil organic matter is tolerant to temperature. Ambio 28: 171–174.

Lukewille A. and Wright R. 1997. Experimentally increased soil temperature causes release of nitrogen at a boreal forest catchment in southern Norway. Global Change Biol. 3: 13–21.

Luo Y., Field C.B. and Mooney H.A. 1994. Predicting responses of photosynthesis and root fraction to elevated CO_2: interaction among carbon, nitrogen and growth. Plant, Cell Environ. 17: 1195–1204.

Luo Y., Wan S., Hui D. and Wallace L.L. 2001. Acclimation of soil respiration to warming in a tall grass prairie. Nature 413: 622–625.

Lüscher A., Hebeisen T., Zanetti S., Hartwig U.A., Blum H., Hendrey G.R., Nösberger J. 1996. Differences between legumes and non-legumes of permanent grassland in their responses to free air carbon dioxide enrichment. In: Körner C. and Bazzaz (eds), Carbon Dioxide, Populations and Communities, Academic Press, San Diego, pp. 287–299.

Lüscher A., Hendry G.R. and Nösberger J. 1998. Long-term responsiveness to free air CO_2 enrichment of functional types, species and genotypes of plants from fertile permanent grassland. Oecologia 113: 37–45.

Lynch A.H. and Wu W. 1999. Impacts of fire and warming on ecosystem uptake in the boreal forest. J. Climate 13: 2334–2338.

Maroco J.P., Edwards G.E. and Ku M.S.B. 1999. Photosynthetic acclimation of maize to growth under elevated levels of carbon dioxide. Planta 210: 115–125.

Matsuki S., Ogawa K., Tanaka A. and Hara T. 2003. Morphological and photosynthetic responses of *Quercus crispula* seedlings to highlight conditions. Tree Physiol. 23: 769–775.

McNulty S.G., Aber J.D., McLellan T.M. and Katt S.K. 1990. Nitrogen cycling in high elevation forests of the northeastern US in relation to nitrogen deposition. Ambio 19: 38–40.

McNulty S.G., Aber J.D. and Boone R.D. 1991. Spatial changes in forest floor and foliar chemistry in spruce-fir forests across New England. Biogeochemistry 14: 13–26.

Medlyn B.E., Badeck F.W., De Pury D.G.G., Barton C.V.M., Broadmeadow M., Ceulemans R., De Angelis P., Forstreuter M., Jach M.E., Kellomäki S., Laitat E., Marek M., Philippot S., Rey A., Strassemeyer J., Laitinen K., Liozon R., Portier B., Roberntz P., Wang K. and Jstbid P.G. 1999. Effects of elevated [CO_2] on photosynthesis in European forest species: a meta-analysis of model parameters. Plant, Cell Environ. 22: 1475–1495.

Medlyn B.E., Barton C.V.M., Broadmeadow M.S.J., Ceulemans R., De Angelis P., Forstreuter M., Freeman M., Jackson S.B., Kellomäki S., Laitat E., Rey A., Roberntz P., Sigurdsson B.D., Strassemeyer J., Wang K., Curtis P.S. and Jarvis P.G. 2001a. Stomatal conductance of forest species after long-term exposure to elevated CO_2 concentration: a synthesis. New Phytol. 149: 247–264.

Medlyn B.E., Barton C.V.M., Broadmeadow M.S.J., Ceulemans R., De Angelis P., Forstreuter M., Freeman M., Jackson S.B., Kellomäki S., Laitat E., Rey A., Roberntz P., Sigurdsson B.D., Strassemeyer J., Wang K., Curtis P.S. and Jarvis P.G. 2001b. Stomatal conductance of forest species after long-term exposure to elevated CO_2 concentration: a synthesis. New Phytol. 149: 247–264.

Melillo J.M., Steudler P.A., Aber J.D., Newkirk K., Lux H., Bowles F.P., Catricala C., Magill A., Ahrens T. and Morrisseau S. 2002. Soil warming and carbon-cycle feedbacks to the climate system. Science 298: 2173–2176.

Miroslavov E.A. and Kravkina I.M. 1991. Comparative analysis of chloroplasts and mitochondria in leaf chlorenchyma from mountain plants grown at different altitudes. Ann. Bot. 68: 195–200.

Morgan J.A., Mosier A.R., Milchunas D.G., LeCain D.R., Nelson J.A. and Parton W.J. 2004a. CO_2 enhances productivity, alters species composition, and reduces forage digestibility of shortgrass steppe vegetation. Ecol. Appl. 14: 208–219.

Morgan J.A., Pataki D.E., Körner C., Clark H., Del Grosso S.J., Grünzweig J.M., Knapp A.K., Mosier A.R., Newton P.C.D., Niklaus P.A., Nippert J.B., Nowak R.S., Parton W.J., Polley H.W. and Shaw M.R. 2004b. Water relations in grassland and desert ecosystems exposed to elevated atmospheric CO_2. Oecologia 140: 11–25.

Morse S.R. and Bazzaz F.A. 1994. Individuals vs stands: differential responses to elevated CO_2 and temperature. Ecology 75: 966–975.

Nadelhoffer K.J., Emmet B.A., Gundersen P., Kjonaaas O.J., Koopsman C.J., Schleppi P., Tietma A. and Wright R.F. 1999. Nitrogen deposition makes a minor contribution to carbon sequestration in temperate forests. Nature 398: 145–148.

Naeem S. and Li S. 1997. Biodiversity enhances ecosystem reliability. Nature 390: 507–509.

National Academy Committee on Abrupt Climate Change 2001. In: Abrupt Climate Change: Inevitable Surprises, Alley R.B., Chair, National Academy Press, Washington, DC.

New England Regional Assessment Group (NERA) 2001. Preparing for a chancing climate: the potential consequences of climate variability and change. New England Overview; U.S. Global Change Research Program, University of New Hampshire, 96 p.

Nielsen E.T. and Orcutt D.M. 1996. Physiology of Plants Under Stress: Abiotic Factors. John Wiley and Sons, New York, p. 689.

Niklaus P.A., Leadley P.W., Stöcklin J. and Körner Ch. 1998. Nutrient relations in calcareous grassland under elevated CO_2. Oecologia 116: 67–75.

Niklaus P.A., Leadley P.W., Schmid B. and Kr Ch. 2001. A long-term field study on biodiversity X elevated CO_2 interactions in grassland. Ecol. Monogr. 71: 341–356.

Norby R.J., Cotrufo M.F., Ineson P., O'Neill E.G. and Canadell J.G. 2001. Elevated CO_2, litter chemistry, and decomposition: a synthesis. Oecologia 127: 153–165.

Norby R.J., Edwards N.T., Riggs J.S., Abner C.H. and Wullschleger S.D. 1997. Temperature-controlled open-top chambers for global change research. Global Change Biol. 3: 259–267.

Norby R.J., Kimball B.A., Kobayashi K. (eds.) 2001b. Rising CO_2 - Future Ecosystems. New Phytol. Special Issue 150(2).

Norby R.J., Long T.M., Hartz-Rubin J.S. and O'Neill E.G. 2000. Nitrogen resorption in senescing tree leaves in a warmer, CO_2-enriched atmosphere. Plant Soil 224: 15–29.

Norby R.J. and Luo L. 2004. Evaluating ecosystem responses to rising CO_2 and warming in a multi-factor world. New Phytol. 162: 281–293.

Nowak R.S., Ellsworth D.S. and Smith S.D. 2004. Functional responses of plants to elevated atmospheric CO_2 - do photosynthetic and productivity data from FACE experiments support early predictions? New Phytol. 162: 253–280.

Park J., Mitchell M.J., McHale P.J., Christopher S.F. and Myers T.P. 2003. Interactive effects of changing climate and atmospheric deposition on N and S biogeochemistry in a forested watershed of the Adirondack Mountains, New York State. Global Change Biology 9: 1602–1619.

Paustian K., Collins H.P. and Paul E.A. 1997. Management controls on soil carbon. In: Paul E.A., Paustian K., Elliott E.T. and Cole C.V. (eds), Soil Organic Matter in Temperate Agroecosystems: Long-term Experiments in North America, CRC Press, Boca Raton, FL, pp. 15–49.

Paustian K., Elliott E.T., Six J. and Hunt H.W. 2000. Management options for reducing CO_2 emissions from agricultural soils. Biogeochemistry 48: 147–163.

Pendall E., Bridgham S., Hanson P.J., Hungate B., Kicklighter D.W., Johnson D.W., Law B.E., Luo Y., Megonigal J.P., Olsrud M., Ryan M.G. and Wan S. 2004. Belowground process response to elevated CO_2 and temperature: a discussion of observations, methods, and models. New Phytol. 162: 311–322.

Peterjohn W.T, Melillo J.M., Steudler P.A., Newkirk K.M., Bowles S.T. and Aber J.D. 1994. Responses of trace gas fluxes and N availability to experimentally elevated soil temperatures. Ecol. Appl. 4: 617–625.

Peterson A.G., Ball J.T., Luo Y., Field C.B., Reich P.B., Curtis P.S., Griffin K.L., Gunderson C.A., Norby R.J., Tissue D.T., Forstreuter M., Rey A.A., Vogel C.S., Cmeal participants 1999. The photosynthesis - leaf nitrogen relationship at ambient and elevated atmospheric carbon dioxide: a meta-analysis. Global Change Biol. 5: 331–346.

Poorter H. 1993. Interspecific variation in the growth response of plants to an elevated ambient CO_2 concentration. Vegetatio 104(105): 77–97.

Poorter H. and Perez-Soba M. 2001. The growth response of plants to elevated CO_2 under non-optimal environmental conditions. Oecologia 129: 1–20.

Poorter H., Roumet C. and Campbell B.D. 1996. Interspecific variation in the growth response of plants to elevated CO_2: a search for functional types. In: Kr C.H. and Bazzaz F.A. (eds), Carbon dioxide, Populations, and Communities, Academic Press, San Diego, pp. 375–412.

Potter C.S., Randerson J.T., Field C.B., Matson P.A., Vitousek P.M., Mooney H.A. and Klooster S.A. 1993. Terrestrial ecosystem production: a process model based on global satellite and surface data. Global Biogeochem. Cycles 7(4): 811–841.

Rahmstorf S. 2002. Ocean circulation and climate during the past 120,000 years. Nature 419: 207–214.

Raich J.W. and Nadelhoffer K.J. 1989. Belowground carbon allocation in forest ecosystems: global trends. Ecology 70: 1346–1354.

Raich J.W. and Potter C.S. 1995. Global patterns of carbon dioxide emissions form soils. Global Biogeochem. Cycles 9: 23–36.

Raich J.W. and Schlesinger W.H. 1992. The global carbon dioxide flux in soil respiration and its relationship to vegetation and climate. Tellus 44: 81–89.

Rastetter E.B. 1996. Validating models of ecosystem response to global change. BioScience 46(3): 190–198.

Read J. 1990. Some effects of acclimation temperature on net photosynthesis in some tropical and extra-tropical Australasian *Nothofagus* species. J. Ecol. 78: 100–112.

Reich P.B., Tilman D., Craine J., Ellsworth D., Tjoelker M.G., Knops J., Wedin D., Naeem S., Bahauddin D., Goth J., Bengtson W. and Lee T.D. 2001a. Do species and functional groups differ in acquisition and use of C, N and water under varying atmospheric CO_2 and N availability regimes? A field test with 16 grassland species. New Phytol. 150: 435–448.

Reich P.B., Knops J., Tilman D., Craine J., Ellsworth D., Tjoelker M., Lee T., Naeem S., Wedin D., Bahauddin D., Hendrey G., Jose S., Wrage K., Goth J. and Bengston W. 2001b. Plant diversity enhances ecosystem responses to elevated CO_2 and nitrogen deposition. Nature 410: 809–812.

Rey A. and Jarvis P.G. 1998. Long-term photosynthetic acclimation to increased atmospheric CO_2 concentration in young birch (*Betula pendula*) trees. Tree Physiol. 18: 441–450.

Ricklefs R.E. 1990. Ecology 3. W.H. Freeman and Company, New York, NY.

Rogers A. and Ellsworth D.S. 2002. Photosynthetic acclimation of *Pinus taeda* (Loblolly Pinte) to long-term growth in elevated pCO_2 (FACE). Plant, Cell, and Environment 25: 851–858.

Rogers A. and Humphries S.W. 2000. A mechanistic evalution of Photosynthetic acclimation at elevated CO_2. Global Change Biol. 6:105.

Running S.W. and Hunt E.R. Jr. 1993. Generalization of a forest ecosystem process model for other biomes, BIOME-BGC and an application for global-scale models. In: Ehleringer J.R. and Field C. (eds), Scaling Processes Between Leaf and Landscape Levels, Academic Press, San Diego, pp. 141–158.

Rustad L.E. and Fernandez I.J. 1998. Experimental soil warming effects on CO_2 and CH_4 flux from a low elevation spruce fir forest soil in Maine, USA. Global Change Biol. 4: 597–607.

Rustad L.E. and Norby R.J. 2002. Effects of increased temperature on terrestrial ecosystems. In: Mooney H.A. and Canadell J. (eds), Encyclopedia of Global Environmental Change, Vol. 2: Biological and Ecological Dimensions of Global Environmental Change, John Wiley & Sons Ltd, Chichester, pp. 575–581.

Rustad L.E., Campbell J., Marion G.M., Norby R.J., Mitchell M.J., Hartley A.E., Cornelissen J.H.C. and Gurevitch J.GCTE-NEWS 2001. A meta-analysis of the response of soil respiration, net N mineralization, and aboveground plant growth to experimental ecosystem warming. Oecologia 126: 543–562.

Saleska S., Shaw M.S., Fischer M., Dunne J., Holman M., Still C., Harte J. 2002a. Plant community composition mediates both large transient decline and predicted long-term recovery of soil carbon under climate warming. Global Biogeochem. Cycles 16(4).

Saleska S., Shaw M., Fischer M., Dunne J., Shaw M., Holman M., Still C., Harte J. 2002b. Carbon-cycle feedbacks to climate change in montane meadows: results from a warming experiment and a natural climate gradient. Global Biogeochem. Cycles 16, 2001gb001573.

Schlesinger W.H. 1997 Carbon balance in terrestrial detritus. Annu. Rev. Ecol. Syst. 8: 51–81.

Shaver G.R., Canadell J., Chapin F.S. III, Gurevitch J., Harte J., Henry G., Ineson P., Jonasson S., Melillo J., Pitelka L. and Rustad L. 2000. Global warming and terrestrial ecosystems: a conceptual framework for analysis. BioScience 50: 871–882.

Shaw M.R., Zavaleta E.S., Chiarello N.R., Cleland E.E., Mooney H.A. and Field C.B. 2002. Grassland responses to global environmental changes suppressed by elevated C. Science 298: 1987–1990.

Sims D.A., Cheng W., Luo Y. and Seeman J.R. 1999. Photosynthetic acclimation to elevated CO_2 in a sunflower canopy. J. Exp. Bot. 50: 645–653.

Six J., Elliott E.T. and Paustian K. 2000. Soil macroaggregate turnover and microaggregate formation: A mechanism for C sequestration under no-tillage agriculture. Soil Biol. Biochem. 32: 2099–2103.

Smith S.D., Husman T.E., Zitzer S.F., Charlet T.N., Houseman D.C., Coleman J.S., Fenstermaker L.K., Seemann J.R. and Nowak R.S. 2000. Elevated CO_2 increases productivity and invasive species in an arid ecosystem. Nature 408: 79–82.

Sterner R.W. and Elser J.J. 2002. Ecological Stoichiometry: The Biology of Elements form Molecules to the iosphere. Princeton University Press, Pinceton, NJ, p. 584.

Stitt M. 1991. Rising CO_2 levels and their potential significance for carbon flow in photosynthetic cells. Plant Cell Environ. 14: 741–762.

Stöcklin J. and Körner Ch. 1998. Interactive effects of elevated CO_2, P-availability, and legume presence on calcareous grassland: results of a glasshouse experiment. Funct. Ecol. 13: 200–209.

Strain B.R., Bazzaz F.A. 1983. Terrestrial plant communities. In: Lemon E. (ed.), CO_2 and Plants: the Response of Plants to Rising Levels of Atmospheric Carbon dioxide, AAAS Selected Symposium 8, Washington, DC, pp. 177-222.

Tian H., Melillo J.M., Kicklighter D.W., McQuire A.D., Moore B. III and Vorosmarty C.J. 1999. Parameters for global ecosystem models. Nature 399: 536.

Tissue D.T., Megonigal J.P. and Thomas R.B. 1997. Nitrogenase activity and N_2 fixation are stimulated by elevated CO_2 in a tropical N_2-fixing tree. Oecologia 109: 28–33.

Tjoelker M.G., Oleksyn J. and Reich P.B. 2001. Modeling respiration of vegetation: evidence for a general temperature-dependent Q_{10}. Global Change Biol. 7: 223–230.

Tjoelker M.G., Reich P.B. and Oleksyn J. 1999. Changes in leaf nitrogen and carbohydrates underlie temperature and CO_2 acclimation of dark respiration in five boreal tree species. Plant, Cell Environ. 22: 767–778.

Updegraff K., Bridgham S.D., Pastor J., Weishampel P. and Harth C. 2001. Response of CO_2 and CH_4 emissions from peatlands to warming and water table manipulation. Ecol. Appl. 11: 311–326.

Vetaas O.R. 2002. Realized and potential climate niches: a comparison of four Rhododendron tree species. J. Biogeogr. 29: 545–554.

Wake C. and Markham A. 2005. Indicators of climate change in the northeast 2005. Clean Aiv-Cool Planet, portsmouter, NH. http://inhale.unh.edu/Publications/wake_2005_indicators.pdf.

Walker B., Kinzig A. and Langridge J. 1999a. Plant attribute diversity, resilience, and ecosystem function: the nature and significance of dominant and minor species. Ecosystems 2: 95–113.

Wand J.K., Tissue D.T., Thomas R.B. and Strain B.R. 1999. Comparative responses of model C3 and C4 plants to drought in low and elevated CO_2. Global Change Biol. 5: 857.

Wolfe D.W., Lakso A.N., Otsuki Y., Poole R.M. and Shaulis N.J. 2005. Climate change and shifts in the spring phenology of three horticultural woody perennials in the Northeastern U.S. Int. J. Biometeorol. 49: 303–309.

Wardle D.A., Bardgett R.D., Klironomos J.N., Setälä H., van der Putten W.H. and Wall D.H. 2004. Ecological linkages between aboveground and belowground biota. Science 304: 1629–1633.

Watling J.R., Press M.C. and Quick W.P. 2000. Elevated CO_2 induces biochemical and ultrastructural changes in leaves of the C4 cereal sorghum. Plant Physiol. 123: 1143–1152.

Webber A.N., Nie G.-Y. and Long S.P. 1994. Acclimation of photosynthetic proteins to rising atmospheric CO_2. Photosynth. Res. 39: 413–425.

Zak D.R., Holme W.E., Finzi A.C., Norby R.J. and Schlesinger W.H. 2003. Soil nitrogen cycling under elevated CO_2: a synthesis of the forest face experiments. Ecol. Appl. 13: 1508–1514.

Zak D.R., Pregitzer K.S., King J.S. and Holmes W.E. 2000. Elevated atmospheric CO_2, fine roots and the resonse of soil microorganisms: a review and hypothesis. New Phytol. 147: 201–222.
Zavaleta E.S., Shaw M.R., Chiariello N.R., Thomas B.D., Cleland E.E., Field C.B. and Mooney H.A. 2003. Grassland responses to three years of elevated temperature, CO_2, precipitation, and N deposition. Ecol. Monogr. 73: 585–604.
Ziska L.H. 1998. The influence of root zone temperature on photosynthetic acclimation to elevated carbon dioxide concentrations. Ann. Bot. 81: 717–721.

Photo. ITEX open top roof in Abisko in winter time (Photograph: N. Å. Andersson).

Plant Ecology (2006) 182:65–77
DOI 10.1007/s11258-005-9031-1

Plant performance in a warmer world: general responses of plants from cold, northern biomes and the importance of winter and spring events

R. Aerts*, J.H.C. Cornelissen and E. Dorrepaal
*Institute of Ecological Science, Department of Systems Ecology, Vrije Universiteit, De Boelelaan 1085, NL-1081 HV Amsterdam, The Netherlands; *Author for correspondence (e-mail: Rien.Aerts@ecology.falw.vu.nl; fax: +31-20-5987123)*

Received 1 September 2004; accepted in revised form 15 December 2004

Key words: Climate warming, Nutrient availability, Peatlands, Phenology, *Sphagnum*, Tundra

Abstract

During the past three decades the Earth has warmed with a rate unprecedented during the past 1000 years. There is already ample evidence that this fast climate warming has affected a broad range of organisms, including plants. Plants from high-latitude and high-altitude sites ('cold biomes') are especially sensitive to climate warming. In this paper we (1) review the response in the phenology of plants, changes in their range and distribution, soil nutrient availability, and the effects on the structure and dynamics of plant communities for cold, northern biomes; and (2) we show, by using data from an ongoing snow and temperature manipulation experiment in northern Sweden, that also winter and spring events have a profound influence on plant performance. Both long-term phenological data sets, experimental warming studies (performed in summer or year-round), natural gradient studies and satellite images show that key phenological events are responsive to temperature increases and that recent climate warming does indeed lead to changes in plant phenology. However, data from a warming and snow manipulation study that we are conducting in northern Sweden show that plants respond differently to the various climatic scenarios that we had imposed on these species and that especially winter and spring events have a profound impact. This indicates that it is necessary to include several scenarios of both summer and winter climate change in experimental climate change studies, and that we need detailed projections of future climate at a regional scale to be able to assess their impacts on natural ecosystems. There is also ample evidence that the range shift of herbs and shrubs to more northern regions is for the vast majority of species mainly caused by changes in the climate. This is in line with the observed 'up-greening' of northern tundra sites. These rapid northern shifts in distribution of plants as a result of climate warming may have substantial consequences for the structure and dynamics of high-latitude ecosystems. An analysis of warming studies at 9 tundra sites shows that heating during at least 3 years increased net N-mineralization from 0.32 ± 0.31 (SE) g N m^{-2} yr^{-1} in the controls to 0.53 ± 0.31 (SE) g N m^{-2} yr^{-1} in the heated plots ($p < 0.05$), an increase of about 70%. Thus, warming leads to higher N availability in high-latitude northern tundra sites, but the variability is substantial. Higher nutrient availability affects in turn the species composition of high-latitude sites, which has important consequences for the carbon and water balance of these systems.

Introduction

The Earth's climate has warmed by approximately 0.6 °C over the past century with two main periods of warming, between 1910 and 1945 and from 1976 onwards. The rate of warming during the latter period has been approximately double that of the first and has been greater than at any other time during the past 1000 years (Houghton et al. 2001). However, as all organisms, plants and plant communities do not respond to approximated global average temperature increases. Rather, they respond to regional temperature increases and these have been shown to be highly spatially heterogeneous. E.g. during the past 30 years northern Europe has warmed by about 0.7–1.0 °C, whereas southern Greenland has experienced cooling of about 0.3–0.7 °C (Walther et al. 2002). Nevertheless, climate scenarios for the 21st century predict further warming with the greatest increase predicted for northern high-latitude sites, with the exception of Greenland, with an estimated increase between 1.0 and 4.5 °C (Houghton et al. 2001). Although the warming that has occurred during the past 30 years is relatively minor compared to the predictions for the coming 50–100 years, there is already ample evidence that these recent climatic changes have affected a broad range of organisms, including plants, with diverse geographical distributions (Walther et al. 2002; Parmesan and Yohe 2003; Thomas et al. 2004).

Plants from high-latitude and high-altitude sites (the 'cold biomes') appear to be sensitive to climate warming (e.g. Wookey et al. 1994, 1995; Parssons et al. 1994, 1995; Callaghan and Jonasson 1995) and as this is the region where the highest temperature increases are expected, the responses of plants from these biomes to climate warming can be regarded as an 'early warning system' for the impacts of climate change on species and ecosystems. The aim of this paper is to review the responses of plants and plant communities from high-latitude and – altitude northern ecosystems to increased temperatures. These responses include a wide range of abiotic and biotic processes. Of the abiotic processes, possible changes in soil nutrient availability have a potentially strong impact as plant growth in most high-latitude sites is nutrient limited (Aerts et al. 1992; Shaver and Chapin 1995; Shaver et al. 2001). As a result, warming-induced changes in soil nutrient availability may affect the structure and dynamics of plant communities. Biotic responses further include changes in phenology and range and distribution. In this paper we concentrate on the responses in (1) the phenology of plants; (2) changes in the range and distribution of plants; (3) soil nutrient availability; and (4) the structure and dynamics of plant communities. We contrast the reported responses to experimental warming studies that are conducted in summer or year-round with some of the main results of a warming and snow manipulation experiment that we are currently conducting in a blanket bog in northern Sweden in which six experimentally induced climate scenarios are included (cf. Dorrepaal et al. 2003; Aerts et al. 2004).

Effects of climate warming on plant phenology

Plant phenology is an important variable in the study of possible effects of climate change on the productivity and distribution of terrestrial vegetation (Walther et al. 2002). Accurate knowledge of the effects of climate change on plant phenology is important, as the presence or absence of a photosynthetically active canopy has dramatic effects on ecosystem processes and on biosphere/atmosphere exchanges (van Wijk et al. 2003). Springtime plant activity in the Arctic is largely initiated by snowmelt and is thus clearly temperature-sensitive. However, the timing of the end of the growing season can be triggered by temperature, photoperiod, genetic constraints and/or internal plant cycles of nutrient use (Shaver and Kummerov 1992; Oberbauer et al. 1998). This implies that climate warming effects on the duration of the growing season are probably most strongly determined by the effects on spring events.

Phenological events such as the timing of leaf bud burst and flowering, are probably the simplest processes that enable us to track the effects of climate change on plant performance. As a result, there is now an overwhelming amount of data available on plant phenological responses to climate warming. These data broadly originate from two types of studies: (1) studies in which key phenological events are recorded over decadal time scales and in which the observed changes are attributed to the climatic warming that has been recorded during the past 30 years; and (2) experimental studies in which air and/or soil temperatures have been increased actively (by heating cables) or passively (by placing Open Top Chambers).

Currently, there are very many long-term phenological observations available for a wide variety of organisms and processes. Meta-analysis is a powerful tool to synthesize the results of these individual studies and to detect more general response patterns. Root et al. (2003) performed a meta-analysis of plant phenology data from northern sites. They split their data into two categories: studies on (amongst others) plant species or species groups between 32 and 49.9° N and between 50 and 72° N. As expected, the estimated mean and SEM of the phenological shifts between 32 and 49.9° N (−4.2 ± 0.2 days) were smaller than between 50 and 72° N (−5.5 ± 0.1 days). Remote sensing data validate these ground observations on larger scales. The Normalized Difference Vegetation Index (NDVI), which is derived from infrared and red Earth surface reflectance, scales with green biomass. NDVI satellite data of the area between 45° N and 70° N showed for 1982–1990 a 8-day shift to an earlier start of the growing season and a delay of 4 days for the declining phase (Myneni et al. 1997). More recent NDVI data (Zhou et al. 2001) suggest that the growing season has become nearly 18 days longer during the past two decades in Eurasia and 12 days longer in North America. This extension of the growing season may have a substantial effect on the amount of carbon sequestered by terrestrial biota. This has e.g. been inferred from the increasing amplitude of the annual oscillations in atmospheric CO_2 between 1960 and 1994 (Keeling et al. 1996).

Despite the remote locations and the harsh climatic conditions, there are many experimental studies on the effects of warming on plant performance at high-latitude and high-altitude sites. These studies are mostly embedded in large international networks, such as the International Tundra Experiment (ITEX) or the Network of Ecosystem Warming Studies (NEWS) of the GCTE-IGBP network. This approach allows generalization and scaling up of the results of individual studies and is therefore a powerful method to analyse the impact of climate change on ecosystem performance (e.g. Arft et al. 1999; Rustad et al. 2001; Parmesan and Yohe 2003). The most recent meta-analysis of phenological data from the ITEX network is presented in Arft et al. (1999) and includes the results from 1 to 4 years of experimental data from 13 different circum-polar sites. Due to practical constraints, these studies include only summer warming or warming year-round. Arft et al. (1999) found that key phenological events such as leaf bud burst and flowering occurred earlier as a result of warming. In addition, quantitative measures of vegetative growth were greatest in the warmed plots in the early years of the experiments, whereas reproductive effort and success increased in later years. Similar results were obtained by Dunne et al. (2003) who combined both experimental warming and natural temperature gradient analysis to study the flowering phenology responses of 11 sub-alpine meadow species in Colorado, USA. The results suggest that the initiation of leaf bud burst and flowering are mainly triggered by temperature, whereas the onset of leaf senescence is more dependent on photoperiod and/or genetic controls. However, in some studies it has been found that warming delays senescence (Molau 1997; Stenström et al. 1997).

So both long-term phenological data sets, experimental warming studies, natural gradient studies and satellite images show that key phenological events are responsive to temperature increases and that recent climate warming does indeed lead to changes in plant phenology. For modelling purposes and up-scaling of results for individual species, it would be helpful if the response of plant phenological events to climate warming could be expressed at a higher hierarchical level than that of individual species. Chapin et al. (1996) developed a hierarchical classification system of plant functional types (PFTs) for arctic plant species, based on environmental gradients and the relative impact of different traits on ecosystem processes. This classification system is mainly growth-form based. In general, there is great similarity in the phenological response within plant functional types (Arft et al. 1999; Dunne et al. 2003). This implies that a classification of response patterns at the PFT level can be a key tool for developing predictive models of plant phenological responses to climate change.

However, a weak point of the studies that have been conduced so far is that they do not include climate scenarios in which winter and spring events are incorporated. This is important as climate scenarios for high latitude areas do not only predict higher summer temperatures, but also larger variation in winter snowfall and winter temperatures (Houghton et al. 2001). So far, most experimental warming studies in (sub-)arctic sites have focussed on the effects of higher summer temperatures on a wide variety of plant, animal and ecosystem responses. However, there are to our knowledge no experimental studies that combine various temperature regimes with altered winter snow fall in a replicated and controlled way.

To fill this gap in our knowledge, we have conducted a climate manipulation experiment in a blanket bog in sub-arctic Sweden near the Abisko Scientific Research Station (68°21′ N, 18°49′ E) since the summer of 2000. At this site, annual precipitation amounts to 320 mm per year with a mean summer temperature of 7 °C and a mean winter temperature of −6 °C. The length of the vegetation season is 130 days (Karlsson and Callaghan 1996). The moss component of this bog is dominated by *Sphagnum fuscum*. The cover of vascular plants is about 25% and mainly consists of *Empetrum hermaphroditum, Betula nana, Rubus chamaemorus, Andromeda polifolia, Calamagrostis lapponica* and *Vaccinium uliginosum*. In this experiment, we manipulate both summer temperatures and winter and spring snow accumulation and temperatures independently (Dorrepaal et al. 2003; Aerts et al. 2004, in which also full details about the design and the statistical analyses can be found). In total we mimic six possible climate scenarios for high-latitude ecosystems (Table 1). Spring and summer warming are established by passive warming using a modified, larger version of the ITEX-open-top chambers (OTCs; see Marion et al. 1997) (Figure 1a). Increased winter snow accumulation is achieved by leaving the OTCs in place to serve as passive snow traps (Figure 1b). Through regular measurements of winter snow depths and continuous recording of air and soil temperatures year-round we are able to quantify the effects of our manipulations on these key abiotic variables and their effect on phenological events. We found that our climate manipulations had significant and realistic effects on air and soil temperatures (Dorrepaal et al. 2003): in winter the OTCs increased the snow thickness two-fold, resulting in 0.5–2.8 °C higher average air temperatures. Spring temperatures in the OTCs increased by 0.7–1.2 °C, whereas summer warming had a maximum effect of 0.9 °C. The data available so far showed no indications of effects of the treatments on soil moisture because, vapour pressure deficit was not affected by the OTCs.

Table 1. Climate treatments used in the experiment.

Treatment	Code	Summer	Winter	Spring
1	AAA	A	A	A
2	ASA	A	S	A
3	ASW	A	S	W
4	WAA	W	A	A
5	WSA	W	S	A
6	WSW	W	S	W

A = ambient, W = warming, S = (passive) snow accumulation.

To study the effects of our climate manipulations on flowering phenology and flower production we chose two species that are relatively easy to monitor: the evergreen dwarf-shrub *Andromeda polifolia* and the herb *Rubus chamaemorus* (Aerts et al. 2004). In the spring and summer period of 2001 and 2002, we determined the numbers of flowers produced and the median flowering date (the date at which 50% of the total number of produced flowers does flower). We were fortunate in that we could compare our treatment effects with the effects of inter-annual variability thanks to a record warm spring and early summer for this region that occurred in 2002. The first year of our phenology study (2001) was a 'normal' year.

In 2001, flowering occurred both in *Andromeda polifolia* and in *Rubus chamaemorus* over a period of 4 weeks, from the end of May until the end of June. In the exceptionally warm year 2002, however, flowering started 1 week earlier and was finished in mid-June. In both species, the median flowering date in 2002 was about 2 weeks earlier than in 2001 (Figure 2). Superimposed on this very substantial inter-annual effect, we found that our climate treatments (except summer warming for *Rubus*) resulted in a 1–4 days earlier median flowering date (Figure 2). Our data show that winter and spring events have a significant impact on reproductive characteristics of high-latitude plant species. Both increased winter snow cover and spring warming led to significantly earlier flowering in both *Andromeda* and *Rubus* (Figure 2). Also the total number of produced flowers was stimulated by spring warming (see Aerts et al. 2004). In addition, we found that the effects of the various climate changes during the year on flowering phenology are additive: warming (be it in spring or in summer) or extra snow in winter lead to earlier flowering and combinations of these treatments lead to the most pronounced effects on flowering phenology (Figure 2).

Figure 1. An overview of part of our experimental climate manipulation site in Abisko in summer (top) and in winter (bottom).

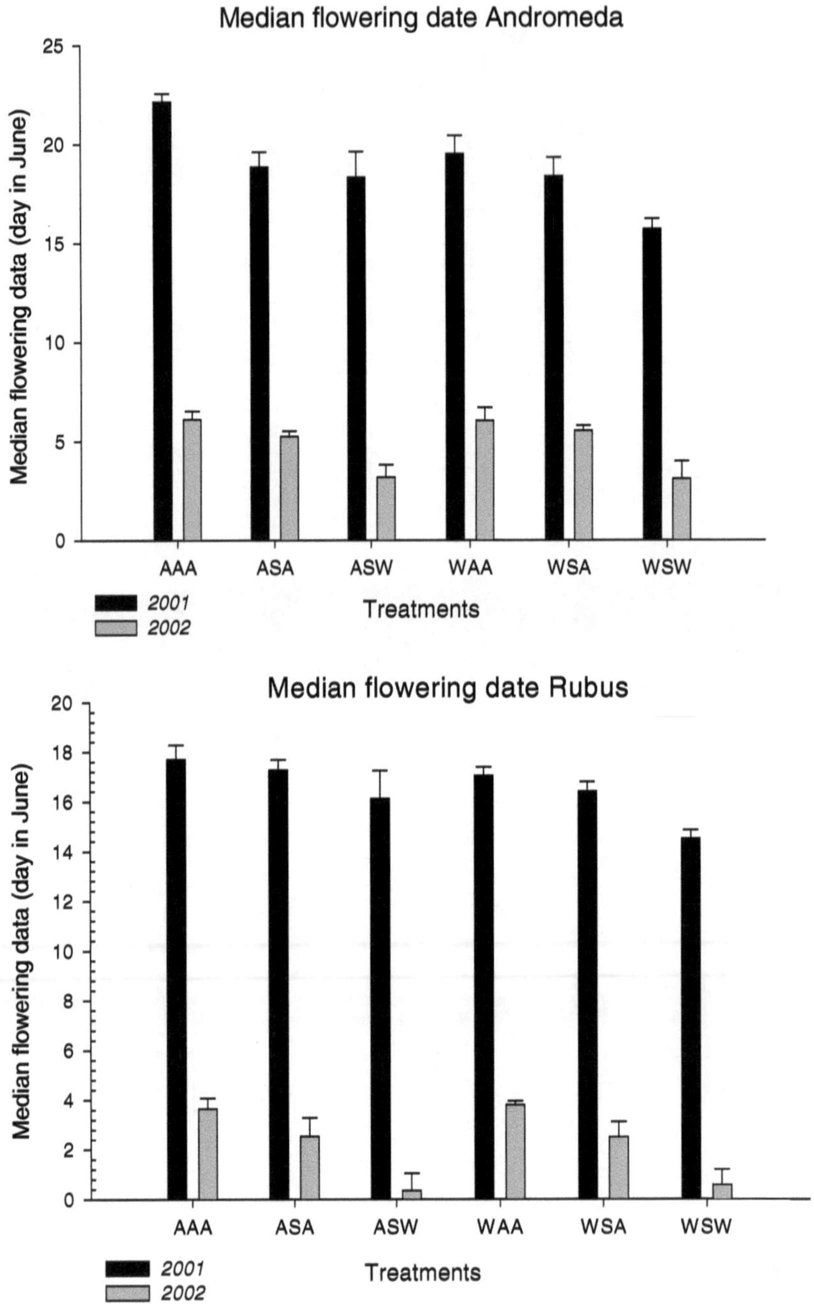

Figure 2. Median flowering date (date at which 50% of the maximum number of flowers has opened) of *Andromeda polifolia* and *Rubus chamaemorus* in the experimental treatments (see Table 1). Error bars are SE ($n = 5$). Reproduction of this figure was kindly permitted by Blackwell Publishing.

These phenological changes may have a substantial impact on plant performance and ecosystem responses. In high-latitude or high-altitude ecosystems, sexual reproduction is mostly increased under more favourable conditions, such as higher temperatures (Wookey et al. 1994, 1995; Molau and Shaver 1997; Dunne et al. 2003). Clonal (=vegetative) growth confers survival potential during unfavourable years, together with the ability to capitalise on nutrient flushes and recycle nutrients internally (Wookey et al. 1995). Investment in sexual reproduction ensures that seed-set is successful during favourable years, even if these occur infrequently. Thus, the balance between vegetative and generative reproduction is important for both long-term survival of plant populations and for maintaining genetic diversity in the population (Callaghan and Jonasson 1995) and it is likely that this balance will shift towards more sexual reproduction in response to climate warming. However, as increased allocation to sexual reproduction usually involves higher nutrient investments (Aerts and Chapin 2000), the long-term responses could become constrained by lack of available nutrients (cf. Callaghan and Jonasson 1995). Since higher temperatures will probably also lead to higher soil nutrient availability (see below), it is difficult to predict what the long-term response will be.

Effects of climate warming on the range and distribution of plants

It is beyond dispute that the range and distribution of many wild species has changed during the past few decades. However, there is less consensus about the causes and their relative importance. Most short-term local changes are not caused by climate change, but by land-use change and by natural fluctuations in the abundance and distribution of species (Parmesan and Yohe 2003). This makes it difficult to detect more general, long-term trends that are caused by climate change. Such underlying trends are often masked by the direct effects of habitat loss. Distribution changes that are caused by climate change are often related to species-specific physiological thresholds of temperature and precipitation tolerance (Woodward 1987), but also to changing biotic interactions (Cornelissen et al. 2001). To the extent that dispersal and resource availability allow, species are expected to track the shifting climate and likewise shift their distributions poleward in latitude and upward in elevation.

Many studies have sought to document these predicted range shifts. Range changes in sedentary organisms, such as plants, follow from the slow processes of population extinctions and colonization. There is now convincing evidence that poleward and upward changes in species ranges have occurred during the 20th century. E.g. the treeline has advanced towards higher altitudes in Europe and New Zealand (Wardle and Coleman 1992; Meshinev et al. 2000; Kullman 2001), shrubs have expanded in formerly shrubless tundra vegetation in Alaska (Sturm et al. 2001), alpine plants in Europe have shown elevational shifts of 1–4 m per decade (Grabherr et al. 1994) and Antarctic plants have shown substantial distribution changes (Kennedy 1995). However, these changes can not unequivocally be attributed to climate change.

As a result of climate warming, non-native species from adjacent areas may cross frontiers and become new elements of the ecosystem. When this concerns species that have crossed long distances, very often human activities have been involved. However, permanent establishment is only possible when the local conditions at the new site are suitable for such invading species. Climate change may lead to improvement of growing conditions for plants and this is probably the cause of the successful spread of (sub-)tropical garden plants into the surrounding countryside of areas with a much colder climate (Dukes and Mooney 1999; Walther 2000). Human interventions, in combination with a warmer climate, are probably also responsible for 50% or more of the higher plant diversity on some remote sub-Antarctic islands during the last two centuries (Smith 1996). The changes in distribution are often asymmetrical with species from lower elevations or latitudes invading faster than resident species are receding upward or poleward. The (probably transient) effect is an increase in species diversity of the considered community as a result of differential rates at which species shift their range.

Parmesan and Yohe (2003) used a number of complementary methods to detect climate-induced shifts in range boundary of a wide variety of organisms. They found that the range shift of herbs and shrubs to more northern regions was for the vast majority of species mainly caused by changes in the climate. This is in line with the observed 'up-greening' of northern tundra sites (Myneni et al. 1997). These rapid northern shifts in distribution of plants as a result of climate warming may have substantial consequences for the structure and dynamics of high-latitude ecosystems.

Factors affecting species distributions interact in complex ways and therefore it is not surprising that simple correlations with temperature changes are not always observed. Proposed patterns of global change include not only predictions about changes in environmental factors such as increased temperature, but also an expectation that changes in the average values of climatic variables will be accompanied by shifts in the probabilities of extreme values occurring (MacGillivray et al. 1995). It is well-known that extreme climatic events may have a disproportional effect on species distributions. Thus, they may be important determinants of the changes in species ranges and distributions. In addition, dispersal capacity varies widely among species. Thus, while a certain species may be able to track and follow decadal warming trends quickly and thereby match the upward and northward shifts of temperature, another species may not.

Effects of climate warming on soil nutrient availability

In most high-latitude ecosystems plant growth is limited by N availability (Aerts et al. 1992; Shaver and Chapin 1995; Shaver et al. 2001). Although soil nutrient availability in high-latitude sites is usually low, these sites may contain considerable amounts of soil nutrients. E.g. boreal soils in an aspen forest at Fairbanks, Alaska contain approximately 300 g N m^{-2} (van Cleve and Alexander 1981). Soil organic matter in northern taiga and tundra ecosystems contains about 95% of the organically bound plant nutrients (Jonasson 1983). However, mineralization of these large stores of nutrients is constrained by low temperatures and by exceptionally high or exceptionally low soil moisture availability (Robinson 2002). This suggests that higher temperatures at high-latitude sites may lead to higher soil nutrient mineralization rates. However, in contrast with N availability, soil warming within realistic limits has probably no effects on soil P availability. The reason for this difference is that detrital N is mostly carbon-bonded whereas detrital P is mostly ester-bonded and the Q10 values for biological processes are much higher than for chemical processes (Aerts and Chapin 2000). A complicating factor is that higher temperatures may also affect soil moisture availability, which in turn also affects N mineralization rates.

The experimental summer warming studies that have investigated the effects on soil nutrient availability have yielded very variable results. Jonasson et al. (1993) measured the effects of higher soil temperatures (up to 2 °C) on soil N and P availability in a sub-arctic heath and a high-altitude fell field near Abisko, northern Sweden. They found no effects of their warming treatments on the net mineralization on N and P in these ecosystems. They suggested that this was due to the strong microbial sink in (sub-) arctic soils for nutrients (cf. Jonasson et al. 1996; Schmidt et al. 1999). Robinson et al. (1995) studied the effects of soil warming (1 °C) on soil N mineralization in a high-arctic polar desert at Svalbard and sub-arctic dwarf shrub heath at Abisko. They also found negligible effects on total net N mineralization. Hartley et al. (1999) observed transient effects of soil warming (5 °C) on soil N mineralization during a 5-year study in a sub-arctic dwarf shrub heath near Abisko. During the first two years of the study soil warming increased soil N mineralization, but this effect had disappeared by the fifth year of the study. This suggests that experimental soil warming only affects the relatively labile soil nutrient pool and that, due to its relatively small size, this pool is depleted within a few years.

However, it should be noted that although the results from individual warming studies may show considerable variation and a lack of statistically significant results, the large-scale response to warming may be more clear-cut. Rustad et al. (2001) performed meta-analysis on the results of warming studies at 32 research sites representing four broadly defined biomes, including high-latitude or -altitude tundra, low tundra, grassland and forest. They also found considerable variation among sites in response to warming. For this meta-analysis, N mineralization data were available for 9 tundra sites. Using the data reported by Rustad et al. (2001), we calculated that for these tundra sites, heating during at least 3 years increased net N-mineralization from 0.32 ± 0.31 (SE) g N m^{-2} yr^{-1} in the controls to 0.53 ± 0.31 (SE) g N m^{-2} yr^{-1} in the heated plots ($p < 0.05$), an increase of about 70%. These results show that warming does indeed lead to higher N availability in high-latitude tundra sites, but the variability is substantial.

Effects of climate warming on the structure and dynamics of ecosystems

The assemblages of species in ecological communities reflect interactions among organisms as well as between organisms and the abiotic environment. Thus, given the influence that climate has on plant phenology, soil nutrient availability and the distribution of organisms there is no doubt that climate change will affect the structure and dynamics of ecosystems in cold biomes. This may also affect the carbon balance of these systems with important feedbacks to the global C cycle (Gorham 1991).

The quick phenological changes that have been observed in response to climate change have the potential to change the relationships that plants have with animal, fungal, and bacterial species that act as pollinators, seed dispersers, herbivores, seed predators and pathogens (Dunne et al. 2003). These changes will have the strongest impact if, the interacting species are influenced by different abiotic factors or if, their relative response to the same factor (e.g. elevated temperatures) is different. Moreover, differences in phenological responses to climate change among plant species of different growth forms may also affect plant–plant interactions such as resource competition. Thus, as climate change not only influences particular plants species, but also has a potentially broad impact on ecosystem interactions it will probably result in significant changes in community structure and ecosystem functioning.

As plant growth is limited by N availability in most high-latitude ecosystems (Aerts et al. 1992; Shaver and Chapin 1995), increased soil N availability may also have a wide variety of effects on the structure and dynamics of northern high-latitude and high-altitude ecosystems. Due to species-specific responses, these changes are difficult to predict at the level of individual species, but show more predictable responses at the level of plant functional types (PFTs; Quested et al. 2003). In general, increased nutrient supply results in an increased abundance of fast-growing species (mostly graminoids, deciduous shrubs and herbaceous species) at the expense of slow growers (mostly evergreens and mosses and lichens; Cornelissen et al. 2001; van Wijk et al. 2004). Press et al. (1998) performed fertilisation experiments in a sub-arctic dwarf shrub heath in northern Sweden and found an 18-fold increase in the abundance of the grass *Calamagrostis lapponica.* Similarly, Robinson et al. (1998) performed a long-term fertilization experiment at a high arctic polar desert at Svalbard and found increased abundance of the deciduous shrub *Salix polaris* and the herb *Polygonum viviparum* and reduced abundance of the evergreens *Dryas octopetala* and *Saxifraga oppositifolia.* Graglia et al. (2001) found that 10 years of nutrient addition in two subarctic dwarf shrubs heaths resulted in a strong increase of the abundance of grasses and a strong decline of cryptogams.

These changes in dominance of PFTs may have profound consequences for the amount of carbon fixed in Net Primary Production (NPP) as NPP usually differs substantially among PFTs, both at a given growth-limiting level of nutrient availability and at a very high level of nutrient availability (Aerts and Chapin 2000). Shaver et al. (2001) added N and P fertilizer to an Alaskan moist tundra over a period of 15 years. The species composition of the vegetation in the fertilized plots changed dramatically from a mix of graminoid, evergreen, deciduous and moss species to strong dominance of the deciduous shrub species *Betula nana.* Concomitant with this change the aboveground biomass and NPP increased by 2.5 times. An important explanation for this dramatic increase in NPP was that *Betula nana* was able to produce much (woody) biomass at a relatively low N investment, thereby outcompeting other species.

Most climate manipulation experiments focus on the performance of vascular plants. However, *Sphagnum* mosses form a major component of northern peatlands and contribute significantly to the sequestration of atmospheric carbon dioxide (Gorham 1991). Over the past millennia, approximately one-third of the total world soil carbon has accumulated in the organic deposits of those peatlands (Gorham 1991), mainly due to the recalcitrant nature of *Sphagnum* litter and the unfavourable conditions for decomposition, such as low temperatures, low pH and a high water table (Johnson and Damman 1993). *Sphagnum* mosses have thus been responsible for the sequestration of large quantities of carbon and play an important role in the global carbon cycle. Climate change is likely to alter the accumulation of carbon by *Sphagnum,* with consequences for the C sink function of northern peatlands.

This is illustrated by data from our research site in northern Sweden, where *Sphagnum fuscum* is the dominant species in the moss layer. The climate manipulations had strong effects on *S. fuscum* (Dorrepaal et al. 2003). Summer warming increased the length increment by 62% in 2001 and by 42% in 2002, whereas there was no effect of the winter snow addition or the spring warming. The summer warming treatment reduced the bulk density of the *Sphagnum* plants, but the winter snow addition treatments and the spring warming had no effect. As a result of the counteracting effects of summer warming on length growth and bulk density, there was only a trend for a positive effect of summer warming on biomass production (Figure 3). However, the dry matter production showed a positive response to the winter and spring treatments with on average an enhanced dry matter production by 33% (Figure 3). These results show that the productivity of *Sphagnum fuscum* shows different responses to the various climatic scenarios that we had imposed on this species and that, as was also the case for the phenology of vascular plants, winter and spring events may be of particular importance.

Not only NPP, but also the amount of C lost through the decomposition pathway will be affected by climate-change induced changes in PFT composition (Hobbie 1996; Aerts and Chapin 2000). Quested et al. (2003) performed a standardized comparative decomposition study in litter beds (cf. Cornelissen 1996) in sub-arctic Sweden with leaf litters of 72 plant species belonging to 9 PFTs. Decomposition rates of the four most abundant PFTs increased in the order: evergreens < graminoids < woody deciduous < herbs. These results suggest that carbon loss through the decomposition pathway increases due to PFT replacement as the PFTs with low decomposition rates are replaced by PFTs with high decomposition rates. The ultimate effect of these PFT changes on ecosystem carbon balance depends on the ratio between the increase in NPP and the increase in carbon loss through litter decomposition. So far, there are no data available to document such changes.

Changes in PFT composition do not only affect NPP and ecosystem carbon balance. The shift from moss-dominated communities towards communities with a higher abundance of vascular plants (cf. van Wijk et al. 2004) is likely also to alter the energy and water exchange between the atmosphere and the soil, which in high-latitude ecosystems are strongly regulated by the moss mat

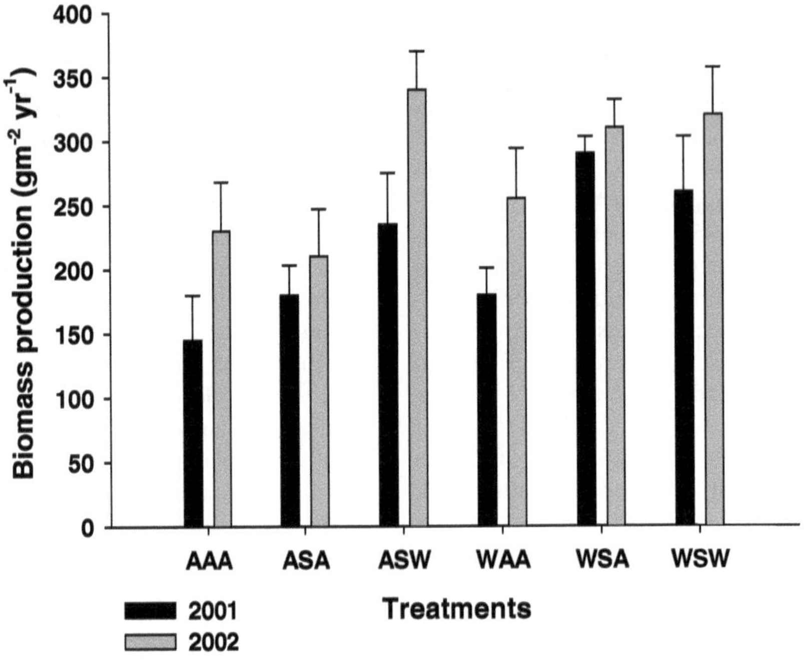

Figure 3. Dry matter production of *Sphagnum fuscum* during the summers of 2001 and 2002 in response to summer and winter climate change scenarios (see Table 1). Error bars are SE ($n = 5$). Redrawing of this figure was kindly permitted by Blackwell Publishing.

(Tenhunen et al. 1992). Van der Wal and Brooker (2004) studied the relation between moss abundance and soil temperatures in arctic tundra at Spitsbergen. They found that the thickness of the moss layer was negatively correlated with soil temperature. Thus, the mosses have a significant impact on the energy balance of the soil. Heijmans et al. (2004) showed that moss evaporation contributes substantially to total ecosystem evapotranspiration. However, this contribution depends strongly on the openness of the vascular plant canopy and on the identity of the moss species. The observed strong influence of habitat type suggests that microclimate is the primary factor determining moss evaporation rates. High moss evaporation rates, which especially occur in *Sphagnum* mosses, suggest a potential cooling effect of mosses. In addition to these properties of the moss mat, it has been shown that high amounts of vascular plant litter instead of a moss mat will probably also suppress seedling emergence (Bosy and Reader 1995). However, in many types of tundra vegetation this effect will not be very important as most species propagate vegetatively in this ecosystem type (Callaghan et al. 1992).

Changes in the dominance of plant functional types may also affect higher trophic levels. Richardson et al. (2002) studied the effects of nutrient addition and warming on plant communities and their insect herbivores in a sub-arctic heath near Abisko, northern Sweden. The treatments resulted in substantial changes in plant community composition with a strong increase in grass biomass and strong decline of the mosses. These changes also resulted in dramatic changes in the community of herbivorous Hemiptera and especially in the families that fed on either bryophytes or grasses. This will undoubtedly have consequences for the overall herbivory pressure, but that was not quantified in this study.

In conclusion, climate warming in cold biomes has a profound impact on a wide range of abiotic and biotic processes. Currently, we have a rather good idea of the effect of a warmer summer climate on these processes, but we clearly need more data on the effects of changes in the winter climate. Further, experimental studies should include several scenarios of both changes in summer and winter climate and that requires detailed projections of future climate at a regional scale to be able to assess their impacts on natural ecosystems. Our climate manipulation experiment in northern Sweden does provide a suitable framework to study the response of high-latitude ecosystems to different summer and winter climate scenarios and we hope that more of such experiments will be started at northern high-latitudes.

Acknowledgements

The experiment described in this paper could not have been performed without the help of Richard van Logtestijn and Sandra Berg. We also thank the staff of the Abisko Scientific Research Station, notably Anders Eriksson for major technical support and Lilian Eriksson, Majlis Kardefeldt and Nils Åke Andersson for the snow depth measurements. This study was financially supported by USF grant 98/24 to RA and a grant of the Royal Swedish Academy of Sciences to ED. The County Administrative Board at Luleå gave permission to perform this experiment in the Abisko National Park.

References

Aerts R. and Chapin F.S. 2000. The mineral nutrition of wild plants revisited: a re-evaluation of processes and patterns. Adv. Ecol. Res. 30: 1–67.

Aerts R., Wallén B. and Malmer N. 1992. Growth-limiting nutrients in *Sphagnum*-dominated bogs subject to low and high atmospheric nitrogen supply. J. Ecol. 80: 131–140.

Aerts R., Cornelissen J.H.C., Dorrepaal E., van Logtestijn R.S.P. and Callaghan T.V. 2004. Effects of experimentally imposed climate scenarios on flowering phenology and flower production of sub-arctic bog species. Global Change Biology 10: 1599–1609.

Arft A.M., Walker M.D. and Gurevitch J. et al. 1999. Response patterns of tundra plant species to experimental warming: a meta-analysis of the International Tundra Experiment. Ecol. Monogr. 69: 491–511.

Bosy J.L. and Reader R.J. 1995. Mechanisms underlying the suppression on forb seedling emergence by grass (*Poa pratensis*) litter. Funct. Ecol. 9: 635–639.

Callaghan T.V. et al. 1992. Clonal plants and environmental change: introduction to the proceedings and summary. Oikos 63: 341–347.

Callaghan T.V. and Jonasson S. 1995. Arctic terrestrial ecosystems and environmental change. Philos. Trans. Roy. Soc. Lond. 352: 259–276.

Chapin F.S., Bret-Harte M.S., Hobbie S.E. and Zhong H. 1996. Plant functional types as predictors of transient responses of arctic vegetation to global change. J. Veg. Sci. 7: 347–356.

van Cleve K. and Alexander V. 1981. Nitrogen cycling in tundra and boreal ecosystems. In: van Clark F.E. and Roswall T. (eds.), Terrestrial Nitrogen Cycles, Ecological Bulletins, Stockholm, pp. 375–404.

Cornelissen J.H.C. 1996. An experimental comparison of leaf decomposition rates in a wide range of temperate plant species and types. J. Ecol. 84: 573–582.

Cornelissen J.H.C., Callaghan T.V., Alatalo J.M., Michelsen A., Graglia E., Hartley A.E., Hik D.S., Hobbie S.E., Press M.C., Robinson C.H., Henry G.H.R., Shaver G.R., Phoenix G.K., Gwynn Jones D., Jonasson S., Chapin F.S. III, Molau U., Neill C., Lee J.A., Melillo J.M., Sveinbjörnsson B. and Aerts R. 2001. Global change and arctic ecosystems: is lichen decline a function of increases in vascular plant biomass? J. Ecol. 89: 984–994.

Dorrepaal E., Aerts R., Cornelissen J.H.C., Callaghan T.V. and van Logtestijn R.S.P. 2003. Summer warming and increased winter snow cover affect *Sphagnum fuscum* growth, structure and production in a sub-arctic bog. Glob. Change Biol. 10: 93–104.

Dukes J.S. and Mooney H.A. 1999. Does global change increase the success of biological invaders?. Trend. Ecol. Evol. 14: 135–139.

Dunne J.A., Harte J. and Taylor K.J. 2003. Subalpine meadow flowering phenology responses to climate change: integrating experimental and gradient methods. Ecological Monographs 73: 69–86.

Gorham E. 1991. Northern peatlands: role in the carbon cycle and probable responses to climatic warming. Ecological Applications 1: 182–195.

Grabherr G., Gottfried M. and Pauli H. 1994. Climate effects on mountain plants. Nature 369: 448.

Graglia E., Jonasson S., Michelsen A., Schmidt I.K., Havström M. and Gustavsson L. 2001. Effects of environmental perturbations on abundance of subarctic plants after three, seven and ten years of treatments. Ecography 24: 5–12.

Hartley A.E., Neill C., Melillo J.M., Crabtree R. and Bowles F.P. 1999. Plant performance and soil nitrogen mineralization in response to simulated climate change in subarctic dwarf shrub heath. Oikos 86: 331–343.

Heijmans M.M.P.D., Arp W.J. and Chapin F.S. 2004. Controls on moss evaporation in a boreal black spruce forest. Glob. Biogeochem. Cycles 18: 1524–1530.

Hobbie S.E. 1996. Temperature and plant species control over litter decomposition in Alaskan tundra. Ecol. Monogr. 6: 503–522.

Houghton J.T., Ding Y. and Griggs D.J. et al. 2001. Climate Change 2001: the Scientific Basis. Cambridge University Press, CambridgeThird IPCC Report.

Jonasson S. 1983. Nutrient content and dynamics in north Swedish shrub tundra areas. Holarctic Ecol. 6: 295–304.

Jonasson S., Havström M., Jensen M. and Callaghan T.V. 1993. In situ mineralization of nitrogen and phosphorus of arctic soils after perturbations simulating climate change. Oecologia 95: 179–186.

Jonasson S., Michelsen M., Schmidt I.K., Nielsen E.V. and Callaghan T.V. 1996. Microbial biomass C, N and P in two arctic soils and responses to addition of NPK fertilizer and sugar: implications for plant nutrient uptake. Oecologia 106: 507–515.

Johnson L.C. and Damman A.W.H. 1993. Decay and its regulation in *Sphagnum* peatlands. Adv. Bryol. 5: 249–296.

Karlsson P.S. and Callaghan T.V. 1996. Plant ecology in the subarctic Swedish Lapland. Ecol. Bull. 45: 1–227.

Keeling C.D., Chin J.F.S. and Whorf T.P. 1996. Increased activity of northern vegetation inferred from atmospheric CO_2 measurements. Nature 382: 146–149.

Kennedy A.D. 1995. Antarctic terrestrial ecosystem response to global environmental change. Ann. Rev. Ecol. Syst. 26: 683–704.

Kullman L. 2001. 20th century climate warming and tree-limit rise in the southern Scandes of Sweden. Ambio 30: 72–80.

MacGillivray C.W., Grime J.P., Band S.R., Booth R.E., Campbell B., Hendry G.A.F., Hillier H., Hodgson J.G., Hunt R., Jalili A., Mackey J.M.L., Mowforth M.A., Neal A.M., Reader R., Rorison I.H., Spencer R.E., Thompson K. and Thorpe P.C. 1995. Testing predictions of the resistance and resilience of vegetation subjected to extreme events. Funct. Ecol. 9: 640–649.

Marion G.M., Henry G.H.R. and Freckman D.W. et al. 1997. Open-top designs for manipulating field temperature in high-latitude ecosystems. Glob. Change Biol. 3(Suppl. 1): 20–32.

Meshinev T., Apostolova I. and Koleva E. 2000. Influence of warming on timberline rising: a case study on *Pinus peuce* Griseb. in Bulgaria. Phytocoenologia 30: 431–438.

Molau U. 1997. Responses to natural climatic variation and experimental warming in two tundra plant species with contrasting life forms: *Cassiope tetragona* and *Ranunculus nivalis*. Glob. Change Biol. 3(Supplement 1): 97–107.

Molau U. and Shaver G.R. 1997. Controls on seed production and seed germinability in *Eriophorum vaginatum*. Glob. Change Biol. 3(Suppl. 1): 80–88.

MyneniR.B. Keeling C.D., Tucker C.J., Asrar G. and Nemani R.R. 1997. Increased plant growth in the northern high latitudes from 1981 to 1991. Nature 386: 698–702.

Oberbauer S.F., Starr G. and Popp E.W. 1998. Effects of extended growing season and soil warming on carbon dioxide and methane exchange of tussock tundra in Alaska. J. Geophys. Res. 103: 29075–29082.

Parmesan C. and Yohe G. 2003. A globally coherent fingerprint of climate change impacts across natural systems. Nature 421: 37–42.

Parsons A.N., Welker J.M., Wookey P.A., Press, M.C., Callaghan T.V. and Lee J.A. 1994. Growth-response of four sub-arctic dwarf shrubs to simulated environmental change. J. Ecol. 82: 307–318.

Parsons A.N., Press M.C., Wookey P.A., Welker J.M., Robinson C.H., Callaghan T.V. and Lee J.A. 1995. Growth responses of *Calamagrostis lapponica* to simulated environmental change in the sub-arctic. Oikos 72: 61–66.

Press M.C., Potter J.A., Burke M.J.W., Callaghan T.V. and Lee J.A. 1998. Responses of a subarctic dwarf shrub heath community to simulated environmental change. J. Ecol. 86: 315–327.

Quested H.M., Cornelissen J.H.C., Press M.C, Callaghan T.V., Aerts R, Trosien F., Riemann P., Gwynn-Jones D., Kondratchuk A. and Jonasson S. 2003. Decomposition of subarctic plants with differing nitrogen economies: a functional role for hemiparasites. Ecology 84: 3209–3221.

Richardson S.J., Press M.C., Parsons A.N. and Hartley S.E. 2002. How do nutrients and warming impact on plant communities and their insect herbivores. A 9-year study from a sub-arctic heath. J. Ecol. 90: 544–556.

Robinson C.H. 2002. Controls on decomposition and soil nitrogen availability at high latitudes. Plant Soil 242: 65–81.

Robinson C.H., Wookey P.A., Parsons A.N., Potter J.A., Lee J.A., Callaghan T.V., Press M.C. and Welker J.M. 1995. Responses of plant litter decomposition and nitrogen mineralisation to simulated environmental change in a high arctic polar semi-desert and a subarctic dwarf shrub heath. Oikos 74: 503–512.

Robinson C.H., Wookey P.A., Lee J.A., Callaghan T.V. and Press M.C. 1998. Plant community responses to simulated environmental change at a high arctic polar semi-desert. Ecology 79: 856–866.

Root T.L., Price J.T., Hall K.R., Schneider S.H., Rosenzweig C. and Pounds J.A. 2003. Fingerprints of global warming on wild animals and plants. Nature 421: 57–60.

Rustad L.E., Campbell J.L. and Marion G.M. et al. 2001. A meta-analysis of the response of soil respiration, net nitrogen mineralization, and aboveground plant growth to experimental ecosystem warming. Oecologia 126: 543–562.

Schmidt I.K., Jonasson S. and Michelsen A 1999. Mineralisation and microbial immobilisation of N and P in arctic soils in relation to season, temperature and nutrient amendment. Appl. Soil Ecol. 11: 147–160.

Shaver G.R. and Kummerov J. 1992. Phenology, resource allocation, and growth of arctic vascular plants. In: Chapin F.S., Jefferies R.L. and Reynolds J.F. (eds.), et al.Arctic Ecosystems in a Changing Climate: an Ecophysiological Perspective, Academic Press, New York, pp. 191–212.

Shaver G.R. and Chapin F.S. 1995. Long-term responses to factorial NPK fertilizer treatment by Alaskan wet and moist tundra sedge species. Ecography 18: 259–275.

Shaver G.R., Bret-Harte M.S., Jones M.H., Johnstone J., Gough L., Laundre J. and Chapin F.S. 2001. Species composition interacts with fertilizer to control long-term vegetation change in tundra productivity. Ecology 82: 3163–3181.

Smith R.I.L. 1996. Introduced plants in Antarctica: potential impacts and conservation issues. Biol. Conserv. 76: 135–146.

Stenström M., Gugerli F. and Henry G.H.R. 1997. Response of *Saxifraga oppositifolia* to simulated climate change at three contrasting latitudes. Glob. Change Biol. 3(Supplement 1): 44–54.

Sturm M., Racine C. and Tape K. 2001. Increasing shrub abundance in the arctic. Nature 411: 546–547.

Tenhunen J.D. et al. 1992. The ecosystem role of poikilohydric tundra plants. In: Chapin F.S. (ed.), et al.Arctic Ecosystems in a Changing Climate: An Ecophysiological Perspective, Academic Press, New York, pp. 213–237.

Thomas C.D., Cameron A., Green R.E., Bakkenes M., Beaumont L.J., Collingham I.C., Erasmus B.F.N., Ferreirade Siqueira M., Grainger A., Hannah L., Hughes L., Huntley B., van Jaarsveld A.S., Midgley G.F., Miles L., Ortega-Huerta M.A., Peterson A.T., Phillips O.L. and Williams S.E. 2004. Extinction risk from climate change. Nature 427: 145–148.

van der Wal R. and Brooker R.W. 2004. Mosses mediate grazer impacts on grass abundance in arctic ecosystems. Funct. Ecol. 18: 77–86.

Walther G.-R. 2000. Climatic forcing on the dispersal of exotic species. Phytocoenologia 30: 409–430.

Walther G.-R., Post E., Convey P., Menzel A., Parmesan C., Beebee T.J.C., Fromentin J.-M., Hoegh-Guldberg O. and Bairlein F. 2002. Ecological responses to recent climate change. Nature 416: 389–395.

Wardle P. and Coleman M.C. 1992. Evidence for rising upper limits of four native New Zealand forest trees. N.Z. J. Bot. 30: 303–314.

Wijk M.T. van, Williams M., Laundre J.A. and Shaver G.R. 2003. Interannual variability of plant phenology in tussock tundra: modelling interactions of plant productivity, plant phenology, snowmelt and soil thaw. Glob. Change Biol. 9: 743–758.

Wijk M.T. van, Clemmensen K.E., Shaver G.R., Williams M., Callaghan T.V., Chapin F.S. and Cornelissen J.H.C. et al. 2004. Long-term ecosystem level experiments at Toolik Lake, Alaska, and at Abisko, Northern Sweden: generalizations and differences in ecosystem and plant type responses to global change. Glob. Change Biol. 10: 105–123.

Woodward F.I. 1987. Climate and Plant Distribution. Cambridge University Press, Cambridge.

Wookey P.A., Welker J.M., Parson A.N., Press M.C., Callaghan T.V. and Lee J.A. 1994. Differential growth, allocation and photosynthetic responses of *Polygonum viviparum* to simulated environmental change at a high arctic polar semi-desert. Oikos 70: 131–139.

Wookey P.A., Robinson C.H., Parsons A.N., Welker J.M., Press M.C., Callaghan T.V. and Lee J.A. 1995. Environmental constraints on the growth, photosynthesis and reproductive development of *Dryas octopetala* at a high arctic polar semidesert, Svalbard. Oecologia 102: 478–489.

Zhou L.M., Tucker C.J., Kaufmann R.K., Slayback D., Shabanov N.V. and Myneni R.B. 2001. Variations in northern vegetation activity inferred from satellite data of vegetation index during 1981 to 1999. J. Geophys. Res.–Atmos. 106: 20069–20083.

Photo. Open-top chambers (OTC's), placed here in a fellfield vegetation on Anchorage Island (67°36′ S), increase the temperature in the vegetation with about 2-3 °C and simulate the temperature increase envisaged in the next decades. In a paired-plot design of a control plot and an OTC, the effects of this temperature increase are studied. Analysis of stable isotope ratio's in various components of the ecosystem can be used to determine sources and pathways of C and N in Antarctic terrestrial ecosystems and the possible changes in the fluxes of this matter due to climate change effects. (photo: Ad Huiskes).

Plant Ecology (2006) 182:79–86
DOI 10.1007/s11258-005-9032-0

Stable isotope ratios as a tool for assessing changes in carbon and nutrient sources in Antarctic terrestrial ecosystems

A.H.L. Huiskes*, H.T.S. Boschker, D. Lud and T.C.W. Moerdijk-Poortvliet
*Netherlands Institute of Ecology, Royal Netherlands Academy of Arts and Sciences, (NIOO-KNAW), Unit for Polar Ecology, P.O. Box 140, 4400 AC, Yerseke, The Netherlands; *Author for correspondence (e-mail: a.huiskes@nioo.knaw.nl)*

Received 1 September 2004; accepted in revised form 15 December 2004

Key words: δ^{13}C, δ^{15}N, *Deschampsia antarctica*, Lichen, Melt water, *Prasiola crispa*, Precipitation

Abstract

Samples of an angiosperm species, nine lichen species and a terrestrial alga, were collected from a variety of Antarctic terrestrial habitats, and were analysed for C and N stable isotope composition. Collections were made along natural gradients, the marine gradient, running from the sea coast inland and the moisture gradient, determined by melt water and precipitation runoff, and running towards the sea coast. Considerable variation in stable isotope ratios was found; δ^{13}C values ranged between −16 and −32‰ and δ^{15}N values between −23 and +23‰ The variation in stable carbon isotope ratios could be attributed in part to species specific differences, but differences in water availability also played a role, as was shown for the terrestrial alga *Prasiola crispa* and the lichen species *Usnea antarctica*. The differences in the isotope ratios of nitrogen could be retraced to the origin of nitrogen: marine or terrestrial. The nitrogen stable isotope ratios were influenced by both the marine gradient from the sea inland and the melt water and precipitation flow running in the opposite direction, towards the sea. This was shown for the lichen species *Turgidosculum complicatulum* and the angiosperm species *Deschampsia antarctica*. The variation in the C and N stable isotope ratios can be used to determine sources and pathways of N and changes in the water availability in Antarctic terrestrial ecosystems. Contrary to earlier reports the use of stable N isotope ratios is possible in this case because of the relative simplicity of the structure of the Antarctic terrestrial ecosystems.

Introduction

Antarctic terrestrial ecosystems are found in the 2% of the continent that is periodically or permanently snow- and ice-free. They increase in complexity from South to North and from the interior towards the coast. In the more extreme situations (at higher latitudes) trophic structure comprises only two levels, primary producers and decomposers. However, also in more benign circumstances, grazing of primary production and predation of herbivores are generally considered to be negligible (Clarke 1985; Convey 1996). The most complex terrestrial ecosystems are found in the Maritime Antarctic (The west of the Antarctic Peninsula and the outlying islands) at low altitudes close to the sea (Smith 1996).

Nutrient input in Antarctic terrestrial ecosystems can be autochtonous, originating from decomposition of organic material *in situ*, but also, depending on the distance to the sea, allochtonous, coming from the marine ecosystem via sea spray and in the excrements of sea birds and sea mammals, or indirectly via precipitation. In this respect the ecosystems resemble island ecosystems (Stapp et al. 1999; Sanchez-Pinero and Polis 2000). Data on the fluxes of carbon and minerals are few (Davis 1981). The relative contribution of the marine ecosystem and of the mineralisation *in situ* to the mineral pool of the terrestrial ecosystems in the Antarctic is not known. The two angiosperms in Antarctica take up most of the minerals by roots, mosses by rhizoids from the substrate and via their aerial parts from runoff water. The substrates for these organisms are protoranker soils or polar 'soils'. These 'soils' are characterised by the absence of stratification in organic horizons but not necessarily by the absence of organic matter, a feature unique to Antarctica (Fogg 1998). Lichens and terrestrial algae, occurring mainly on scree or rock faces, take up minerals via their thalli, from what moisture sources are available: precipitation, melt water, but also sea spray, mist and fog.

The use of stable isotopes at natural abundance levels provides a powerful approach for understanding food web – and environmental interactions in ecology (Peterson and Fry 1987; Robinson 2001; Dawson et al. 2002). Isotopic compositions of elements, such as C and N, change in predictable ways, during their course through the biosphere, which makes them ideal tracers of the pathways and the origin of these elements. Robinson (2001) discusses the limited use of δ^{15}N as a tracer for N sources. We hypothesize that it is possible to use δ^{15}N as a tracer for N in simple systems, such as the Antarctic terrestrial ecosystem, where the presence of herbivores and carnivores is negligible, and the organic matter produced enters directly the decomposition chain. Even more so, because the potential sources for N have discrete differences in δ^{15}N.

Isotope ratios are expressed in the delta notation

$$\delta X_{\mathrm{std}} = (R_{\mathrm{sam}}/R_{\mathrm{std}} - 1)1000(‰)$$

in which δX_{std} (δ^{13}C or δ^{15}N) is the isotope ratio in delta units relative to a standard (PDB for ^{13}C and atmospheric nitrogen for ^{15}N) and R is the ratio ^{13}C/^{12}C or ^{15}N/^{14}N.

Nitrogen of marine origin has different δ^{15}N values as compared to that of terrestrial origin (Peterson and Fry 1987) and this has been used successfully as source markers for organic matter of marine origin vs. terrestrial origin (e.g. Peters et al. 1978). In order to investigate the source of N into the Antarctic terrestrial ecosystem we assessed the stable isotope ratios of N in the various primary producers of this ecosystem.

As primary producers dominate the terrestrial Antarctic ecosystem, fractionation of CO_2 by the different autotrophic groups of organisms determines largely the δ^{13}C of the organic carbon in the system. However, the exact source of C is difficult to determine as the carbon isotope ratios of the different autotrophic groups show considerable overlap (Galimov 2000). These diffences in carbon isotope ratios are in part species specific (Máguas et al. 1993; Galimov 2000), but can for lichens and algae also be attributed to the relationship between the carboxylation rate and the liquid phase diffusion rate of CO_2 (Máguas et al. 1993; Lange et al. 1988).

Increased melt water run-off may therefore not only alleviate the input of marine N into the ecosystem but, when the liquid phase diffusion of CO_2 becomes increasingly rate limiting and the internal CO_2 pressure increases, the ^{13}C content of the photosynthates increases and less negative δ^{13}C values are the result (Lange et al. 1988).

In this paper, we want to show that differences in water availability in Antarctic terrestrial ecosystems can therefore be detected by differences in the stable isotope ratios of both C and N.

By sampling contrasting habitats, we are able to assess the influence of temperature and water availability and the influence of the marine ecosystem on the terrestrial ecosystem. We focus in this study on the stable isotope ratios of the angiosperm *Deschampsia antarctica*, of lichen species, and of the terrestrial alga *Prasiola crispa* ssp. *antarctica* as they illustrate our hypothesis best.

Materials and methods

Study area

The coastal zones of the Léonie Islands Archipelago (67°36′ S, 68°15′ W), to which Léonie Island and Anchorage island belong, are rocky. More inland, up to ca. 50 m of altitude, many small rock terraces are present, interspersed with rock formations and scree slopes. In depressions of the coastal rock and on the rock terraces pockets of protoranker soils may be found: soils with no structure and of very limited organic content. Here the angiosperm species *Deschampsia antarctica* and the terrestrial alga *Prasiola crispa* ssp. *antarctica* are found. Some of these terraces may be wet for prolonged periods, due to stagnating melt water. The lichens occur on rocky substrates or on scree. Mineral supply is allegedly originating from decomposition, rock weathering and/or of marine origin: nesting Skua's (*Catharacta mccormicki*, *Catharacta antarctica*, and their hybrids), precipitation, and sea spray are the main vectors of minerals of marine origin.

Samples of the Antarctic angiosperm species *Deschampsia antarctica*, nine lichen species, and the terrestrial alga *Prasiola crispa* ssp. *antarctica*, were collected from Léonie Island and Anchorage Island (both in the Léonie Islands Archipelago), in a number of contrasting habitats (coastal–non-coastal, wet–dry, ornitocoprophylic or not influenced by bird guano) (Table 1). The lichen *Usnea antarctica* was also collected from more southerly areas, on Charcot Island (70° S, 75° W) and at Mars Oasis (Alexander Island, 71° S 68° W) from inland rocks. Details about the collection sites are given in Table 2.

Collection and analysis of the samples

Fresh samples were collected in the field and cleaned of adhering debris, by the use of a pair of tweezers and by brushing them gently with tissue paper. They were put in vials and subsequently stored and transported to the Netherlands at −20 °C. Prior to analysis, the samples were freeze-dried (for about 78 h) and powdered in a ball mill. A subsample of known weight was used for further analysis.

Particulate matter of marine origin was sampled in triplicate by filtration of 20 l of surface sea water (about 50 m from the shore) over a pre-combusted (400 °C, 4 h) glass fibre filter on board an inflatable craft, used for diving operations. Within 3 h the filters were freeze dried and subsequently stored and transported to the Netherlands at −20 °C. A section of the filter was cut out and used for further analysis. For the analysis of organic matter of terrestrial origin soil samples

Table 1. Samples collected for this study.

Group	Species	Substratum/Habitat/ Collection area
Angiosperm	*Deschampsia antarctica*	'soil' pockets/wet to dry rock terraces, inland to coastal/ Léonie Island, Anchorage island
Terrestrial alga	*Prasiola crispa* ssp. *antarctica*	places with prolonged water availability/inland (rock terraces, fellfields) to coastal (depressions of coastal rocks)/ Léonie island, Anchorage Island
Lichens	*Pseudephebe minuscula*	rocks/not influenced by birds, intermittent moisture availability, inland/ Léonie Island, Anchorage Island
	Rhizoplaca melanophthalma	rocks/moderate influence by birds, intermittent moisture availability, inland to coastal/Léonie Island; Anchorage Island
	Stereocaulon alpinum	scree/not influenced by birds, intermittent moisture availability, inland/Léonie Island
	Turgidosculum complicatulum	rocks/influenced by birds, intermittent moisture availability, inland to coastal/Léonie Island
	Umbilicaria antarctica	rocks/indirectly influenced by birds via melt water, frequent moisture availability, inland/Anchorage Island, Léonie Island
	Umbilicaria decussata	rocks/indirectly influenced by birds via melt water, intermittent moisture availability, inland to coastal/Léonie Island
	Umbilicaria umbilicarioides	rocks/indirectly influenced by birds via melt water, intermittent moisture availability, inland to coastal/Léonie Island
	Usnea antarctica	rocks/none to moderate influence by birds, intermittent moisture availability, inland to coastal/Charcot Island, Mars Oasis, Anchorage Island, Léonie Island
	Usnea subantarctica	rocks/none to moderate influence by birds, intermittent moisture availability, inland/ Léonie Island

Table 2. Collection sites.

Site notation	Substratum	Moisture availability	Distance to sea (m)	Altitude a.s.l. (m)	Animal influence
Léonie Island coast	Rocky, with depressions containing 'soil'	Intermittent: precipitation, local snow melt, sea water	1–10	5	Resting sea birds
Léonie Island rock outcrop	Rocky, with fellfield and 'soil'	Intermittent: precipitation and local snow melt	50	10	7 Skua nests present
Léonie Island rocky slope	Boulders and scree, pockets of 'soil'	Intermittent: precipitation, local snow melt	50	25–30	5 Skua nests present
Léonie Island terrace	Terrace with boulders and scree	Prolonged: precipitation, melt water runoff, stagnating water	100	40	No Skua nests
Léonie Island scree	Scree slope with boulders	Intermittent: precipitation, local snow melt	100	40	No Skua nests
Léonie Island fellfield	Fellfield: boulders, scree, pockets of 'soil'	Prolonged: Melt water runoff, locally stagnating, precipitation	150	60	±20 Skua nests
Anchorage Island coast	Rocky, with depressions containing 'soil'	Intermittent: precipitation, local snow melt, sea water	1–10	5	Resting sea birds, sea mammals
Anchorage Island terrace	Terrace with boulders and scree	Prolonged: precipitation, melt water runoff, stagnating water	80	20	Skua nests in close vicinity
Anchorage Island scree	Scree slope with boulders	Intermittent: precipitation, local snow melt	100	25	No Skua nests
Anchorage Island fellfield	Fellfield: boulders, scree, pockets of 'soil'	Prolonged: Melt water runoff, locally stagnating, precipitation	150	30	±20 Skua nests
Anchorage Island high rock	Rocky	Intermittent: Steep rock, only precipitation	150	35	No
Anchorage Island wet rock	Rocky	Prolonged: in melt water stream	100	30	No
Charcot island inland rock	Rocky	Occasional local snow melt	±50	20	No
Mars Oasis inland rock	Rocky	Occasional local snow melt	>1 km	>100	No

were collected at a number of places near the collection sites for the plant material. The samples were treated in a similar way as the plant samples.

The carbon and nitrogen isotopic composition of the samples was determined using a Fisons NA-1500 element analyzer coupled on line via a Finnigan conflo II interface, to a Finnigan MAT Delta S isotope ratio mass spectrometer (IRMS). Inorganic carbon was removed by adding 30% HCl (suprapure) to the subsample to be measured (Nieuwenhuize et al. 1994).

Results

The $\delta^{13}C$ and $\delta^{15}N$ values of the samples of the terrestrial alga *Prasiola crispa* ssp. *antarctica* from various habitats, shown in Figure 1, have a comparatively large variation in both the carbon and nitrogen isotope ratios. In coastal and wet habitats $\delta^{13}C$ and $\delta^{15}N$ values are higher as compared to the values obtained from samples from more inland and drier habitats.

In Figure 2, $\delta^{13}C$ and $\delta^{15}N$ ratios of nine lichen species are plotted, collected from various habitats. Both the $\delta^{13}C$ and the $\delta^{15}N$ ratio's show a large variation. With respect to the carbon isotope ratios there are two species distinctly separated from the other seven species: *Stereocaulon alpinum*, which has a more depleted carbon isotope ratio than the majority of the lichen species studied, and*Turgidosculum complicatulum* which has a less depleted carbon isotope ratio than the remainder of the lichen species. *T. complicatulum* has also the highest $\delta^{15}N$ ratios, but there is a wide range in the latter ratios.

With respect to N isotope ratios, $\delta^{15}N$ values in *Usnea antarctica* range from 5 to −20‰ while the other species show mostly positive $\delta^{15}N$ values, with the exception of *Umbilicaria decussata* showing isotope ratios of N, ranging from +8 to −5‰, with one extreme value of −17.32‰.

In Figure 3 $\delta^{15}N$ values of *Deschampsia antarctica* are presented. The $\delta^{13}C$ ratio of *D. antarctica* ranges between −28 and −31‰ which is in the range of C3 plants in lower latitudes, but on the lower side of this range. The $\delta^{15}N$ ratios form a gradient from 8‰ in plants growing close to the coast, to 16‰ in plants growing in the highest terrace sampled. We added in this figure also the results of the lichen *Turgidosculum complicatulum* as the range in $\delta^{15}N$ ratios shown in Figure 2 have a similar relationship with the distance to the sea. Regression lines were calculated with the STATISTICA programme. In this figure we also plotted the $\delta^{15}N$ ratios of particulate organic material

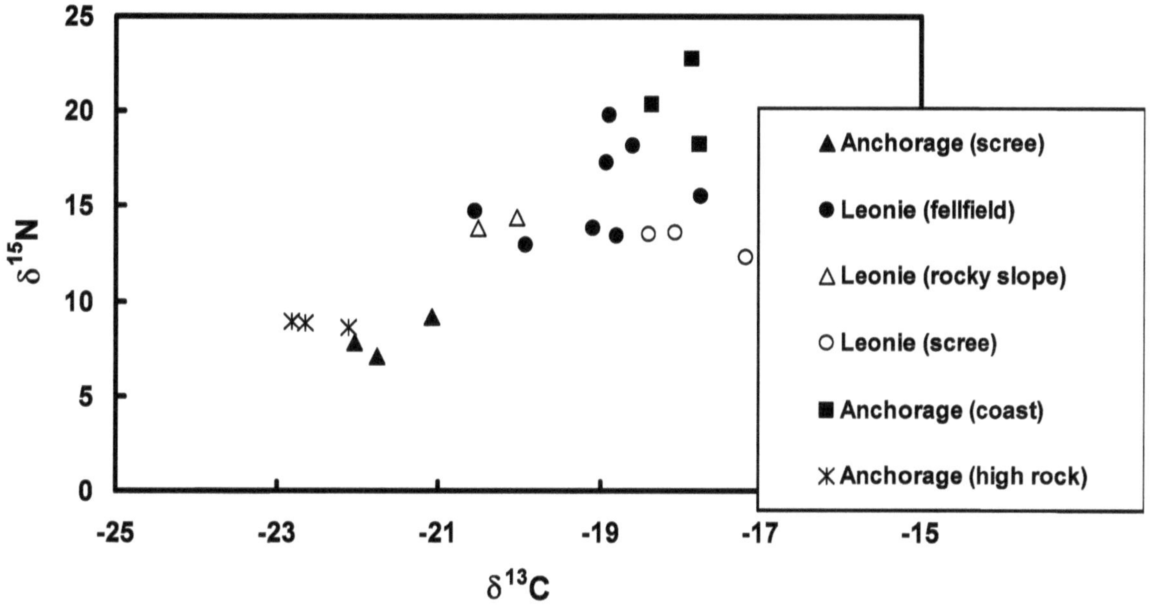

Figure 1. $\delta^{15}N$ ratios of organic material of the terrestrial alga *Prasiola crispa* ssp. *antarctica* plotted against the $\delta^{13}C$ ratio of this material.

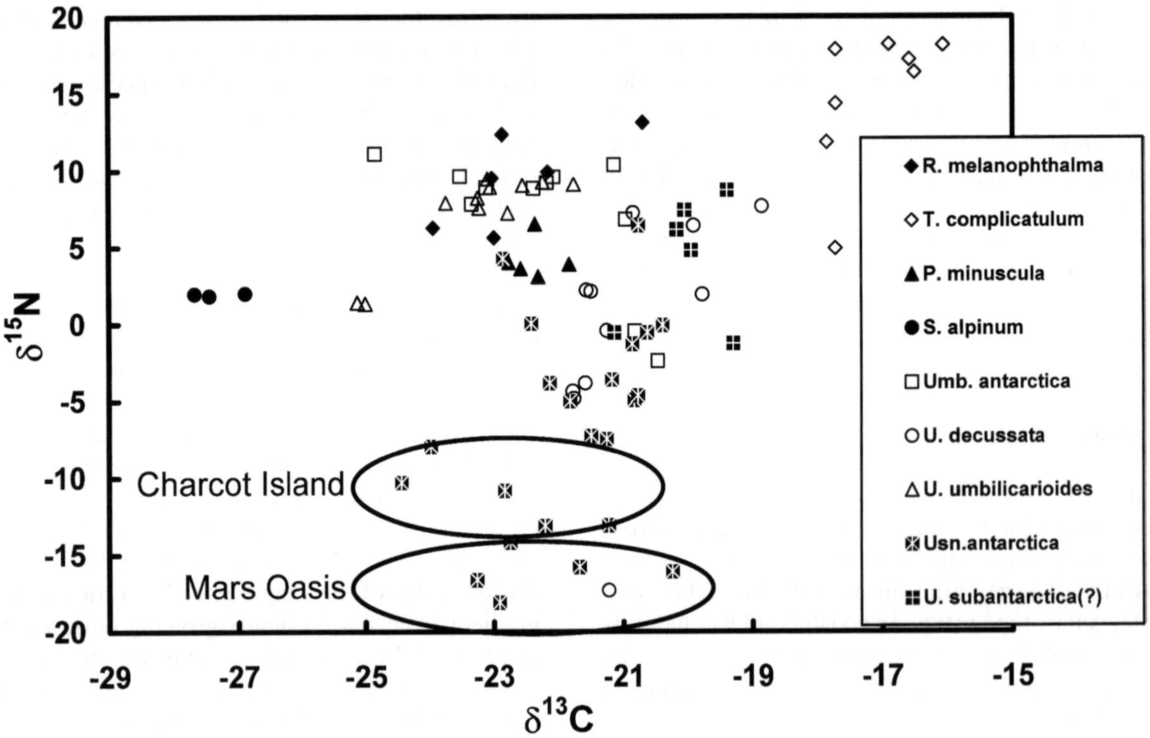

Figure 2. δ^{15}N ratios of organic material of the lichen species *Rhizoplaca melanophthalma, Turgidosculum complicatulum, Pseudephebe minuscula, Stereocaulon alpinum, Umbilicaria antarctica, Umbilicaria decussata, Umbilicaria umbilicarioides, Usnea antarctica*, and *Usnea* cf. *subantarctica* against the δ^{13}C ratio of this material.

collected in sea water and of particulate organic matter in substrates in the PC site, which is the site we sampled furthest away from the sea.

Discussion

Galimov (2000) reported on the carbon isotope composition of Antarctic photoautotrophic organisms (angiosperms, mosses, lichens and the terrestrial alga *Prasiola crispa* ssp. *antarctica*). Although these samples were collected in the South Shetland and South Orkney Islands at 7° of latitude lower than where most of our collections were made, the δ^{13}C ratios were quite similar. He concluded that the variation in carbon isotope composition was species-controlled, rather than a result of variations in environmental factors. These conclusions were corroborated by Máguas et al. (1993), who concluded that 90% of the variation in carbon isotope ratios in the lichen species they studied was species-specific and it depended on the nature of the symbiosis what the specific carbon isotope composition would be. They showed that lichen associations with a green alga together with a cyanobacterium as photobionts were more depleted in δ^{13}C, resembling that of angiosperms with a C3 photosynthesis. Lichens with only a cyanobacterium as a photobiont had less depleted δ^{13}C ratios, while lichens with green algae as photobiont were the least depleted in ^{13}C.

These conclusions are largely corroborated by our results. They explain the relatively low δ^{13}C ratio of *Stereocaulon alpinum* (Figure 2) which has both a green alga and a cyanobacterium as photobionts, resembling the δ^{13}C ratio of C3 plants (e.g. *D. antarctica* with δ^{13}C ratios between −28 and −31‰) while at the other end of the scale we find *Turgidosculum complicatulum*, a lichen species found in the coastal habitat, which has *P. crispa* (a green alga) as a photobiont. This species shows similar δ^{13}C values as the carbon isotope composition of its photobiont in the coastal habitat (Figure 1).

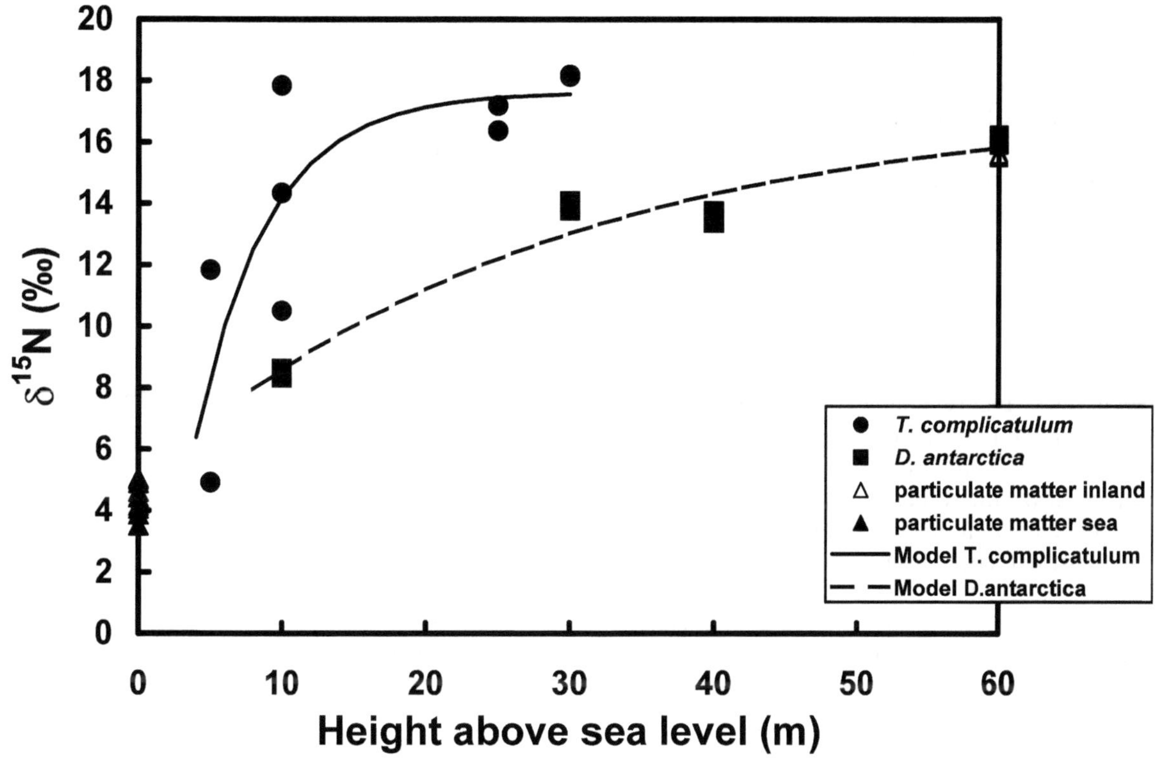

Figure 3. δ^{15}N ratios of organic material of the angiosperm *Deschampsia antarctica* (filled squares) and the lichen *Turgidosculum complicatulum* (filled circles) plotted against the height above sea level (hsl) of the collection sites. δ^{15}N ratios of particulate matter are shown as triangles. The model used for calculating the regression lines was: $\delta^{15}N = a + b \times (1\text{-exp}(\text{-c} \times \text{hsl}))$.

However, the C and N stable isotope ratios of samples cannot be ascribed entirely to species-specific differences. Figure 1 for instance, depicts the stable carbon isotope ratios of *P. crispa* sampled in different habitats, and shows that samples from the wet PC site on Léonie island and the wet coastal zone on Anchorage island are less depleted in ^{13}C than e.g. those from the scree slopes and high rocks on Anchorage island (further away from the sea and moistened intermittently by precipitation). This corroborates with the conclusion drawn by Lange et al. (1988) that, when the liquid phase diffusion of CO_2 becomes increasingly rate limiting (e.g. by stagnating melt water) and the internal CO_2 pressure increases, the ^{13}C content of the photosynthates increases and less negative δ^{13}C values are the result. In dry areas the organisms use atmospheric CO_2 directly, which results in a more depleted δ^{13}C ratio. The δ^{13}C ratios of lichen species from different habitats (Figure 2) support this: *Usnea antarctica* samples from the more continental Antarctic (and therefore more arid) habitats Mars Oasis and Charcot Island are more depleted in ^{13}C then e.g. *Usnea* spp. samples from the maritime Léonie Island.

The δ^{15}N values of *Usnea antarctica* vary considerably between −20 and +10‰, with the lowest values found in the more continental sites. We hypothesize that this could be explained by Erskine et al. (1998) who show that vegetation close to birds nests (as is the case on Léonie Island), may derive N (with a positive δ^{15}N ratio) directly from this source (via melt water), while vegetation further away (at Charcot Island or Mars Oasis) derives N from atmospheric NH_3 (with a negative δ^{15}N ratio) volatilized from bird guano and precipitated with snow. The δ^{15}N ratio, gives us information about the existence of a direct link between the N source and the organisms utilising it, and hence about the water availability.

The δ^{15}N values in Figure 3 show clearly that close to the sea, the δ^{15}N ratio resembles that of particulate organic matter in the sea, while further away from the sea, the ratio resembles that of particulate organic matter of terrestrial origin. This is the case for the angiosperm *D. antarctica* which takes up N mainly by its roots from the soil, but also for the lichen *T. complicatulum* and for the alga *P. crispa* (Figure 1), which take up N from water (melt water, precipitation or sea spray).

The results presented in this paper show that water (e.g. melt water or precipitation) plays an important role in the C and N supply of photoautotrophic organisms in Antarctic terrestrial ecosystems and that the stable isotope ratios of C and N depict this water availability. For instance, the curves in Figure 3 show a steeper curve for *T. complicatulum* than for the angiosperm *D. antarctica*. *T. complicatulum* takes its N from melt water running down the slope, which is only close to the sea appreciably 'diluted' with N from sea water.

Robinson (2001) discussed the limited use of $\delta^{15}N$ as a tracer for N in ecosystem studies. With this study we have shown that in simple ecosystems, such as the Antarctic terrestrial ecosystem, with discrete N isotope ratios of the sources (marine and terrestrial) it is possible to use $\delta^{15}N$ as a tracer for N sources, even if the distances between source and sink are large.

In the Antarctic Peninsula region climate change is allegedly responsible for increased precipitation and increased snow melt (and hence for larger amounts of melt water being present). As these vectors have an effect on the $\delta^{13}C$ and $\delta^{15}N$ values of the photoautotrophic organisms, stable isotope ratios can be used to detect changes in water availability in the terrestrial ecosystem, and can be utilized therefore as indicators for climate change.

Acknowledgements

This study was executed with a grant from the Netherlands Antarctic Programme (NAAP, Grant No. 851.20.012), administered by the Netherlands Organisation for Scientific Research (NWO). We thank J. Nieuwenhuize, Y. Van der Maas, and P. Van Breugel for analysing the stable isotope ratios. The hospitality and the logistic support of the personnel of Rothera Research Station (British Antarctic Survey) is gratefully acknowledged. We dedicate this paper to Jelte Rozema on the occasion of becoming a full professor at the Vrije Universiteit in Amsterdam. Jelte is a long-time friend and a collaborator in many joint projects of the Vrije Universiteit and the Netherlands Institute of Ecology. We keep pleasant memories on the time that some of us collaborated with him in field research at Léonie Island.

References

Clarke A. 1985. Food webs and interactions: an overview of the Antarctic ecosystem. In: Bonner W.N. and Walton D.W.H. (eds.), Key Environments; Antarctica, Pergamon Press, Oxford, pp. 329–350.

Convey P. 1996. The influence of environmental characteristics on the life history attributes of Antarctic terrestrial biota. Biol. Rev. 71: 191–225.

Davis R.C. 1981. Structure and function of two Antarctic terrestrial moss communities. Ecol. Monogr. 51: 125–143.

Dawson T.E., Mambelli S., Plamboeck A., Templer P.H. and Tu K.P. 2002. Stable isotopes in plant ecology. Annu. Rev. Ecol. Syst. 33: 507–559.

Erskine P.D., Bergstrom D.M., Schmidt S., Stewart G.R., Tweedie C.E. and Shaw J.D. 1998. Subantarctic Macquarie Island – a model ecosystem for studying animal-derived nitrogen sources using ^{15}N natural abundance. Oecologia 117: 187–193.

Fogg G.E. 1998. The Biology of Polar Habitats. Oxford University Press, Oxford.

Galimov E.M. 2000. Carbon isotope composition of Antarctic plants. Geochim. Cosmochim. Acta 64: 1737–1739.

Lange O.L., Green T.G.A. and Ziegler H. 1988. Water status related photosynthesis and carbon isotope discrimination in species of the lichen genus *Pseudocyphellaria* with green or blue-green photobionts and in photosynbiodemes. Oecologia 75: 494–501.

Máguas C., Griffiths H., Ehleringer J. and Serôdio J. 1993. Characterization of photobiont associations in lichens using carbon isotope discrimination techniques. In: Ehleringer J., Hall A. and Farquhar G. (eds.), Stable Isotopes and Plant Carbon–Water Relations, Academic Press, New York, pp. 201–212.

Nieuwenhuize J., Maas Y.E.M. and Middelburg J.J. 1994. Rapid analysis of organic carbon and nitrogen in particulate materials. Mar. Chem. 45: 217–224.

Peters K.E., Sweeney R.E. and Kaplan I.R. 1978. Correlation of carbon and nitrogen stable isotope ratios in sedimentary organic matter. Limnol. Oceanogr. 23: 598–604.

Peterson B.J. and Fry B. 1987. Stable isotopes in ecosystem studies. Ann. Rev. Ecol. Syst. 18: 293–320.

Robinson D. 2001. $\delta^{15}N$ as an integrator of the nitrogen cycle. Trends Ecol. Evol. 16: 153–162.

Sanchez-Pinero F. and Polis G.A. 2000. Bottom-up dynamics of allochthonous input: direct and indirect effects of seabirds on islands. Ecology 81: 3117–3132.

Smith R.I.L. 1996. Terrestrial and freshwater biotic components of the Western Antarctic Peninsula. Antarct. Res. Ser. 70: 15–59.

Stapp P., Polls G.A. and Pinero F.S. 1999. Stable isotopes zreveal strong marine El Nino effects on island food webs. Nature 401: 467–469.

Photo. Rice cultivation in the Philippines. Photograph by Peter Verburg.

Plant Ecology (2006) 182:89–106
DOI 10.1007/s11258-005-9033-z

Upscaling regional emissions of greenhouse gases from rice cultivation: methods and sources of uncertainty

Peter H. Verburg[1,*], Peter M. van Bodegom[2], Hugo A. C. Denier van der Gon[1]
Aldo Bergsma[1] and Nico van Breemen[1]
*[1]Laboratory of Soil Science and Geology, Wageningen University, PO Box 37, 6700 AA, Wageningen, The Netherlands; [2]Institute of Ecological Science, Free University, De Boelelaan 1085, 1081 HV, Amsterdam, The Netherlands; *Author for correspondence (e-mail: peter.verburg@wur.nl; phone: +31-317-485208; fax: +31-317-482419)*

Received 1 September 2004; accepted in revised form 15 December 2004

Key words: Climate change, Methane emission, Rice paddy fields, Uncertainty, Upscaling

Abstract

One of the important sources of greenhouse gases is the emission of methane from rice fields. Methane emission from rice fields is the result of a complex array of soil processes involving plant-microbe interactions. The cumulative effects of these processes at the level of individual plants influence the global atmospheric composition and make it necessary to expand our research focus from small plots to large landscapes and regions. However, present extrapolations ('upscaling') are tenuous at best because of methodological and practical problems. The different steps taken to calculate regional emission strengths are discussed and illustrated by calculations for a case-study in the Philippines. The applicability of high quality, process-based, models of methane emission at the level of individual plants is limited for regional analysis by their large data requirements. Simplified models can be used at the regional level but are not able to capture the complex emission situation. Data availability and model accuracy are therefore often difficult to match. Other common sources of uncertainty are the quality of input data. A critical evaluation of input data should be made in every upscaling study to assess the suitability for calculating regional emissions. For the case-study we show effects of differences in input data caused by data source and interpolation technique. The results from the case-study and similar studies in literature indicate that upscaling techniques are still troublesome and a cause of large uncertainties in regional estimates. The results suggest that some of the stumbling blocks in the conventional upscaling procedure are almost impassable in the near future. Based on these results, a plea is made for meso-level measurements to calibrate and validate upscaling methods in order to be better able to quantify and reduce uncertainties in regional emission estimates.

Introduction

Local human activities can lead to global changes in climate. Anthropogenic activities leading to climate change are those related to fossil fuel production and burning, to forestry and agriculture, to waste disposal, and to ozone depleting chemical (ODC) manufacture and use (Wilbanks and Kates 1999). These activities emit trace gases like CO_2, CH_4, N_2O, or Ozone Depleting Compounds (ODCs) that accrue in the atmosphere and increase radiative forcing, thereby warming the surface of the earth. One of the important sources of greenhouse gas emissions is the emission of methane from rice fields (Intergovernmental Panel on Climate Change 1997). Worldwide emission from rice has been extrapolated from reports from China, India, Vietnam, Korea, and the Philippines to be from 21 to 30 teragrams per year (1 teragram = 10^{12} g) (Sass et al. 2002b). These values are less than several estimates since 1981, but still represent a globally significant source. As we move into the future, rice grain production must increase to feed an increasing population, while at the same time, methane emissions from irrigated rice agriculture need to be reduced to help stabilize the global climate. Thus, the relationship between rice grain yield and the emission of methane from irrigated rice fields emerges as a major scientific and policy issue (Sass and Cicerone 2002).

Methane emission from rice fields is the result of a complex array of soil processes involving plant microbe interactions. Flooding rice fields promotes anaerobic fermentation of carbon sources supplied by the rice plants and other incorporated organic substrates. Methane thus produced may be partly oxidised under influence of oxygen released by the rice plant. The majority of the remaining methane is again released through the rice plant into the atmosphere. Plant growth and agricultural management are therefore linked to future climate change. Indirectly, climate may affect methane emission through effects on rice growth as suggested by various authors (Sass et al. 1991; Denier van der Gon et al. 2002).

The cumulative effects of the plant-microbe interactions leading to the emission of methane make it necessary to expand our research predictions from small plots of land to large landscapes and regions. However, present extrapolations are tenuous at best because of methodological and practical problems. Expanding the scale of analysis, decreases the integrity of individual units, and increases the variability among them (Levin 1993). Woodmansee (1989) argues that an appropriate analysis of trace gas emissions should be based on flux rates of these gases determined at their sources, that is, at the location where the biological activity responsible for their formation is prevalent. That location is, at least for most agricultural sources of greenhouse gas emissions, at the soil pore and individual bacterium, bacterial colony or soil enzyme scale of spatial resolution. Aggregate emissions for landscapes, regions or the whole globe are obtained by 'upscaling'. This 'scale-up' or 'bottom-up' paradigm is the idealised 'first principles' approach attempted by most natural science studies (Root and Schneider 1995). Empirical observations made at small scales are used to determine possible mechanistic associations or 'laws of nature' that are then extrapolated to predict larger scale responses. If this knowledge gained and tested on small systems is to be applied to larger ones, rules must be found by which the fine-scale information can be scaled and applied to coarser-scale phenomena. Often, fine-scale relations/processes are applied directly, or with minor changes, to describe the properties of coarser-scale aggregates. The problem with this approach is that the aggregate does not generally behave the same way as the fine-scale components from which it is constituted, because of feedbacks within the system and non-linear system behaviour (O'Neill 1979).

Large numbers of flux measurements at detailed scales within wetland rice fields or under laboratory conditions have greatly increased our understanding at the scale of individual plants of the processes controlling methane emission (Khalil et al. 1998b; Minami et al. 1994). The results of these studies reveal a huge variation of flux rate values mostly depending on soil, climate, and water, nutrient management and plant growth. Therefore, the first attempts to provide regional/global estimates based on a simple multiplication of a locally measured methane flux and the area of paddy soils (e.g. Cicerone and Shetter 1981; Wassmann et al. 1993), are unrealistic. Later improvements of the method started with distinguishing different types of water management and multiplying rice areas by management specific emission factors (Neue et al. 1990; Intergovernmental Panel on Climate Change 1997). Other improvements include methods that are based on differences in primary production or the amount of carbon returned to the rice soil during the rice crop cycle (Bachelet et al. 1995). Because it is acknowledged that emission factors cannot capture the large number of variables explaining differences in rice-growing environments, there is a joint effort to replace emission factors by process or semi-empirical models (Cao et al. 1995; Huang et al. 1998a; Van Bodegom et al. 2001; Cai et al. 2003). These models are used to calculate the emission source strength for the specific conditions in the different mapping units that can be distinguished within the studied region (Li et al. 2004). This paper discusses the various upscaling problems for a case-study on regional methane emissions from wetland rice fields in the Philippines. This paper will draw special attention to the different steps made during such upscaling procedures and illustrates why and where uncertainties arise. Based on this analysis recommendations will be made for the improvement of future regional emission estimation procedures.

Methods and data

Upscaling procedure and sources of uncertainty

The upscaling methodology is divided into a number of methodological steps (Figure 1), described in this section. The uncertainties that arise during this procedure are indicated and quantified for the case study region addressed in this paper.

Step 1 encompasses the definition of the factors that determine emissions and the quantification of underlying processes. This is often done by process-based modelling. Process-based models are frequently based on the analysis of processes and emission measurements at the local scale. A first category of uncertainties are introduced because of misinterpretation of processes, model abstractions and inaccuracies in the model definition. These inaccuracies are assessed by validating the model with field measurements.

Ideally, a detailed process-based model should form the basis for the calculation of regional fluxes. However, data availability is limiting at the regional scale, hindering the application of a process-based model. A simpler model is often needed. Such a simplified model relates emission strengths to available, key variables using simplified process relations or empirical equations. Simplified models based on empirical equations might suffer from a lack of causality. Therefore, simplified models that are derived from detailed process models are generally more valid for upscaling studies (Williams et al. 1997; Huang et al. 1998b). These simplifications induce new uncertainties in the actual calculation of emission strengths because of the less complete system

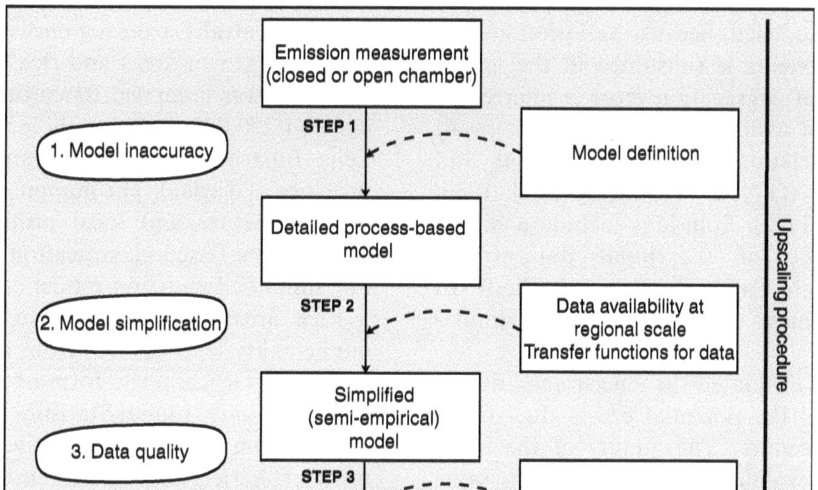

Figure 1. Procedure followed for upscaling of small-scale emission measurements to regional scale emission estimates (right side of figure) and the uncertainties induced during this procedure (left side).

description. We have assessed this source of uncertainty by comparing the results of a detailed process-based model with the results of a simplified model that was selected to make best use of the data available in the case study.

Step 3 involves the calculation of emission strengths for the individual spatial units distinguished within the region of interest with the simple emission model. During these calculations, errors can arise from data inaccuracies, or as a consequence of inaccurate transfer functions that generate the required data out of available data. Another source of error is related to the spatial data resolution. Because of the impracticality of handling large numbers of fine-scale components individually, they are generally lumped into an aggregated component and treated collectively. For example, instead of calculating the emission source strength for every single 1 m^2 within the region, more or less homogeneous areas are lumped into aggregated components, called patches, which are often represented as mapping units. The fine-scale model is often applied directly, or with minor changes, to describe the properties of these coarse-scale aggregates. This procedure causes errors. A coarse-scale model, assembled from fine-scale relationships, can be inaccurate even when the underlying, fine-scale processes are well understood and can themselves be adequately modelled, because variation among fine-scale components is subsumed in the aggregate. This type of aggregation error is referred to as the 'fallacy of averages' and is a direct result of non-linear relationships between input and output variables (O'Neill 1979; King et al. 1989; Rastetter et al. 1992). Solutions include a higher spatial resolution of the input data (more patches) or Monte-Carlo simulation methods. In this study we have used Monte-Carlo simulations.

We have both evaluated the inaccuracies due to data quality and the potential errors due to the aggregation procedure. The quality of the input parameters was evaluated by comparing the same variable as derived from different sources for the study area. Rice areas were derived from national statistical surveys and from the interpretation of radar images, respectively. At the same time we have tested the influence of different methods to characterise soil texture variability based on a soil map. Aggregation errors were explored by calculating emissions with and without taking within-unit variability of soil and yield into account with the help of Monte-Carlo simulation methods.

Study area and local case-studies

The study area is located in the Central Luzon plain in the Philippines, not far north of Manila (Figure 2). The study area covers around 800,000 ha and is one of the most important rice growing areas of the Philippines bounded by mountain ranges. Within the study area two smaller, local, case-study areas are distinguished for which additional data were available. The first area (Muñoz area) measures about 20,000 hectare located in the province of Nueva Ecija and is representative for the irrigated, intensively used rice-growing areas in the region. The second study area for which detailed data are available is the Victoria area which is representative for the rainfed rice-growing conditions in the region and is located in Victoria municipality, Tarlac province (IRRI 1992, 1995).

Data

For the central Luzon region we used soil texture data, rice grown areas and rice yields. A soil texture map was compiled based on 4 provincial soil maps (1:250,000 scale for the provinces of Nueva Ecija, Bulacan and Pampanga and 1:50,000 for the province of Tarlac). The mapping units were based on soil texture and local properties (Figure 3). This texture based classification was the input of our simplified emission model described below.

Rice areas and yields were available at the municipality level ($n = 95$) from both the National Statistical Office in the form of the 1991 Agricultural Census (National Statistics Office 1994a, b, c, d) and from the provincial offices of the Department of Agriculture. Although the NSO statistics are probably more reliable because of more advanced processing and double-checking of the data, they do not distinguish between the dry and wet season and between rainfed and irrigated rice-cultivation systems as did the statistics from the provincial offices. These are, however, very important distinctions for estimating methane

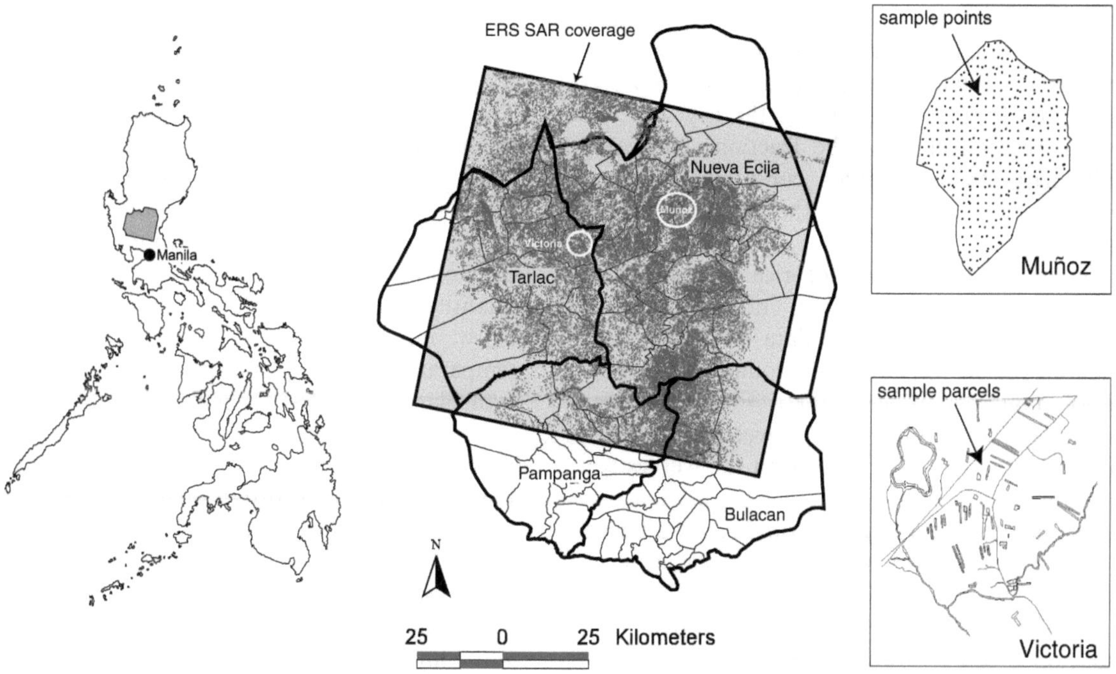

Figure 2. Location of the study region. Left: within the Philippines; Middle: map of the four provinces in the study area with municipality boundaries and area covered by the ERS-SAR image (black: rice area); Right: detailed maps of the two sites with more detailed sample data.

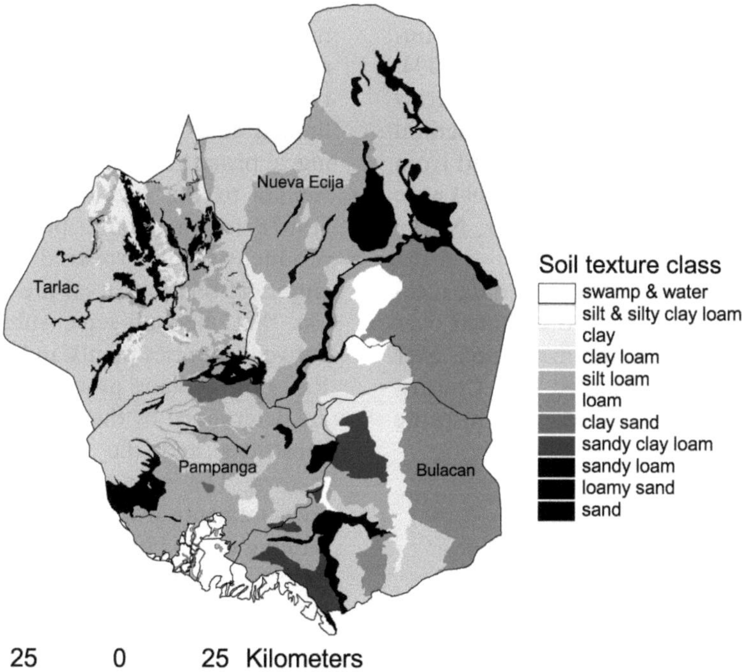

Figure 3. Soil texture map of the study area.

emissions (Yagi et al. 1996). Therefore, we have chosen to combine the two data sources by dividing the total rice area reported in NSO statistics by the fraction grown in the dry and wet season, respectively and the fractions cultivated under irrigated and rainfed conditions reported in provincial statistics of the Department of Agriculture.

Higher resolution data of rice location were obtained from a supervised interpretation of multi-temporal ERS-SAR data for the area (Figure 2) using very limited ground-truth information (van der Woerd 2000). ERS-SAR area data could be derived for the wet season only, consequently all analyses focus on the wet season emission.

For the detailed analysis in the Muñoz area we used an extensive data set of soil properties, crop management and yields compiled by Oberthür et al. (1996, 1999). Soil data in this area were based on a detailed, quasi-systematic soil survey, made in 1993. A regular 750×750-m grid with 341 sample locations was used. In addition, samples were taken from 43 randomly selected fields. Chemical and physical analysis was performed for all samples. For this study we have used soil texture, organic C and Fe contents of the soil. Crop sampling and farm management practices were recorded for a random selection of the soil sampling sites during the dry and wet seasons of 1994. Variables describing yield, total biomass, straw production and seedling treatment method (transplanting/direct seeding) were extracted from this data-base. Daily climate records for 1994 were recorded at the climate station in Muñoz, Nueva Ecija (Climate Unit 1995).

For the Victoria area, we used yield data measured on approximately 80 fields distributed over the different environments found in the area over the period 1990–1996. The Rainfed Rice Consortium of the International Rice Research Institute collected these data (IRRI 1995).

Models

Detailed process model (step 1 in Figure 1)
The detailed process-based model used in this study is described by Van Bodegom et al. (2001). Two compartments, a rhizosphere and a bulk soil, are distinguished in this model. To calculate methane emissions, the model contains simplified process-based descriptions of methane production, transport and oxidation for each compartment. The model can also be used for situations in which the rice yield is not optimal (Van Bodegom et al. 2001). Among the most important variables needed for the simulations are the soil reducible iron content and the soil organic carbon content. Other controlling variables are rice variety, rice yield, quantity of NO_3 or SO_4 containing fertilisers, straw input, length of growing period, daily temperature, seedling treatment (direct seeding or transplanting) and water management.

Simplified semi-empirical model (step 2 in Figure 1)
Based on their own process-based model (Huang et al. 1998b), Huang et al. (1998a, b) derived a simplified model for methane emission that predicts methane emissions with limited information which makes the model applicable to large areas with limited data sets. The relation between methane production and rice growth and development was given much emphasis in the process-based model. The influence of the soil environment was characterised by the relative effect of soil texture on methane production/emission (Huang et al. 1997) and linked to soil sand content. The simplified model was derived from the detailed model by generalising the process-equations and adding empirically derived relations to replace relations that cannot be quantified because of the limited data availability at regional scales following a procedure for model simplification that is similar to the one followed by Williams et al. (1997) for a model predicting gross primary productivity of terrestrial ecosystems. The semi-empirical relations used are based on the hypothesis that methanogenic substrates are primarily derived from rice plants and added organic matter. Rates of methane production in flooded rice soils are determined by the availability of methanogenic substrates and the influence of environmental factors. The model was validated against field measurements from various regions of the world (Huang et al. 1998b). The simplified model needs information on growth duration, grain yield, soil texture (% sand), average temperature and a rice variety index as inputs; these are variables that are generally more easily available than the set of inputs needed by the detailed process model. Huang et al. (1998a) used the model to make a regional

emission estimate for China. The model does not distinguish rainfed rice systems from irrigated systems, therefore we adjusted calculated emissions for rainfed rice by 40% according to the IPCC guidelines (Intergovernmental Panel on Climate Change 1997).

Model results were compared for approximately 65 sites within the Muñoz area for which all necessary data needed by both the detailed model by Van Bodegom et al. (2001) and the simplified model by Huang et al. (1998a, b) were available for the dry and wet season of 1994.

Aggregation procedure (step 3 in Figure 1)

A number of different, common aggregation procedures are compared to calculate the regional methane flux for the Central Luzon region.

Procedure 1 is based on the calculation of the region-wide average value of the variables determining methane emission, followed by the calculation of the average seasonal emission for the area with the simplified model of Huang et al. (1998a, b). This average emission strength was multiplied by the rice harvested area as derived from the ERS-SAR image interpretation. The average parameter values were calculated based on a weighted-averaging procedure using the relative rice area in the soil-mapping unit (soil texture) and municipality (yield). Sand percentages were calculated by taking the average value of the range of sand percentages belonging to the texture class in the legend of the soil map (Figure 3; (Soil Survey Staff 1995)).

Procedure 2 simulates the emissions for all individual mapping units (n$\sim$700) resulting from an overlay of the soil map and the municipality map. For every mapping unit a unique combination of yield, fraction rainfed rice cultivation and soil texture was determined. The specific methane emission strength for all individual mapping units was determined with the simplified emission model. These emissions were multiplied with the rice area derived from the ERS-SAR image interpretation for the mapping unit, after which the total emission for the study area was calculated. Sand percentages were calculated following procedure 1.

Procedure 3 uses a sophisticated approach for upscaling that aims at minimising aggregation errors modulated by spatial heterogeneity that inherently remains within the mapping units (Rastetter et al. 1992). King et al. (1989) developed this method by using a multivariate set of frequency distributions as input rather than lumped parameters related to a spatially explicit subdivision of land units. Monte Carlo techniques were used to integrate solutions of the simulation model across the geographic region of interest (Figure 4).

Frequency distributions of methane emission for all mapping units based on assumed frequency distributions for yields and soil texture composition were generated using Monte Carlo simulations with the simplified model. Frequency distributions for yield variability within a typical mapping unit were derived from the two case-study areas for irrigated rice yields and rainfed rice yields, respectively. It was assumed that all mapping units have similar variability in yield around the reported average yield. Variability in sand percentage was derived from the high-resolution soil sampling available in the Muñoz area. Again we assumed that this variability is representative for the whole study region. After a frequency distribution for methane emission was generated for all mapping units and distribution functions were fitted, we have again used a Monte Carlo procedure to integrate these results into a regional emission frequency distribution. This was done by consequent sampling from the frequency distributions of the emission strength of the individual mapping units and multiplying the sampled emission strengths by the rice area in the mapping unit.

Procedure 4 uses a method similar to procedure 3, only now we assume a different distribution for the variability in soil texture within the mapping units. Instead of using the variability of sand percentage in the Muñoz area, a uniform distribution of sand percentage between the values valid for the texture class based upon the soil texture triangle was assumed.

Results

Model accuracy (uncertainty 1 in Figure 1)

The detailed process model of Van Bodegom et al. (2001) was validated against measurements of a closed flux-chamber experiment conducted in Maligaya, which is within the Muñoz study area in Nueva Ecija province (Corton et al. 1995).

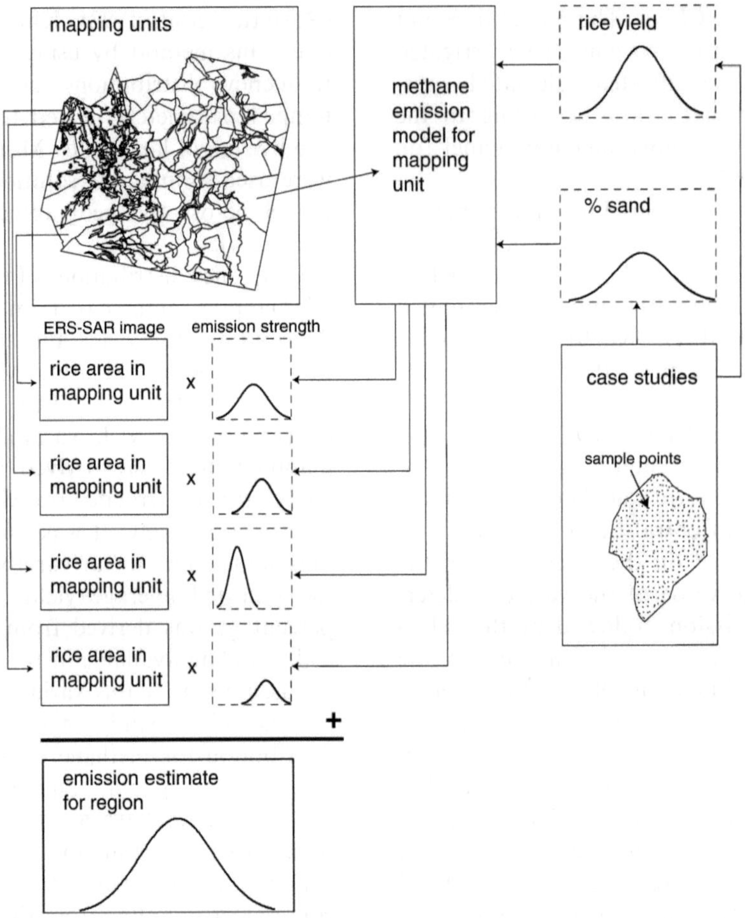

Figure 4. Upscaling procedure with Monte-Carlo simulations (procedure 3 in text). For each mapping unit a frequency distribution is generated based on frequency distributions of rice yield and soil texture with a simplified emission model. Emission source strengths for the individual mapping units are multiplied with the rice area within the mapping unit and integrated into a region emission estimate.

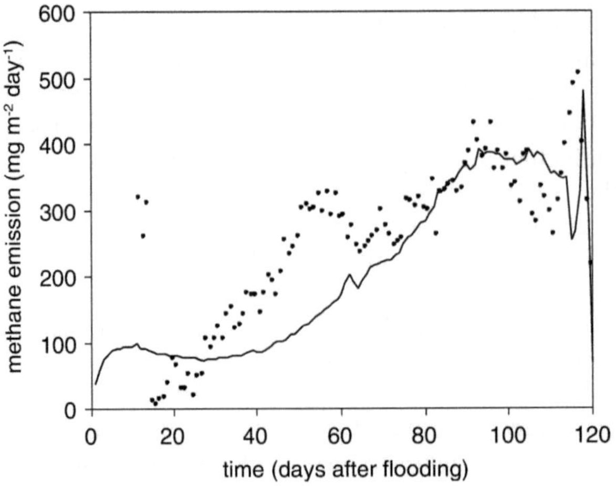

Figure 5. Validation of model performance (line) for measured values in Maligaya, Nueva Ecija Provinde (dots).

Measured and modelled emissions were available on a daily basis. Overall performance of the simulation model is reasonable and not significantly different from the measurements (Figure 5). In spite of some overestimation and underestimation during parts of the growing season, the total seasonal emission is well captured. The model was also validated for other experimental sites across Asia. For 9 sites spread over East Asia an average deviation of 7% in seasonal methane emission was found, indicating that the major processes driving methane emissions from rice field were well captured by the model (see Van Bodegom et al. (2001) for a full description of the validation).

Model simplification (uncertainty 2 in Figure 1)

Figure 6 shows, for each of the 65 sample points in the Muñoz area, the emission calculated by the detailed process-model and the semi-empirical model for the dry season and the wet season of 1994. Although the order of magnitude of the emissions calculated by the models is similar, there is little correspondence between the results of the two models.

The difference between the simulation results of the models of Huang et al. (1998a) and van Bodegom et al. (2001) is strongly related to the sand and organic carbon content of the soil. Given the observation that the estimate of methane emissions is sensitive to the description of mineralisation, as was shown before by van Bodegom et al. (2000) it is most likely that the deviation between the model results is caused by the difference in the description of the mineralisation process. Huang et al. (1998b) use a correlation between methane emissions and sand content. This is because the fact that sand content is a proxy for soil mineralisation (see e.g. (Parton et al. 1987)). The higher the sand content, the less organic matter is protected and thus the more can be mineralised. Van Bodegom et al. (2001) directly used a semi-empirical model to calculate mineralisation. This semi-empirical model relates organic carbon to mineralisation; the higher organic carbon content, the more mineralisation occurs. In addition to these factors, it is important to understand what happens to the mineralised carbon. Huang et al. (1998b) assume that the produced carbon converted into methane is a function of the redox potential. The redox potential is however taken as a fixed function of time, independent of soil type. Van Bodegom et al. (2001) allow changes in the time dependent function of Eh by using the total reducible iron content as an input parameter. Iron is the most important alternative electron acceptor in soils and is thus the dominant factor determining the soil Eh. By incorporating iron instead of Eh, it is possible to distinguish soils with a slow and fast change in Eh. The higher the iron content and the lower the soil C content, the slower the decrease of Eh and thus the smaller the methane production.

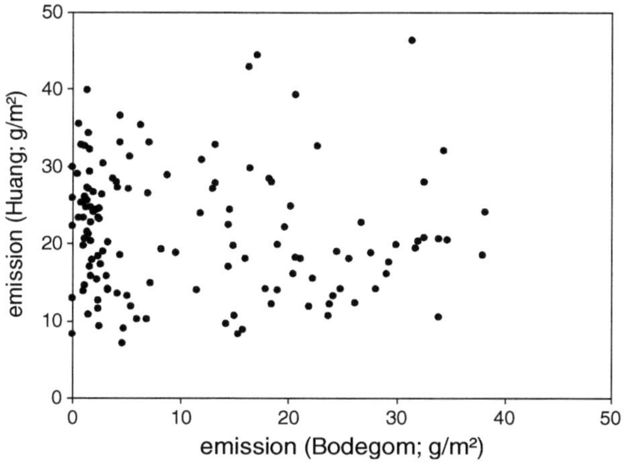

Figure 6. Methane emission for approx. 65 sites for both the wet and dry season calculated with the process-based simulation model of Van Bodegom et al. 2001 and the simplified model of Huang et al. 1998b, respectively.

Data quality (uncertainty 3 in Figure 1)

The rice area in the wet season determined by supervised classification of the ERS-SAR image is shown in Figure 2. A comparison between the rice area derived from the ERS-SAR image and the rice area given in the statistical surveys, shown in Figure 7, shows large deviations between the two sources for a number of municipalities. In general, the rice grown area as identified by ERS-SAR is smaller than the area reported in statistics (almost 30% for the total study area). There can be several causes for this difference:

- Inconsistency in statistical sources; statistics by the National Statistical Office differ considerably from those by the Bureau of Agricultural Statistics (Philrice/BAS 1995) and statistics by the provincial offices of the Department of Agriculture.
- Differences between the years of reporting in statistics (1991) and the ERS-SAR image (1995). However, in time series of the Bureau of Agricultural Statistics, there is no major difference in total rice area between these years.
- Interpretation and classification problems of the ERS-SAR image. Many rice fields are relatively small and irrigation canals and dykes are abundant. These small elements can cause an underestimation of the rice area in the ERS-SAR interpretation. More fieldwork might improve the interpretation of the ERS-SAR image.

It is clear that the regional emission calculated with the rice area of the ERS-SAR is lower than the emission calculated with the statistical data.

Aggregation errors (uncertainty 4 in Figure 1)

The results of the four different upscaling procedures are shown in Table 1. Emissions differ only slightly among the methods used. Also the different frequency distributions for soil texture (procedures 3 and 4) make little difference in total regional emission. The spatial distribution of average emission source strengths for irrigated and rainfed rice fields respectively, as calculated with procedure 3, is shown in Figure 8. Spatial variability within the region is large. Patterns of emission source strength for irrigated and rainfed rice correspond to a large degree because of similar soil texture, a correlation between irrigated and rainfed yields and the assumed constant reduction of emissions in rainfed vs. irrigated fields.

Discussion

Model selection and simplification

Model validation indicated that we were capable to simulate the processes leading to methane emissions from rice fields to a reasonable degree with the detailed process model. Other models that were based on process knowledge have led to comparable good performances (Khalil et al. 1998a; Matthews et al. 2000; Cai et al. 2003). These models have proven to be relevant for assessments of the conditions that might lead to a reduction of emissions (Frolking et al. 2004) and shape management recommendations. However, these models require detailed data that are not available at regional scales. Therefore, we are not always able to use this type of process models for

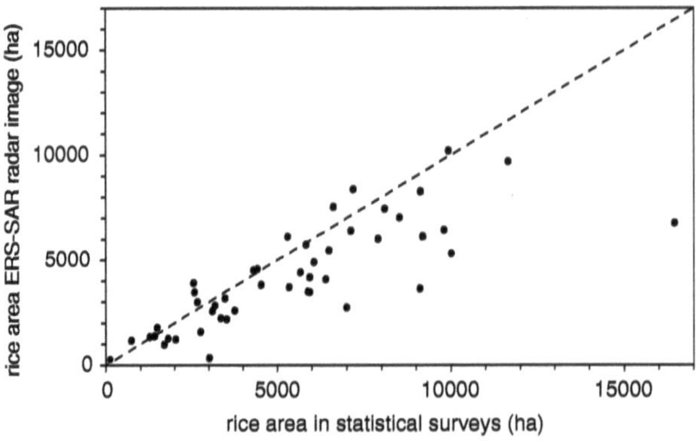

Figure 7. Rice area for municipalities within the study area derived from statistics and radar image interpretation respectively.

Table 1. Total methane emission calculated for the Central Luzon region based upon different upscaling methodologies or the IPCC methodology (Intergovernmental Panel of Climate Change 1997).

Upscaling procedure	Emission (10^6 kg per season) (standard deviation between brackets)	Aggregation procedure
Procedure 1		
Rainfed	13.5	One simulation based on weighted averages of soil/management parameters
Irrigated	44.0	
Total	57.5	
Procedure 2		
Rainfed	13.6	Simulation for each mapping unit (soil/management) based on mapping unit information followed by summation of emissions
Irrigated	43.6	
Total	57.1	
Procedure 3		
Rainfed	13.4 (0.4)	Simulation for each mapping unit (soil/management) based on mapping unit parameters and Monte-Carlo simulation of within unit variability; followed by summation
Irrigated	43.6 (1.5)	
Total	57.0 (1.5)	
Procedure 4		
Rainfed	13.5 (0.2)	Same as procedure 3 with different assumptions of within unit variability
Irrigated	43.2 (1.2)	
Total	56.8 (1.2)	
IPCC		
Rainfed	13.0–15.6	Use of emission factors (IPCC 1995)
Irrigated	40.2–48.2	
Total	53.2–63.8	

upscaling procedures. In case of the discussed process-based model information is needed on soil iron and organic carbon contents that are unavailable for the large (800,000 ha) study area. The selection of an appropriate simplified model turns out to be a very critical step within the upscaling process. In this study, the model by Huang et al. (1998a, b) was selected because it made full use of the available data for the study region. The model was validated for a number of independent methane measurements in the USA, Italy, Indonesia, the Philippines and China. So, based on this information it was assumed, as is common practice in upscaling studies, that this model was appropriate for the upscaling exercise. Results from the Muñoz case study area, nested within the larger study area of Central Luzon, indicate that there is no correspondence between the detailed model and the simplified model. As is shown in Table 2, the average emission for the 65 sample points in the Muñoz area is about twice as high when calculated with the simplified model. Because validation by measurements is not possible (65 measurement sites within a region would be too expensive) we can only conclude that the choice of model has an important influence on the obtained results. This makes model selection a large source of uncertainty for the regional estimate and the generated spatial emission pattern (e.g. Figure 8).

Also in other studies it has been indicated that a model validated under a range of circumstances does not necessarily lead to good predictions in other conditions. An example is given by a region-wide validation of the DNDC model that was developed and validated mainly based on the cropping practices and soil conditions in the U.S. and China (Frolking et al. 2004). The validation over a wide range of locations indicated that, in spite of good simulation results in many locations, poor results were obtained for a number of other locations, mostly attributed to local farming practices and soil conditions for which the model was not validated in earlier instants (Cai et al. 2003).

Although model comparison and validation might facilitate the selection of an appropriate model for a certain region, it might not necessarily reduce the uncertainty in emission estimates. A considerable amount of variability in emissions may be driven by natural stochasicity in biogeochemical processes, limiting the ability of researchers to constrain the estimates at this level of analysis (Sass et al. 2002a).

Data quality

Data quality is critical to any upscaling procedure. The comparison between various sources of land

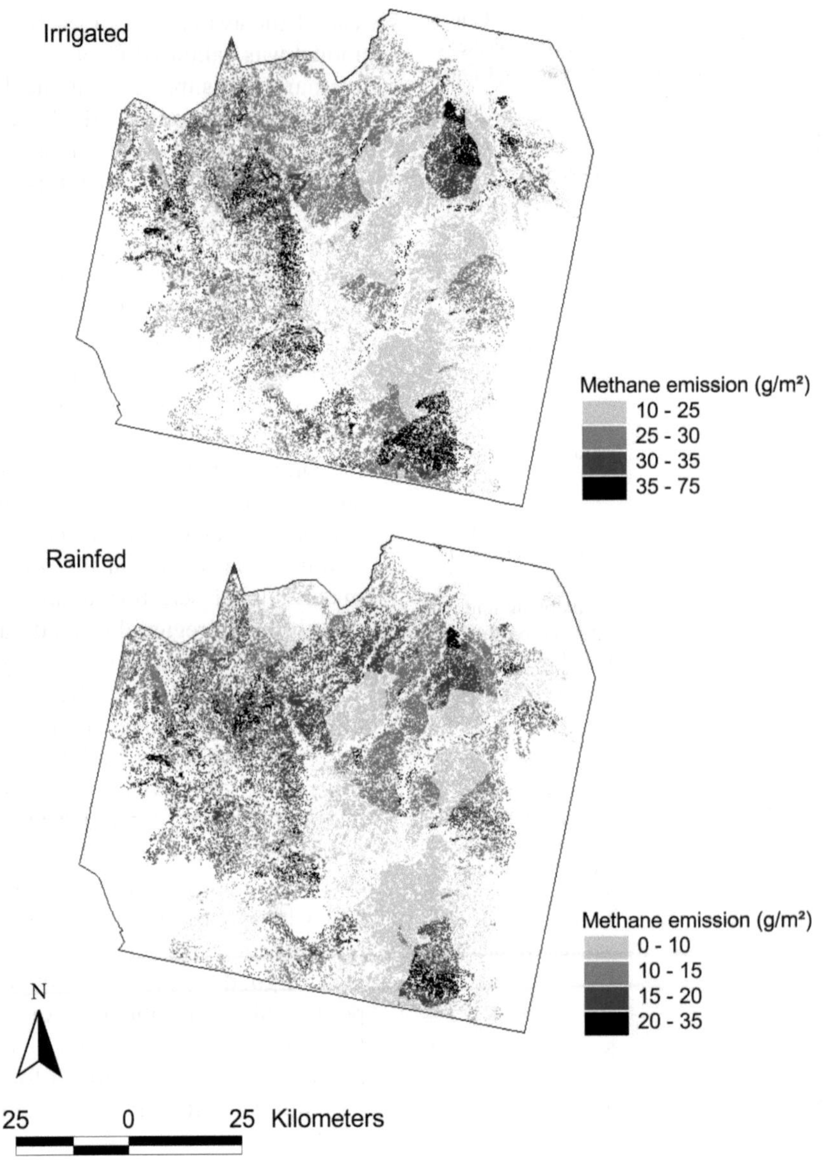

Figure 8. Spatial variability in source strength of methane emission (g/m^2) for the study region for irrigated and rainfed rice fields respectively; rice area is derived from ERS-SAR image interpretation, no distinction is made between irrigated and rainfed rice area.

use statistics and remote sensing information illustrated that the parameter 'rice area', which is generally assumed to be the most straightforward parameter, is subject to large uncertainties. Census reports based on farm sampling are often subject to classification and report problems and do not always distinguish between cropping seasons, while that is needed for greenhouse gas emission inventories. Large uncertainties in statistical data are also well known for other countries. In China, one of the largest contributors to the global methane emission from wetland rice fields, surveys of rice area differ by about 40%, causing an enormous uncertainty in total methane emission (Smil 1999; Verburg and Denier van der Gon 2001). Also the use of radar images is not without uncertainties as classification can still be problematic (Le Toan et al. 1997; van der Woerd 2000). The advantage of using the detailed maps produced by radar images is the high spatial

Table 2. Comparison of average emission for Muñoz case study for two models.

Season (1994)	Average of 65 simulations and standard deviations	Simulation based on average input values
Bodegom		
Dry season	10.41 (10.6)	4.27
Wet season	12.07 (11.4)	3.22
Huang		
Dry season	22.77 (9.0)	23.02
Wet season	20.60 (7.4)	20.65

The average emission is calculated based upon the average of 65 individual simulations or based upon one simulation using average input parameters.

resolution, which allowed us to make better use of the spatial detail of other parameters. Statistical information is mostly limited to administrative units, highest resolution being municipalities in the case of the Philippines, which tend to stretch across several soil units. Therefore, no proper linkage can be obtained between municipality data and soil properties because a breakdown of the rice area within the municipality is not possible. For example, some municipalities cover both lowland area and mountainous area in the considered study area. Using municipality statistics for the rice area would induce large errors in emission calculations because of the incorporation of soil data from these mountainous areas.

There are also uncertainties involved with the use of soil maps. Different options exist for translating soil texture classes into a sand percentage and estimating the within-unit variability. The two procedures evaluated in this paper (procedures 3 and 4) yielded similar results. Other uncertainties lie in the reliability of the soil maps used. Oberthür et al. (1996, 1999) assessed the quality of the soil map we used for the Muñoz area by comparing the soil textures delineated on the soil map with sampled soil data and interpolated samples using Thiessen polygon and kriging techniques. They concluded that these reconnaissance soil maps poorly represent the variability in soil texture and generally have a low accuracy. Van Bodegom et al. (2002a) assessed the effect of this type of data quality issues for methane emissions for the island of Java, Indonesia. The authors calculated methane emissions based on respectively the FAO soil map connected to an international database of soil profiles (Bachelet and Neue 1993; Batjes 1995) for soil chemical properties (Fe and organic C) and compared this to emissions created independently using soil data derived from a kriging interpolation of 555 (top-)soil samples of rice soils taken throughout Java. Resulting total emission estimates, calculated with the process-based model of Van Bodegom et al. (2001), differed by 43%, an alarming difference. Reasons for this difference include: (1) Rice soils differ from non-rice soils as a result of waterlogging and oxidation-reduction sequences, which alter the topsoil chemical composition. Therefore the topsoil chemical properties of a soil type grown to rice for many years will differ from the topsoil composition of that same soil type under forest, pasture or grown to a non-rice crop. (2) Topsoil properties play a negligible role in the criteria for soil type definition and the within-soil type variation of topsoil properties may be large. Therefore it can by no means be guaranteed that linking topsoil chemical properties from reference profiles of soil types collected at many different sites to a soil map of a specific region captures, or even correlates to, the actual topsoil composition. (3) Mapping units of soil maps are mostly based on geomorphological and morphological properties. Therefore, the mapping unit delineations often have little relevance for chemical properties. Within-mapping unit variance of these properties might therefore be equal to the variability between mapping units. In spite of these considerations it is common practice to use soil maps for upscaling purposes (Cao et al. 1998; Ven and Tempel 1994) but should, based on these results, be done with great care. Lack of alternative data is often a major constraint, but this study illustrates that using the 'best available' does by no means guarantee that the data are appropriate to generate reliable emission estimates.

Aggregation errors

We assessed the effects of within-unit variability in yield and soil texture on the calculated, regional, methane emission. Monte Carlo techniques are an appropriate means to reduce aggregation effects and approximate the within-unit variability. We found that taking within-unit variability into account only resulted in slightly different results with a relatively small standard deviation (compare procedures 2 and 3 in Table 1). Even the calculation that disregards all variability within the study area (procedure 1) resulted in a similar result. This similarity is caused by the relatively linear behaviour of the simplified emission model over the range of parameter values occurring in the study area, which reduces aggregation errors. The use of a more detailed, non-linear, process model would probably have resulted in much larger scaling errors. The use of a detailed model was not possible for the region as a whole because of data limitations. However, for the smaller Muñoz area we were able to use both the simplified and the detailed model. A simulation was made with average parameter values for the area based on the 65 sample points and compared to the average emission based on simulations for all the individual sample points. Table 2 presents the results. The detailed model indeed shows a very large aggregation error, whereas the aggregation error within the simplified model is very small. This indicates that, although the simplified model is insensitive to aggregation errors, aggregation errors can seriously affect the regional estimate depending on the non-linearity of the processes involved. The choice of model influences the extent of aggregation error that is reflected in the results. Confidence intervals presented are therefore not necessarily confidence intervals based on the actual aggregation error but are instead model specific, determining the applicability and validity of the models.

Other upscaling studies report relatively low aggregation errors: for methane emissions in Java an aggregation error of approx. 10% was found (Van Bodegom et al. 2002b) while for nitrous oxide emissions in Costa Rica a similar 10% error was found (Plant 2000). Both studies were conducted with detailed process-based models.

Other scaling issues

Summation of source strengths for individual spatial units (e.g., mapping units) is invalid when lateral interactions exist in the response of the individual patches, i.e. if feedbacks at a higher aggregation level occur (O'Neill 1988). Oxidation of atmospheric methane at the soil interface or within aboveground rice biomass could cause such a feedback. However, methane oxidation in rice paddy soils hardly occurs at atmospheric methane concentrations (Hanson and Hanson 1996) while for oxidation within a rice canopy no significant methane oxidation could be detected (Van Bodegom et al. 2002a). All lateral interactions can thus be neglected, as was previously assumed by Aselmann (1989).

All upscaling issues addressed in this paper refer to spatial scaling issues. However, also from a temporal point of view, scaling issues need to be taken into account: yields and climatic conditions change yearly. In our study we have tried to use the data available for 1995, but sometimes had to supplement the database with data from other years, causing another inaccuracy in the input data. On longer time-scales the emission of methane is subject to changes in rice area and management (Denier van der Gon 2000). Land use changes can affect emissions and should be taken into account when using emission inventories to formulate emission reduction policies (Khalil et al. 1993; Verburg and Denier van der Gon 2001).

Implications for upscaling research

The sources of uncertainty in regional estimates of methane emissions from rice fields distinguished in this study are interconnected: the choice of a more accurate simulation model might involve the need for parameters that only can be obtained with high uncertainties or large costs. Model selection also influences the aggregation error (Table 2) as it reflects only model behaviour. The real aggregation error depends on the non-linearity involved with the processes causing methane emissions. Monte-Carlo techniques are an appropriate means to reduce aggregation errors. Reduced uncertainties and more realistic confidence intervals will result if model behaviour is closer to reality and reliable estimates of the within-unit variability can be made.

Most upscaling studies do not account for all these different types of uncertainties but simply report the regional estimate obtained. Such an estimate is often legitimised through the comparison of different calculation methods for methane emission. Sass et al. (1999) compared 21 studies which utilise various biological, chemical, and physical factors to calculate a total annual methane flux from China. In spite of the differences in methods used to obtain these estimates, they are impressively similar, with an average methane emission estimate of 13.0 ± 3.3 Tg methane yr^{-1}. This similarity does not necessarily demonstrate that we are able to estimate the value of the emission level very accurately. Instead, it only indicates that the simplified models all direct us to similar, average seasonal fluxes. Other uncertainties are not assessed in any of the methods reviewed (Sass et al. 1999): they are all based on rice area data derived from official statistics, which are known to underestimate the actual rice area by approx. 40% (Smil 1999), and none of the methods takes into account aggregation errors. For our case-study we found a striking similarity in the emissions calculated with the simplified model and a straightforward application of IPCC emission factors (Table 1). Again the same rice area was used for both estimates and methane fluxes are subject to large uncertainties for both methods. Li et al. (2004) assessed uncertainties in greenhouse gas emissions for China based on variability in county-level soil conditions. Although the range of emissions reported in their paper captures some of the potential uncertainties in aggregate emission estimates, these do not address all other uncertainties that might influence the emission estimate.

Conclusions

This paper has shown, through a series of calculations and analyses for a case study in the Philippines that (conventional) upscaling of greenhouse gas emissions from rice paddies involves a large number of methodological pitfalls and uncertainties. A reduction of the uncertainties involved with upscaling cannot solely be achieved by a further improvement of our knowledge of the processes leading to methane emissions. The complexity and variability of methane emissions is well studied and a large number of measurements have been made. However, the complexity and number of interacting processes involved leads to inaccuracies when simplified models are used. So, at present, less data demand goes hand-in-hand with a reduced ability to capture the existing and observed variability. Even though simplified models might capture the average emission level correctly, these methods are unsuitable to predict the spatial pattern of emissions, which is needed for target-oriented mitigation strategies. Obtaining data for more detailed, process-based models will reduce model errors at the cost of increased data inaccuracies. Especially the use of soil chemical data, almost essential for methane emission calculations, remains a major source of uncertainty because of the high sampling requirements for reliable data interpolation. At the moment it seems impossible to meet the data requirements of detailed, process-based models. The results of this study indicate that uncertainties in regional emission estimates can largely be reduced by investment in regional to continental scale database. Such data will not only reduce the uncertainty in the data but, at the same time, allow the use of models that are better able to capture the processes of methane emission.

This study has indicated that uncertainty in regional emission estimates is determined by context specific variability at different levels of analysis and errors in data sources, models and aggregation procedure. Therefore, it is not possible to make general statements of the total uncertainty in emission estimates. Often, published uncertainty ranges only reflect the uncertainty in a number of parameters at the scale of application, disregarding the influence of uncertainty at higher spatial resolutions that might influence the regional estimates through non-linear aggregation.

Methods to validate upscaling estimates are urgently needed. Validation is restricted as a result of the present difficulty to measure emissions at regional scales that could provide the data to validate our upscaling methodologies such as aircraft and or tethered balloon gas flux measurement and box model methods (Hanna 1982; Liu et al. 2000), which will certainly help to improve upscaling efforts. One alternative means to obtain regional emissions strengths are downscaling methods that calculate regional methane emissions with inverse modelling using atmospheric transport models (Heimann and Kaminski 1999). Such a downscaling procedure would be able to yield independent regional estimates of regional methane emissions (Denier van der Gon et al. 2000). However, the present limitation of this method is the spatial resolution, which is $8° \times 10°$ (Houweling et al. 1999), much larger than the study area presented in this paper.

If meso-scale methods and downscaling procedures are integrated with upscaling procedures, our estimates of regional fluxes may be improved. Such an integration could be made along the line of the procedure proposed by Root and Schneider (1995) called Strategic Cyclical Scaling that combines upscaling and downscaling methods in a cyclic procedure. In case of methane emissions, the approach could be used to select appropriate simplified models that are valid for the larger scale regions but still underpinned by causal relationships. Investment in further development of mesoscale measurements and downscaling methods will enable such innovative scaling procedures that might be successful in stepping beyond the uncertainties and pitfalls of (traditional) upscaling methods.

Acknowledgements

We would like to thank all people that helped with the collection of statistical information and soil maps in the Philippines, especially Jocy Bajita (PhilRice) and Dr S.P. Kam (IRRI). We thank Thomas Oberthür and Achim Dobermann for making their Muñoz data set available, To Phuc Tuong and Annie Boling for providing the Victoria data set and the UPRICE project (van der Woerd 2000) for the radar interpretation of the area.

References

Aselmann I. 1989. Global-scale extrapolation: a critical assessment. In: Andreae M.O., Schimel D.S. and Robertson G.P. (eds), Exchange of Trace Gasses Between Terrestrial Ecosystems and The Atmosphere. Report of the Dahlem Workshop, Berlin 1989, February 19–24. Wiley, Chichester.

Bachelet D., Kern J. and Tolg M. 1995. Balancing the rice carbon budget in China using spatially-distributed data. Ecol. Modelling 79: 167–177.

Bachelet D. and Neue H.U. 1993. Methane emissions from wetland rice areas of Asia. Chemosphere 26: 219–237.

Batjes N.H. 1995. A homogenized soil data file for global environmental research: a subset of FAO, ISRIC and NRCS profiles (Version 1.0). 95/10, ISRIC, Wageningen.

Cai Z.C., Sawamoto T., Li C.S., Kang G., Boonjawat J., Mosier A., Wassmann R. and Tsuruta H. 2003. Field validation of the DnDC model for greenhouse gas emissions in East Asian cropping systems. Global Biogeochem. Cyc. 17(4): 1107. doi:101029/2003GB002046, 2003.

Cao M.K., Dent J.B. and Heal O.W. 1995. Modeling methane emissions from rice paddies. Global Biogeochem. Cyc. 9: 879–897.

Cao M.K., Gregson K. and Marshall S. 1998. Global methane emission from wetlands and its sensitivity to climate change. Atmos. Environ. 32: 3293–3299.

Cicerone R.J. and Shetter J.D. 1981. Sources of atmospheric methane: Measurements in rice paddies and a discussion. J. Geophys. Res. 86: 7203–7209.

Climate Unit I. 1995. Compilation of Weather Data for 1994, Climate Unit, Agronomy, Plant Physiology, and Agroecology Division, International Rice Research Institute, Los Banos, Philippines.

Corton T.M., Bajita J.B., Asis C.A. Jr. and Pamplona R. 1995. Methane emission from an irrigated Phillipine paddy field subjected to several fertilizer treatments. Philippine J. Crop Sci. 20(1): 39–55.

Denier van der Gon H.A.C. 2000. Changes in CH_4 emissions from rice fields from 1960 to the 1990s. I. Impacts of modern rice technology. Global Biogeochem. Cyc. 14: 61–72.

Denier van der Gon H.A.C., Houweling S., Van Bodegom P.M., Verburg P.H. and Van Breemen N. 2000. Combining upscaling and downscaling of methane emissions from rice fields: methodologies and preliminary results. Nutrient Cyc. Agroecosyst. 58: 285–301.

Denier van der Gon H.A.C., Kropff M.J., van Breemen N., Wassmann R., Lantin R.S., Aduna E., Corton T.M. and van Laar H.H. 2002. Optimizing grain yields reduces CH4 emissions from rice paddy fields. Proceedings of the National Academy of Sciences 99(19): 12021–12024.

Frolking S., Li C.S., Braswell R. and Fuglestvedt J. 2004. Short- and long-term greenhouse gas and radiative forcing impacts of changing water management in Asian rice paddies. Global Change Biol. 10: 1180–1196.

Hanna S.R. 1982. Handbook on Atmospheric Diffusion. Technical Information Center, Department of Energy, U.S.

Hanson R.S. and Hanson T.E. 1996. Methanotrophic bacteria. Microbiol. Rev. 60: 439–471.

Heimann M. and Kaminski T. 1999. Inverse modelling approaches to infer surface trace gas fluxes from observed atmospheric mixing ratios. In: Bouwman A.F. (ed.), Approaches to Scaling of Trace Gas Fluxes in Ecosystems. Developments in Atmospheric Science 24, Elsevier, Amsterdam.

Houweling S., Kaminski T., Dentener F.J., Lelieveld J. and Heimann M. 1999. Inverse modeling of methane sources and sinks using the adjoint of a global transport model. J. Geophys. Res. 104(D21): 26137–26160.

Huang Y., Sass R.L. and Fisher F.M. 1997. Methane emission from Texas rice paddy soils. 1. Quantitative multi-year dependence of CH_4 emissions to soil, cultivar and grain yield. Global Change Biol. 3: 479–489.

Huang Y., Sass R.L. and Fisher F.M. 1998a. Model estimates of methane emission from irrigated rice cultivation of China. Global Change Biol. 4: 809–821.

Huang Y., Sass R.L. and Fisher F.M. 1998b. A semi-empirical model of methane emission from flooded rice paddy soils. Global Change Biol. 4: 247–268.

Intergovernmental Panel of Climate Change 1997. Revised 1996 IPCC Guidelines for National Greenhouse Gas Inventories. Workbook, OECD, Paris.

IRRI 1992. Agrohydrologic and socioeconomic characterization of the rainfed lowland rice research consortium site at

Victoria, Tarlac, Philippines, Program report for 1992. International Rice Research Institute, Manila, pp. 66–68.
IRRI 1995. Program Report for 1995. International Rice Research Institute, Manila, Philippines.
Khalil M.A.K., Rasmussen R.A. and Shearer M.J. 1998a. Effects of production and oxidation processes on methane emissions from rice fields. J. Geophys. Res. 103: 25233–25239.
Khalil M.A.K., Rasmussen R.A., Shearer M.J., Dalluge R.W., Ren L.X. and Duan C.L. 1998b. Measurements of methane emissions from rice fields in China. J. Geophys. Res. 103: 25181–25210.
Khalil M.A.K., Rasmussen R.A., Wang M.X. and Ren L. 1993. Methane sources in China: historical and current emissions. Chemosphere 26: 127–142.
King A.W., Johnson A.R., O'Neill R.V. and DeAngelis D.L. 1989. Using ecosystem models to predict regional CO_2 exchange between the atmosphere and the terrestrial biosphere. Global Biogeochem. Cyc. 3: 337–361.
Le Toan T.F., Ribbens F., Wang L.-F., Floury N., Ding K.-H., Kong A., Fujita M. and Kurosu T. 1997. Rice crop mapping and monitoring useing ERS-1 data based on experiment and modelling results. IEEE T. Geosci. Remote Sens. 35(1): 41–56.
Levin S.A. 1993. Concepts of Scale at the Local Level. In: Ehleringer J.R. and Field C.B. (eds.), Scaling Physiological Processes; Leaf to Globe. Physiological Ecology, Academic Press, San Diego, pp. 7–20.
Li C.S., Mosier A., Wassmann R., Cai Z.C., Zheng X.H., Huang Y., Tsuruta H., Boonjawat J. and Lantin R. 2004. Modeling greenhouse gas emissions from rice-based production systems: sensitivity and upscaling. Global Biogeochem. Cyc. 18, GB1043. doi:101029/2003GB002045.
Liu J.G., Zhang Y.H., Shao K.S., Shao M., Zeng L.M., Lu S.H., Slanina S. and Denier van der Gon H.A.C. 2000. Estimation of regional methane emission from rice fields using simple atmospheric diffusion models. Nutrient Cyc. Agroecosyst. 58(1–3): 303–310.
Matthews R., Wassmann R. and Arah J. 2000. Using a crop/soil simulation model land GIS techniques to assess methane emissions from rice fields in Asia. I. Model development. Nutr. Cycl. Agroecosyst. 58(1–3): 141–159.
Minami K., Mosier A. and Sass R.L. 1994. CH_4 and N_2O: Global emissions and controls from rice fields and other agricultural and industrial sources. Nat Inst Agro-Environm Sci, Tsukuba, Japan.
National Statistics Office 1994a. 1991 Census of Agriculture, Volume 2: Final report (by location of the farm). Bulacan, Republic of the Philippines.
National Statistics Office 1994b. 1991 Census of Agriculture, Volume 2: Final report (by location of the farm). Nueva Ecija, Republic of the Philippines.
National Statistics Office 1994c. 1991 Census of Agriculture, Volume 2: Final report (by location of the farm). Pampanga, Republic of the Philippines.
National Statistics Office 1994d. 1991 Census of Agriculture, Volume 2: Final report (by location of the farm). Tarlac, Republic of the Philippines.
Neue H.U., Becker-Heidmann P. and Scharpenseel H.W. 1990. Organic matter dynamics, soil properties and cultural practices in rice lands and their relationship to methane production. In: Bouwman E.D. (ed.), Soils and the Greenhouse Effect. John Wiley and Sons, pp. 457–466.
Oberthür T., Dobermann A. and Neue H.U. 1996. How good is a reconnaissance soil map for agronomic purposes?. Soil Use Manag. 12: 33–43.
Oberthür T., Goovaerts P. and Dobermann A. 1999. Mapping soil texture classes using field texturing, particle size distribution and local knowledge by both conventional and geostatistical methods. Eur. J. Soil Sci. 50(3): 457–479.
O'Neill R.V. 1979. Transmutation across hierarchical levels. In: Innis G.S. and O'Neill R.V. (eds.), Systems Analysis of Ecosystems, International Cooperative Publishing House, Fairlands, Md., pp. 59–78.
O'Neill R.V. 1988. Hierarchy theory and global change. In: Rosswall T., Woodmansee R.G. and Risser P.G. (eds.), Scales and Global Change., John Wiley and Sons, New York, pp. 29–45.
Parton W.J., Schimel D.S., Cole C.V. and Ojima D.S. 1987. Analysis of factors controlling soil organic matter levels in the Great Plains grasslands. Soil Sci. Soc. Am. J. 51: 1173–1179.
Philrice/BAS 1995. Provincial Rice Statistics (1995). Philrice, Maligaya, Philippines.
Plant R.A.J. 2000. Regional analysis of soil-atmosphere nitrous oxide emissions in the Northern Atlantic zone of Costa Rica. Global Change Biol. 6(6): 639–653.
Rastetter R.A., King A.W., Cosby B.J., Hornberger G.M., O'Neill R.V. and Hobbie J.E. 1992. Aggregating fine-scale ecological knowledge to model coarse-scale attributes of ecosystems. Ecol. Applications 2: 55–70.
Root T.L. and Schneider S.H. 1995. Ecology and climate: research strategies and implications. Science 269: 334–341.
Sass R.L. and Cicerone R.J. 2002. Photosynthate allocations in rice plants: food production or atmospheric methane, invited commentary. Proceedings of the National Academy of Sciences 99: 11993–11995.
Sass R.L., Fisher F.M., Ding A. and Huang Y. 1999. Exchange of methane from rice fields: National, regional, and global budgets. J. Geophys. Res. 104(D21): 26943–26951.
Sass R.L., Fisher F.M., Turner F.T. and Jund M.F. 1991. Methane emission from rice fields as influenced by solar radiation, temperature and straw incorporation. Global Biogeochem. Cyc. 5: 335–350.
Sass R.L., Andrews J.A., Ding A. and Fisher F.M. 2002a. Spatial and temporal variability in methane emissions from rice paddies: implications for assessing regional methane budgets. Nutrient Cyc. Agroecosyst. 64(1–2): 3–7.
Sass R.L., Mosier A. and Zheng X.H. 2002b. Introduction and summary: International Workshop on Greenhouse Gas Emissions from Rice Fields in Asia. Nutrient Cyc. Agroecosyst. 64(1–2): ix–xv.
Smil V. 1999. China's agricultural land. China Quart. 158: 414–429.
Soil Survey Staff (1995). Soil Survey Manual, Bureau of Plant Industry, Soils and Agricultural Engineering, USDA.
Van Bodegom P.M., Leffelaar P.A., Stams A.J.M. and Wassmann R. 2000. Modelling methane emissions from rice paddies: variability, uncertainty and sensitivity analysis of processes involved. Nutrient Cyc. Agroecosyst. 58(1–3): 231–248.
Van Bodegom P.M., Verburg P.H. and Denier van der Gon H.A.C. 2002a. Upscaling methane emissions from rice paddies: problems and possibilities. Global Biogeochem. Cy. 16(1): Art. No. 1014.

Van Bodegom P.M., Verburg P.H., Stein A., Adiningsih S. and Denier van der Gon H.A.C. 2002b. Effects of interpolations and data resolution on methane emission estimates from rice paddies. Ecol. Environ. Stat. 9: 5–26.

Van Bodegom P.M., Wassmann R. and Metra-Corton T.M. 2001. A process-based model for methane emission predictions from flooded rice paddies. Global Biogeochem. Cy. 15: 247–263.

van der Woerd H. 2000. Upscaling Methane Emissions from Wetland Rice Fields (UPRICE); Feasibility study on employing radar data for the Philippines – Final report to the BCRS on Phase-1 – Proejct 4.2/AP-09, Institute for Environmental Studies, Amsterdam.

Ven T.v.d. and Tempel P. 1994. ISIS 4.0, ISRIC Soil Information System. 14, ISRIC, Wageningen.

Verburg P.H. and Denier van der Gon H.A.C. 2001. Spatial and temporal dynamics of methane emissions from agricultural sources in China. Global Change Biol. 7(1): 31–49.

Wassmann R., Schütz H., Papen H., Rennenberg H., Seiler W., Aiguo D., Renxing S., Xingjian S. and Minxing W. 1993. Quantification of methane emissions from Chinese rice fields (Zhejiang Province) as influenced by fertiliser treatment. Biogeochemistry 20: 83–101.

Wilbanks T.J. and Kates R.W. 1999. Global changes in local places: How scale matters. Climatic Change 43: 601–628.

Williams M., Rastetter R.A., Fernandes D.N., Goulden M.L., Shaver G.R. and Johnson L.C. 1997. Predicting gross primary productivity in terrestrial ecosystems. Ecol. Appl. 7: 882–894.

Woodmansee R.G. 1989. Ecosystem Processes and Global Change. In: Rosswall T., Woodmansee R.G. and Risser P.G. (eds.), Scales and Global Change, Wiley, New York, pp. 11–27.

Yagi K., Tsuruta H., Kanda K. and Minami K. 1996. Effect of water managment on methane emission from a Japanese rice paddy field: automated methane monitoring. Global Biogeochem. Cyc. 10: 255–267.

Photo 1. Two of the eight metal frames with six UV-lamps each in Adventdalen, Spitsbergen. The vegetation can be classified as middle arctic tundra typical of valley floors along the West Coast of Spitsbergen. The vegetation is dominated by *Salix polaris*, *Saxifraga oppositifolia*, *Bistorta vivipara* and several mosses including *Sanionia uncinata*. The lichen cover is relatively sparse, with *Peltigra didactyla* and *Peltigra rufescent* dominating.

Plant Ecology (2006) 182:109–118
DOI 10.1007/s11258-005-9034-y

Effects of enhanced UV-B radiation on nitrogen fixation in arctic ecosystems

Bjørn Solheim[1,*], Matthias Zielke[1,2], Jarle W. Bjerke[1,3] and Jelte Rozema[4]
*[1]Department of Biology, Faculty of Science, University of Tromsø, N-9037 Tromsø, Norway; [2] University Centre in Svalbard, N-9171 Longyearbyen, Norway; [3]Norwegian Institute for Nature Research, The Polar Environmental Centre, N-9296 Tromsø, Norway; [4]Department of Systems Ecology, Institute of Ecological Science, Climate Center Vrije Universiteit, De Boelelaan 1087, 1081 Amsterdam, HV, The Netherlands; *Author for correspondence (e-mail: bjorn.solheim@ib.uit.no)*

Received 1 September 2004; accepted in revised form 15 December 2004

Key words: Bryophytes, Cyanobacteria, Cyanolichens, Nitrogen fixation, The Arctic, UV-B radiation

Abstract

Recent global climate models predict a further significant loss of ozone in the next decades, with up to 50% depletion of the ozone layer over large parts of the Arctic resulting in an increase in ultraviolet-B radiation (UV-B) (280–315 nm) reaching the surface of the Earth. The percentage of total annual ecosystem N input due to biological nitrogen fixation by cyanobacteria might be as high as 80% and the contribution to total annual N uptake by plants up to 20%. A possible reduction of nitrogen fixation raises serious concerns about already nutrient impoverished plant communities. This review shows that nitrogen fixation by moss-associated cyanobacteria in arctic vegetation was dramatically reduced after six years of exposure to enhanced UV-B radiation. In subarctic vegetation, nitrogen fixation activity of moss-associated cyanobacteria was not affected by 6 years of enhanced UV-B radiation. However, a 50% increase of summer precipitation resulted in a 5- to 6-fold increase in activity. Long-term effects of UV-B radiation on nitrogen fixation activity have been examined only in two lichens, giving contrasting results. *Peltigera aphthosa* (L.) Willd., having external cephalodia, experienced a significant reduction, whereas *Peltigera didactyla* (With.) J.R. Laudon, having cyanobacteria in the photobiont layer below the upper cortex, did not experience any changes due to radiation regimes. The difference is probably related to the location of the cyanobacteria. While the *Nostoc* cells are protected by the fungal, melanized upper cortex in *P. didactyla*, they are exposed and unprotected in *P. aphthosa*, and their own synthesis of UV-B absorbing compounds appears to be low. Under certain environmental conditions, an increasing UV-B radiation will dramatically affect nitrogen fixation in arctic tundra vegetation, which in turn may have severe influence on the nitrogen budget in these environments. Further long-term studies are necessary to conclude if these effects are temporal and how concurrent climatic changes will influence the nitrogen balance of the ecosystem.

Introduction

Objective and structure of article

In this review our objective is to discuss how elevated levels of UV-B radiation might change the nitrogen input into arctic ecosystems. The main factors for nitrogen input in the ecosystem is biological nitrogen fixation by cyanobacteria, which in turn is affected by the extent of the predicted increase in UV-B radiation. Possible changes in UV-B radiation reaching the surface of the Earth and the importance of biological nitrogen fixation for the nitrogen input in Arctic ecosystems are dealt with in the introduction. Then the response of cyanobacteria to UV-radiation and the effects on nitrogen fixation are reviewed. Especially the effects of UV-B radiation in long-term experiment based on enhanced UV-B radiation by UV-lamps in the field are discussed before reaching a conclusion.

Changes in UV-B radiation reaching the surface of the earth

The depletion of the stratospheric ozone layer, caused by anthropogenic release of halocarbons into the atmosphere, was reported about three decades ago (Molina and Rowland 1974). Since then the concurrent increase in harmful ultraviolet-B radiation (UV-B) (280–315 nm) reaching the Earth's surface has been dramatic (e.g. McKenzie et al. 2003). This may be particularly true for polar regions (McKenzie et al. 2003) where, due to low solar angles and a thick layer of ozone, ecosystems have existed under low levels of UV-B for millennia. The long lifetimes of ozone-depleting gases combined with the compounding influence of trophospheric warming and increased water vapour in the stratosphere (Jones and Shanklin 1995; Shindell et al. 1998a; Kirk Davidoff et al. 1999) imply that, despite reductions in the emission of ozone-depleting gases, continued and possibly increasing ozone depletion will remain in the North for the foreseeable future. Moreover, recent global climate models predict a further significant loss of ozone in the next decades, with up to 50% depletion of the ozone layer over large parts of the Arctic (Taalas et al. 2000). UV-B-radiation induces damage of DNA and proteins of plants (reviewed in Caldwell et al. 2003), fungi (reviewed in Caldwell et al. 2003; Hughes et al. 2003) and bacteria (Vincent and Neale 2000). Enhancement of UV-B radiation has wide-ranging impacts on decomposition and nutrient cycling (Zepp et al. 1998). Johnson et al. (2002) studied how soil microorganisms in a subarctic habitat respond to enhanced UV-B, elevated CO_2 and a combination of both. They concluded that a long-term enhancement of UV-B alone or combined with increased CO_2 significantly changed the C:N ratio of the microbial biomass and this could alter the availability of nitrogen. Another source of nitrogen for the ecosystem is the annual input from biological nitrogen fixation. Since nitrogen often is the limiting nutrient for primary production in the Arctic, any change in nitrogen availability will have substantial consequences for the ecosystem.

Importance of nitrogen fixation in arctic ecosystems

Nitrogen fixing symbioses between soil bacteria and higher plants like legumes or actinorhizal plants have been reported for subarctic and alpine tundra, but such symbioses are not found in the Arctic and Antarctic (Henry and Svoboda 1986). Nitrogen fixation by heterotrophic soil bacteria is either limited or absent because of low soil temperature and insufficient nutrient availability (Jordan et al. 1978; Smith 1985; Christie 1987). Virtually all biological nitrogen fixation in polar regions originates from cyanobacteria (Alexander 1974; Kallio 1974; Lennihan et al. 1994; Solheim et al. 1996). Cyanobacteria in the Arctic exist as free-living organisms or closely associated with other organisms such as bryophytes or as a component of lichens. Although the absolute values for nitrogen fixation in polar regions are generally lower than the ones measured in temperate and tropical environments, the contribution to the total nitrogen input to terrestrial polar ecosystems is major (Croome 1973; Alexander 1974; Alexander et al. 1974; Henry and Svoboda 1986; Lennihan and Dickson 1989; Chapin and Bledsoe 1992). Cyanobacteria are often found in great variety in the High Arctic, and in several habitats they are not only the dominant microbial phototrophs and diazotrophs, but constitute most of the microbial ecosystem biomass and productivity (Vincent 2000). However, cyanobacterial nitrogen fixation is constrained by environmental conditions and therefore shows wide spatial and temporal variations (Chapin and Bledsoe 1992). Based on available published data Chapin and Bledsoe (1992) calculated annual nitrogen fixation rates in different arctic and alpine ecosystems to be within a range of 19–255 mg $m^{-2}\,yr^{-1}$ with a mean of 127 mg $m^{-2}\,yr^{-1}$. Calculated as percent of total annual ecosystem N input the range was between 25 and 82% with a mean of 52.4% and as percent of total annual N plant uptake the range was between 0.7 and 20.5% with a mean of 7.1%. Nitrogen fixation in the Antarctic has been estimated from 10 to 192 mg N m^{-2} yr^{-1} (Vincent 1988). Inputs by rain and snow are generally less than 30 mg m^{-2} yr^{-1} in the Arctic (Barsdate and Alexander 1975). Nitrogen fixation contributes to the ecosystem on an average with four times more nitrogen than atmospheric deposition, and in the most productive areas of the Arctic the contribution is even higher. Still with a few exceptions, such as the ornithogenic tundra (tundra highly influenced by deposition of fecal matter of birds), the amount of nitrogen available in the soil is one of the major factors limiting plant growth in the Arctic (Shaver and Chapin III 1980; Nadelhoffer et al. 1992).

Nitrogen fixation by free-living cyanobacteria compensates for the lack of nitrogen in arctic soils (Lennihan et al. 1994) and permits the subsequent colonization of these habitats by other microorganisms and higher plants (Bliss and Gold 1994). In addition to the ability of fixing nitrogen from the air, cyanobacteria also fix carbon through photosynthesis. These two properties make them to pioneer colonizers of soils of newly exposed grounds, such as zones of glacial retreat or areas emerged from the sea (Tedrow 1977; Van Coppenolle et al. 1995; Bolch et al. 1996), where nitrogen can hardly be measured (Vincent 2000).

The first plants to be established in the succession are usually bryophytes, and in most arctic ecosystems the vegetation is dominated by mosses (Elvebakk 1994) many harbouring epiphytic nitrogen fixing cyanobacteria (Solheim et al. 2004; for review see Solheim and Zielke 2002). Solheim et al. (1996) described biological nitrogen fixation on Spitsbergen, the largest island in Svalbard, and found that cyanobacteria in association with mosses were by far the most important source of nitrogen in vegetated areas. In a study of nitrogen fixation in different vegetation types on Spitsbergen (Zielke et al. 2002; Solheim et al. 2005) the importance of moss-associated cyanobacteria was confirmed. Further, Dodds et al. (1995) stated that moss-associated cyanobacteria provide 2–58% of the nitrogen input to arctic ecosystems and 42–84% in antarctic moss vegetation. Nitrogen fixation by moss-associated cyanobacteria has also been found to be of importance in subarctic and alpine regions (Solheim and Zielke 2002) and in boreal forests (DeLuca et al. 2002).

Cyanobacterial lichens play a central role in polar ecosystems since they contribute significantly to the nitrogen economy. Lichens can be classified as bipartite and tripartite based on the number of symbionts with two or three symbionts, respectively. Cyanobacteria are the primary photobiont of several pan-arctic species from the genera *Peltigera*, *Placynthium*, *Collema*, *Leptogium*, *Lempholemma*, *Massalongia*, *Pannaria* and *Fuscopannaria*. Tripartite lichens have a green algal primary photobiont and cyanobacteria in isolated external or internal structures called cephalodia. All species of the genera *Stereocaulon*, *Solorina*, *Psoroma* and *Placopsis* and some species of *Nephroma* and *Peltigera* are cephalodiate. Most of these genera have a foliose or fruticose growth form, and are among the lichens with the highest biomass and productivity in the Arctic. Thus, their fixation of atmospheric nitrogen gives a highly important input of nitrogenous compounds, which becomes available to associated organisms through leaching or fragmentation and decomposition of lichens. The symbiotic cyanobacteria in arctic lichens are mostly regarded as *Nostoc* and *Stigonema*, with the latter having lower nitrogen fixation measured as acetylene reduction activity than the former (Nash III 1996).

The actual input of N fixed by lichens to the nitrogen economy in cold environments is still uncertain, because there is a number of potentially serious errors associated with these calculations (Nash III 1996). Estimates from boreal and alpine ecosystems indicate annual inputs from 0.2 to 40.0 kg N ha^{-1} (Kallio 1974; Crittenden 1975; Alexander et al. 1978; Vitousek et al. 1987). Despite the extreme variation in the estimates, hardly anyone refuses that cyanolichens are important for the nitrogen economy in polar ecosystems, which are frequently nitrogen-limited (see e.g. Longton 1988; Chapin and Bledsoe 1992; Nash III 1996; Crittenden 2000).

Moisture, temperature and light are the most important factors for the regulation of nitrogen fixation in lichens. Nitrogen fixation shows the same diurnal cycle as gross photosynthesis (Nash III and

Olafsen 1995), and it may well be that nitrogen fixation is dependent upon photosynthetates (Nash III 1996). Although maximum nitrogen fixation is reached at temperatures between 10 and 20 °C (Huss-Danell 1977), measurable nitrogen fixation has been detected at subzero temperatures (Kallio et al. 1972). Thus, arctic lichens may fix nitrogen even during shorter periods in winter.

Responses of cyanobacteria to UV-radiation

Being photoautotrophic organisms, solar radiation is one of the most important factors determining the growth of cyanobacteria in their natural habitats. Consequently, cyanobacteria are in most ecosystems exposed to sunlight, including UV-B. Cyanobacteria have several mechanisms to respond to UV-B stress and mitigate the UV-B induced damages (He and Häder 2002; Kumar et al. 2003; reviewed by Sinha et al. 2001). Two of them can be classified as biochemical defence mechanisms: (i) the production of UV-B-absorbing substances such as mycosporine-like amino acids and scytonemin, or the ability to produce a mucilaginous sheath that screens out some of the UV-B, and (ii) the production of agents that neutralize the highly toxic reactive oxygen species caused by UV-B. Finally, cyanobacteria possess mechanisms, such as photoreactivation, to repair DNA by splitting cyclobutane dimers and to repair damaged photosynthetic apparatus. Light-independent repair of DNA by nucleotide excision also takes place (Kumar et al. 2003).

During the last decade an increasing number of studies has investigated the effect of enhanced UV radiation, especially UV-B, on different types of subarctic and arctic ecosystems (reviewed in Hessen 2002) as well as single types of microorganisms (Mancinelli and White 2001 and quoted references). Most of these studies focused on the effects of enhanced UV-radiation on either entire ecosystems or the physiology of organisms. Very little is known about how the effects on single organisms become transferred to the ecosystem level. There is an extensive literature on cyanobacteria and UV-radiation (for review see Castenholz and Garcia-Pichel 2000), but most of the work deals with free-living organisms. Despite the ecological importance of cyanolichens, only a few studies have focused on general effects of enhanced UV-B radiation on these organisms, and in particular on effects on nitrogen fixation activity. Although a dozen UV-B effect studies on lichens have been undertaken in polar or alpine ecosystems (see review by Bjerke 2003), most of them dealt with green-algal species. In the summary of results from various European UV experiments, Björn et al. (1997) noted that no UV-B induced morphological damage was found in the two foliose, tripartite lichens *Nephroma arcticum* (L.) Torss. and *Peltigera aphthosa*, but it was not stated for how long they had been exposed and which morphological characters that were examined. Sass and Vass (1998) exposed the same lichens to extreme levels of UV-B radiation for up to 35 h. Measurements of chlorophyll fluorescence and thermoluminescence indicated that the lichens were remarkably tolerant to the extreme radiation, which kills vascular plants within a day.

A short-term experiment at the Antarctic Peninsula using different filters to modify the incident UV radiation included the tripartite lichen *Stereocaulon alpinum* Laurer (Huiskes et al. 2001). No significant effects on the chlorophyll fluorescence of this species, or any of the other examined plants and lichens could be detected.

The widespread, foliose, old-growth forest lichen *Lobaria pulmonaria* (L.) Hoffm. with cyanobacteria in internal cephalodia has been used in several studies on the effects of light stress, and two of them involved UV-B radiation. Gauslaa and Solhaug (2001) and Solhaug et al. (2003) found that dark brown melanins that accumulate in the upper cortex of the lichen significantly reduce the cortical transmittance of UV-B radiation and that the synthesis of these pigments is induced by UV-B radiation and water. UV-B radiation did not affect the chlorophyll fluorescence or the net photosynthetic CO_2 uptake in *L. pulmonaria* exposed to solar radiation for 3 weeks (Solhaug et al. 2003). In another experiment *N. arcticum* was exposed to enhanced UV-B radiation at Abisko, Swedish Lapland (68.35° N, 18.82° E, 360 m a.s.l.) for one field season (80 days). Enhanced UV-B radiation caused a significant reduction in maximal photosystem II efficiency, but no changes in the concentration of UV-absorbing, cortical and medullary phenolics (Bjerke et al. 2005).

Effects of UV-B on nitrogen fixation

In the comprehensive literature on cyanobacteria and UV-radiation very few studies deal with the effects on nitrogen fixation. This is surprising taking into consideration the importance of the process and the fact that it has been known for more than 20 years that UV-B in non-lethal doses severely inhibited nitrogen fixation in the cyanobacterium *Anabaena flos-aquae* whereas other physiological activities were unaffected (Newton et al. 1979). Babu et al. (1998) reported that enhanced UV-B caused a significant reduction of biomass and overall productivity of cultures of *Anabaena*, *Nostoc*, and *Scytonema* as a result of impaired photosynthesis and nitrogen fixation, and Rai et al. (1998) showed an interactive effect of UV-B and heavy metals on metabolism and nitrogen fixation in cultures of *Anabaena*. Sinha (1996), Sinha et al. (2001) and Kumar et al. (2003) investigated the effect of enhanced UV-B on several physiological and biochemical processes in various cyanobacteria species in laboratory experiments. They found significant inhibition of growth, significant reduction in CO_2 uptake, RuBISCO activity and heterocyst differentiation, and a complete inhibition of nitrogenase when cyanobacteria cultures were exposed to increased UV-B. The simultaneous exposure of both UV-B and visible light, however, did reverse some of these effects. This finding and the fact that several studies were performed with doses of UV-B which were unnaturally high compared to the irradiances present in solar radiation (Newton et al. 1979; Quesada et al. 1995) should make us cautious when extrapolating results from laboratory experiments to natural environments.

Physical and biochemical protection against UV-B mediated for example by other members of a cyanobacterial mat community (Sheridan 2001), soil and dust (Cockell et al. 2003), or leaves of the host plant, may also play an important role for nitrogen fixation potential in natural environments.

The effects of enhanced UV-B radiation on biological activity of organisms are dependent upon the relationship between damage and repair (Stapleton 1999). Photochemical damage of nucleic acids and proteins in cells are temperature independent while the repair mechanisms, needing gene activation and protein synthesis, only function at temperatures and humidity conditions required for normal cell functions (Takeuchi et al. 1993). UV inhibition of growth of mat-forming cyanobacteria from the Antarctic increased linearly with decreasing temperature, consistent with the hypothesis that the activity represents a balance between temperature-indedependent photochemical damage and temperature-dependent biosynthetic repair (Roos and Vincent 1998). This balance is dependent upon sufficient water for cellular activities. Potentially, this makes microorganisms in the Arctic especially vulnerable to UV-B damage at low temperatures in early spring before and during snowmelt, and later in the season due to low cellular activity caused by lack of available water (Dickson 2000).

Long-term exposure of tundra and heath vegetation to enhanced UV-B radiation, effects on moss-associated cyanobacteria

Two recent field experiments have studied the effect of long-term exposure of sub-arctic and arctic vegetation on enhanced UV-B radiation. Vegetation from a high arctic site in Adventdalen, Svalbard (78.17° N, 16.00° E) completely dominated by the moss *Sanionia uncinata* (Hedw.) was sampled in 1998, 1999 (Solheim et al. 2002), 2001 (Zielke 2004) and 2003 (this paper) representing 3, 4, 6 and 8 years of enhanced UV-B generated by UV-lamps placed in the field on the vegetation during the growing period equivalent to a 15% ozone depletion (Johanson et al. 1995) (Photo 1). Ambient and enhanced biologically effective UV-B (UV-BBE) radiation, on a clear day in early July, at this latitude are 3.0 and 3.8 kJ m^{-2} day^{-1}. Solheim et al. (2002) found significant decreases of acetylene reduction activity of 46% and 55% after 3 and 4 years of exposure, respectively, and Zielke (2004) found a significant reduction of 28% after 6 years of exposure to enhanced UV-B compared with ambient UV. However, no significant differences between samples treated with enhanced UV-B and control samples could be found after 8 years (Figure 1).

In a similar experiment in a subarctic environment in Abisko, Northern Sweden, samples of the moss *Hylocomium splendens* (Hedw.), which had been treated with enhanced UV-B radiation during the growing season equivalent to a 15% ozone

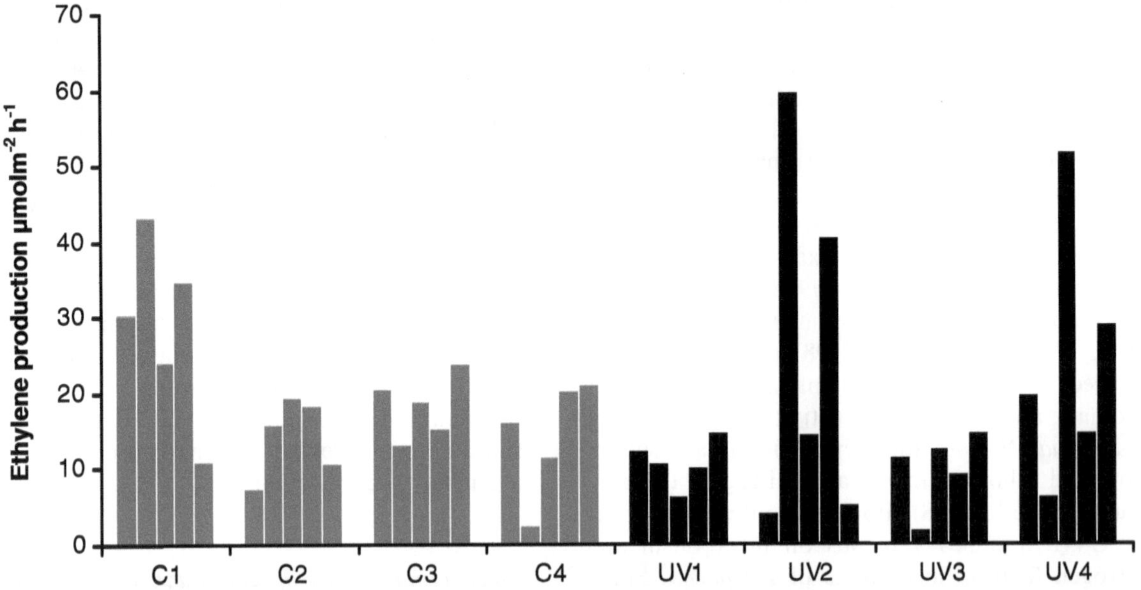

Figure 1. Nitrogen fixation expressed as acetylene reduction activity (μmol ethylene produced m^{-2} h^{-1}). Grey bars from C1 to C4 represent activity in five individual samples of vegetation dominated by *Sanionia uncinata* from each control plot (ambient UV-B radiation), and black bars from UV1 to UV4 represent the same for vegetation treated with enhanced UV-B radiation for 8 years.

depletion (Johanson et al. 1995) for seven years in combination with increased precipitation showed no significant differences in acetylene reduction activity response to enhanced UV-B, but a 5- to 6-fold higher activity was found in vegetation receiving 50% higher summer precipitation. There were no interactive effects of the two treatments (Solheim et al. 2002). The natural and enhanced biologically effective UV-B (UVBBE) radiation, on a clear day in early July, at this latitude are 4.6 and 5.8 kJ m^{-2} day^{-1}.

Long-term exposure of tundra and heath vegetation to enhanced UV-B radiation, effects on lichenized cyanobacteria

The first long-term study on a lichen was the one by Solheim et al. (2002) on *P. aphthosa*, from heath vegetation at Abisko (see above), which had been treated for 8 years with enhanced UV-B radiation simulating a 15% ozone depletion. A highly significant 46% reduction in acetylene reduction activity was detected, whereas thalli exposed for not more than 11 weeks did not experience any reduction. Bjerke et al. (2003) measured acetylene reduction activity rates in the bipartite cyanolichen *P. didactyla* from Adventdalen, which had been exposed to enhanced UV-B radiation, also simulating a 15% ozone depletion, for five years, and they could not find any isolated effects of UV-B radiation. Some morphological and chemical characters in this species and in *Peltigra rufescens* (Weiss) Humb. were also investigated, and enhanced UV-B radiation did not lead to significant differences from control levels, although there was a trend towards smaller thalli under enhanced UV-B. Warming, however, caused significant increases in the concentrations of the two depsides methyl gyrophorate and gyrophoric acid, both which may serve as antiherbivoral agents. This increase was not found in plots treated with both warming and UV-B. Thus, enhanced UV-B inhibited the stimulatory effects of warming.

Discussion

In all the long term studies discussed in this paper the UV-B radiation was enhanced by UV-lamps in the field to a level equivalent to 15% depletion of the ozone layer. This simulates a realistic scenario based on the latest climate models.

Large variation in acetylene reduction activity of cyanobacteria in different seasons and/or years might be due to the interaction of UV-B radiation, temperature and water content. These factors may affect both the cellular activity, including nitrogen fixation, and the relationship between photochemical damage and the cells repair of nucleic acids and proteins. Thus, already small variations in water content in the vegetation might lead to large differences in cellular activity. However, the reduction in acetylene reduction activity of cyanobacteria associated with *S. uncinata* under enhanced UV-B radiation gave a significant reduction of the nitrogen input into the ecosystem after 3 and 4 years of exposure. Though the reduction in nitrogen fixation activity found after 6 years, it was still significant. The dramatic reduction in nitrogen fixation in cyanobacteria associated with *S. uncinata* the first years of enhanced UV-B radiation followed by less reduction and finally no significant difference can be due to adaptation of the cyanobacterial community to higher UV-B doses or more favourable conditions for DNA repair at the end of the period.

In contrast to the results with *S. uncinata* in Spitsbergen, no significant differences in nitrogen fixation were observed between control and UV-B plots containing cyanobacteria associated with *H. splendens* at the sites in Abisko. The impacts of UV-B on higher plant growth and cyanobacterial physiology can be cumulative and long-term. The difference in UV-B impact between *S. uncinata* in Adventdalen and *H. splendens* at Abisko might be due to thicker snow cover in spring and more favourable conditions for DNA repair at the warmer and more humid sites at Abisko.

Few long-term responses to increased UV-B radiation have been studied in lichens, possibly due to the very slow growth of such organisms. Nevertheless, cumulative impacts on growth and physiological integrity may be expected in lichens. This might be especially true for nitrogen fixing lichens as the cyanobacterial partner *Nostoc* occurs on the most exposed part of the lichen and the growth rate of the microsymbiont is dependent on the overall growth rate of the lichen. The effects of UV-B radiation on nitrogen fixation activity have been examined only in two lichens, giving contrasting results. *P. aphthosa*, having external cephalodia, experienced a significant reduction, whereas *P. didactyla*, having cyanobacteria in the photobiont layer below the upper cortex, did not experience any changes due to radiation regimes. The difference is probably related to the location of the cyanobacteria. While the *Nostoc* cells are protected by the melanized upper cortex in *P. didactyla*, they are dependent upon own protection strategies in *P. aphthosa*. Although the cephalodia of *P. aphthosa* produce small amounts of the UV-absorbing compound scytonemin (Buskens 2002) the concentrations are probably too low to render sufficient protection against UV-B radiation over long time spans. This difference in cyanobacterial location within the lichen thallus may also reflect the differences in habitat preferences of the two species, with *P. aphthosa* mostly growing on the forest floor thereby avoiding high radiation intensities, and *P. didactyla* growing on soil in open conditions without shading from adjacent vegetation. The reduction of nitrogen fixation in *P. aphthosa* appears to be a long-term cumulative effect as three months exposure to UV-B radiation did not reduce the lichens potential for nitrogen fixation (Solheim et al. 2002).

The short-term UV-B experiments lasting for one field season or less failed to detect any significant effects on the cyanolichens (Solheim et al. 2002), except for the reduction in chlorophyll fluorescence of *N. arcticum* (Bjerke et al. 2005). These results may indicate either that cyanolichens are not susceptible to enhanced UV-B radiation or that the experiments were of too short duration (Aphalo 2003). Most likely, the latter is the case, since one of the two long-term experiments showed severe effects (Solheim et al. 2002), and since cyanolichens generally have low light compensation levels (Demmig-Adams et al. 1990; Rai 1990).

Conclusion

Studies so far raise serious concerns about already nutrient-impoverished arctic plant communities. Especially considering ongoing and future ozone depletion (Shindell et al. 1998b) and the negative impacts of UV-B on other nutrient processes such as mycorrhizal infection (Klironomos and Allen 1995; van de Staaij et al. 1999), decomposition

(Gehrke et al. 1995; Paul et al. 1999; Moody et al. 2001) and biogeochemical cycling (Zepp et al. 1998). Under certain environmental conditions, an increasing UV-B radiation will dramatically affect nitrogen fixation in arctic tundra vegetation, which in turn may have severe influence on the nitrogen budget in these environments. Further long-term studies are necessary to conclude if these effects are temporal and how concurrent climatic changes will influence the balance between DNA damage and repair.

Acknowledgements

The facilities of enhanced UV-B radiation of the vegetation were started under the European Community contract EV5V-CT910032, and we are grateful for the support of scientists involved and being able to continue UV-B-treatment at the Adventdalen site since 1998. The logistic help of Dr Sigmund Spjelkavik is greatly appreciated and Prof Dr A.S. Blix, Department of Arctic Biology, University of Tromsø, Norway, is greatly acknowledged for permission for storage of UV equipment and for mains power at the Adventdalen site.

References

Alexander V. 1974. A synthesis of the IBP Tundra biome, circumpolar study of nitrogen fixation. In: Holding A., Heal J.O.W., MacLean S.F. and Flanagan P.W. (eds.), Soil Organisms and Decomposition in Tundra, Tundra Biome Steering Committee, Stockholm, pp. 109–121.

Alexander V., Billington M. and Schell D. 1974. The influence of abiotic factors on nitrogen fixation rates in the Barrow, Alaska, arctic tundra. Rep. Kevo Subarctic Res. Station 11: 3–11.

Alexander V., Billington M. and Schell D.M. 1978. Nitrogen fixation in Arctic and Alpine tundra. In: Tieszen L.L. (ed.), Vegetation and Production Ecology of an Alaskan Arctic Tundra, Springer Verlag, New York, pp. 539–558.

Aphalo P.J. 2003. Do current levels of UV-B radiation affect vegetation? Importance of long-term experiments. New Phytol. 160: 273–275.

Babu G.S., Joshi P.C. and Viswanathan P.N. 1998. UVB-induced reduction in biomass and overall productivity of cyanobacteria. Biochem. Biophys. Res. Commun. 244: 138–142.

Barsdate K.J. and Alexander V. 1975. The nitrogen balance of arctic tundra: Pathways, rates and environmental implications. J. Environ. Quality 4: 111–117.

Bjerke J.W. 2003. Arctic-alpine lichens and global change: how do ultraviolet-B radiation and warming affect secondary metabolism, morphological characters and physiological processes? Doctoral thesis, Department of Biology, Faculty of Science. University of Tromsø, Tromsø.

Bjerke J.W., Zielke M. and Solheim B. 2003. Long-term impacts of simulated climatic change on secondary metabolism, thallus structure and nitrogen fixation activity in two cyanolichens from the Arctic. New Phytol. 159: 361–367.

Bjerke J.W., Gwynn-Jones D. and Callaghan T.V. 2005. Effects of enhanced UV-B radiation in the field on the concentration of phenolics and chlorophyll fluorescence in two boreal and arctic-alpine lichens. Environ. Exp. Bot. 53: 139–149.

Björn L.O., Callaghan T.V., Johnsen I., Lee J.A., Manetas Y., Paul N.D., Sonesson M., Wellburn A.R., Coops D., Heide-Jorgensen H.S., Gehrke C., Gwynn-Jones D., Johanson U., Kyparissis A., Levizou E., Nikolopoulos D., Petropoulou Y. and Stephanou M. 1997. The effects of UV-B radiation on European heathland species. Plant Ecol. 128: 252–264.

Bliss L.C. and Gold W.G. 1994. The patterning of plant coastline – implications for succession. Can. J. Bot. 72: 1095–1107.

Bolch C.J.S., Blackburn S.I., Neilan B.A. and Grewe P.M. 1996. Genetic characterization of strains of cyanobacteria using PCR-RFLP of the *cpcBA* intergenic spacer and flanking regions. J. Phycol. 32: 445–451.

Buskens A. 2002. The tracing of UV-B absorbing compounds in pollen from *Alnus glutinosa*, and the effects of natural and/or enhanced ultraviolet radiation on a moss-cyanobacterial community (1) and an arctic tundra vegetation community (2). Doctoral thesis, Department of Systems Ecology, Vrije Universiteit, Amsterdam.

Caldwell M.M., Ballaré C.L., Bormann J.F., Flint S.D., Björn L.O., Teramura A.H., Kulandaivelu G. and Tevini M. 2003. Terrestrial ecosystems, increased solar ultraviolet radiation and interactions with other climatic change factors. Photochem. Photobiol. Sci. 2: 29–38.

Castenholz R.W. and Garcia-Pichel F. 2000. Cyanobacterial responses on UV-radiation. In: Whitton B.A. and Potts M. (eds.), The Ecology of Cyanobacteria: Their Diversity in Time and Space, Kluwer Academics Publishers, Dordrecht/London/Boston, pp. 591–611.

Chapin D.M. and Bledsoe C.S. 1992. Nitrogen fixation in arctic plant communities. In: Chapin F.S. III, Jefferies R.L., Reynolds J.F., Shaver G.R., Svoboda J. and Chu E.W. (eds.), Arctic Ecosystems in a Changing Climate: An Ecophysiological Perspective, Academic Press Inc., San Diego, pp. 301–319.

Christie P. 1987. Nitrogen in two contrasting Antarctic bryophyte communities. J. Ecol. 75: 73–93.

Cockell C., Rettberg P., Horneck G., Scherer K. and Stockes M.D. 2003. Measurements of microbial protection from ultraviolet radiation in polar terrestrial microhabitats. Polar Biol. 26: 62–69.

Crittenden P.D. 1975. Nitrogen fixation by lichens on glacial drift in Iceland. New Phytol. 74: 41–49.

Crittenden P.D. 2000. Aspects of ecology of mat-forming lichens. Rangifer 20: 127–139.

Croome R.L. 1973. Nitrogen fixation in the algal mats on Marion Island. S. Afr. Tydskrif vir Antarkiese Navorsing 3: 64–67.

DeLuca T.H., Zackrisson O., Nilsson M.-C. and Sellstedt A. 2002. Quantifying nitrogen-fixation in feather moss carpets of boreal forests. Nature 419: 917–920.

Demmig-Adams B., Adams W.W. III, Green T.G.A., Czygan F.C. and Lange O.L. 1990. Differences in the susceptibility to light stress in two lichens forming a phycosymbiodeme, one partner possessing and one lacking the xanthophyll cycle. Oecologia 84: 451–456.

Dickson L.G. 2000. Constraints to nitrogen fixation by cryptogamic crusts in a polar desert ecosystem, Devon Island, N.W.T., Canada. Arctic Antarctic Alpine Res. 32: 40–45.

Dodds W.K., Gudder D.A. and Mollenhauer D. 1995. The ecology of Nostoc. J. Phycol. 31: 2–18.

Elvebakk A. 1994. A survey of plant associations and alliances from Svalbard. J. Veg. Sci. 5: 791–802.

Gauslaa Y. and Solhaug K.A. 2001. Fungal melanins as a sun screen for symbiotic green algae in the lichen Lobaria pulmonaria. Oecologia 120: 462–471.

Gehrke C., Johanson U., Callaghan T.V., Chadwick D. and Robinson C.H. 1995. The impact of enhanced ultraviolet-B radiation on litter quality and decomposition processes in *Vaccinium* leaves from the Subarctic. OIKOS 72: 213–222.

He Y.-Y. and Häder D.P. 2002. UV-B-induced formation of reactive oxygen species and oxidative damage of the cyanobacterium *Anabaena* sp.: protective effects of ascorbic acid and *N*-acethyl-L-cysteine. J. Photochem. Photobiol. B: Biol. 66: 115–124.

Henry G.H.R. and Svoboda J. 1986. Dinitrogen fixation (acetylene reduction) in High Arctic sedge meadow communities. Arctic Alpine Res. 18: 181–187.

Hessen D.O. 2002. UV radiation and arctic ecosystems. In: Hessen D.O. (ed.), Ecological Studies, Springer Verlag, Heidelberg.

Hughes K.A., Lawley B. and Newsham K.K. 2003. Solar UV-B radiation inhibits the growth of Antarctic terrestrial fungi. Appl. Environ. Microbiol. 69: 1488–1491.

Huiskes A.H.L., Lud D. and Moerdijk-Poortvliet T.C.W. 2001. Field research on the effects of UV-B filters on terrestrial antarctic vegetation. Plant Ecol. 154: 77–86.

Huss-Danell K. 1977. Nitrogen fixation by *Stereocaulon paschale* under field conditions. Can. J. Bot. 55: 585–592.

Johanson U., Gehrke C., Björn L.O., Callaghan T.V. and Sonesson M. 1995. The effects of enhanced UV-B radiation on the growth of dwarf shrubs in a subarctic heathland. Funct. Ecol. 9: 713–719.

Johnson D., Campbell C.D., Lee J.A., Callaghan T.V. and Gwynn-Jones D. 2002. Arctic microorganisms respond more to elevated UV-B than to CO_2. Nature 416: 82–83.

Jones A.E. and Shanklin J.D. 1995. Continued decline of total ozone over Halley, Antarctica, since 1985. Nature 376: 409–411.

Jordan D.C., McNicol P.J. and Marshall M.R. 1978. Biological nitrogen fixation in the terrestrial environment of a high Arctic ecosystem (Truelove Lowland, Devon Island, N.W.T.). Can. J. Microbiol. 24: 643–649.

Kallio P. 1974. Nitrogen fixation in subarctic lichens. Oikos 25: 194–198.

Kallio P., Suhonen S. and Kallio S. 1972. The ecology of nitrogen fixation in *Nephroma arcticum* and *Solorina crocea*. Rep. Kevo Subarctic Res. Station 9: 7–14.

Kirk-Davidoff D.B., Hintsa E.J., Anderson J.G. and Keith D.W. 1999. The effect of climate change on ozone depletion through changes in stratospheric water vapour. Nature 402: 399–401.

Klironomos J.N. and Allen M.F. 1995. UV-B-mediated changes on bellow-ground communities associated with the roots of *Acer saccharum*. Func. Ecol. 9: 923–930.

Kumar A., Tyagi M.B., Singh N., Tyagi R., Jha P.N., Sinha R.P. and Häder D.P. 2003. Role of white light in reversing UV-B-mediated effects in the N_2-fixing cyanobacterium *Anabaena* BT2. J. Photochem. Photobiol. B: Biol. 71: 35–42.

Lennihan R., Chapin D.M. and Dickson L.G. 1994. Nitrogen fixation and photosynthesis in high arctic forms of *Nostoc commune*. Can. J. Bot. 72: 940–945.

Lennihan R. and Dickson L.G. 1989. Distribution, abundance and physiological aspects of *N. commune* in a high arctic ecosystem. J. Phycol. (Suppl.) 25: 16.

Longton R.E. 1988. The Biology of Polar Bryophytes and Lichens. Cambridge University Press, Cambridge.

Mancinelli R.L. and White M.R. 2001. Inhibition of denitrification by ultraviolet radiation. Adv. Space Res. 26: 2041–2046.

McKenzie R.L., Björn L.O., Bais A. and Ilyasd M. 2003. Changes in biologically active ultraviolet radiation reaching the earth's surface. Photochem. Photobiol. Sci. 2: 5–15.

Molina M.J. and Rowland F.S. 1974. Stratospheric sink for chlorofluoromethanes: chlorine atom-catalysed destruction of ozone. Nature 249: 810–812.

Moody S.A., Paul N.D., Björn L.O., Callaghan T.V., Lee J.A., Manetas Y., Rozema J., Gwynn-Jones D., Johanson U., Kyparissis A. and Oudejans A.M.C. 2001. The direct effects of UV-B radiation on *Betula pubescens* litter decomposing at four European field sites. Plant Ecol. 154: 29–36.

Nadelhoffer K.J., Giblin A.E., Shaver G.R. and Linkins A.E. 1992. Microbial processes and plant nutrient availability in arctic soils. In: Chapin F.S. III, Jefferies R.L., Reynolds J.F., Shaver G.R., Svoboda J. and Chu E.W. (eds.), Arctic Ecosystems in a Changing Climate, Academic Press, Inc., San Diego, pp. 281–300.

Nash T.H. III 1996. Nitrogen, its metabolism and potential contribution to ecosystems. In: Nash T.H. III (ed.), Lichen Biology, Cambridge University Press, Cambridge, pp. 121–135.

Nash T.H. III and Olafsen A.G. 1995. Climate change and the ecophysiological response of arctic lichens. Lichenologist 27: 559–565.

Newton J.W., Tyler D.D. and Slodki M.E. 1979. Effect of ultraviolet-B (280 to 320 nm) radiation on blue green algae (cyanobacteria), possible biological indicators of stratospheric ozone depletion. Appl. Environ. Microbiol. 37: 1137–1141.

Paul N.D., Callaghan T.V., Moody S.A., Gwynn-Jones D., Johanson U. and Gehrke C. 1999. UV-B impacts on decomposition and biogeochemical cycling. In: Rozema J. (ed.), Stratospheric Ozone Depletion. The effects of enhanced UV-B radiation on terrestrial ecosystems, Backhuys Publishers, Leiden, pp. 117–133.

Quesada A., Mouget J.L. and Vincent W.F. 1995. Growth of Antarctic Cyanobacteria under Ultraviolet-radiation – UVA Counteracts UVB Inhibition. J. Phycol. 31: 242–248.

Rai A.N. 1990. Cyanobacterial–fungal symbioses: the cyanolichens. In: Rai A.N. (ed.), CRC Handbook of Symbiotic Cyanobacteria, CRC Press, Boca Raton, pp. 9–41.

Rai L.C., Tyagi B., Rai P.K. and Mallick N. 1998. Interactive effects of UV-B and heavy metals (Cu and Pb) on nitrogen and phosphorus metabolism of a N_2-fixing cyanobacteria *Anabaena doliolum*. Environ. Exp. Bot. 39: 221–231.

Roos J.C. and Vincent W.F. 1998. Temperature dependence of UV radiation effects on Antarctic cyanobacteria. J. Phycol. 34: 118–125.

Sass L. and Vass I. 1998. Characterization of UV-B tolerance in lichens by photosystem II electron transport measurements. In: Garab G. (ed.), Photosynthesis: Mechanisms and Effects, Kluwer Academic Publishers, Dordrecht, pp. 2381–2384.

Shaver G.R. and Chapin F.S. III 1980. Response to fertlization by various plant growth forms in an Alaskan tundra: Nutrient accumulation and growth. Ecology 61: 662–675.

Sheridan R.P. 2001. Role of ultraviolet radiation in maintaining the three-dimensional structure of cyanobacterial mat community and facilitating nitrogen fixation. J. Phycol. 37: 731–737.

Shindell D.T., Rind D. and Lonerang P. 1998a. Climate change and the middle atmosphere. Part IV: ozone response to doubled CO_2. J. Climate 11: 895–918.

Shindell D.T., Rind D. and Lonerang P. 1998b. Increased stratospheric ozone loses and delayed recovery owning to increasing greenhouse-gas concentrations. Nature 392: 589–592.

Sinha R.P. 1996. Effects of UV radiation on certain physiological and biochemical processes in cyanobacteria. J. Photochem. Photobiol. B: Biol. 32: 107–113.

Sinha R.P., Klisch M., Gröniger A. and Häder D.P. 2001. Responses of aquatic algae and cyanobacteria to solar UV-B. Plant Ecol. 154: 221–236.

Smith V.R. 1985. Heterotrophic acetylene reduction in soils at Marion Island. In: Siegfried W.R., Condy P.R. and Laws R.M. (eds.), Antarctic Nutrient Cycles and Food Webs, Springer-Verlag, Berlin, pp. 185–191.

Solhaug K.A., Gauslaa Y., Nybakken L. and Bilger W. 2003. UV-induction of sun-screening pigments in lichens. New Phytol. 158: 91–100.

Solheim B., Endal A. and Vigstad H. 1996. Nitrogen fixation in Arctic vegetation and soils from Svalbard, Norway. Polar Biol. 16: 35–40.

Solheim B., Johanson U., Callaghan T.V., Lee J.A., Gwynn-Jones D. and Björn L.O. 2002. The nitrogen fixation potential of arctic cryptogram [*sic.*] species is influenced by enhanced UV-B radiation. Oecologia 133: 90–93.

Solheim B., Wiggen H., Røberg S. and Spaink H.P. 2004. Associations between arctic cyanobacteria and mosses. Symbiosis 37: 169–187.

Solheim B. and Zielke M. 2002. Associations between cyanobacteria and mosses. In: Rai A.N., Bergman B. and Rasmussen U. (eds.), Cyanobacteria in Symbiosis, Kluwer Academic Publishers, Dordrecht Boston London, pp. 137–152.

Solheim B., Zielke M., Spjelkavik S. and Olsen R.A. 2005. Seasonal variations in nitrogen fixation in high arctic vegetation, Svalbard, Norway. Arct. Antarct. Alp. Res. 37: 372–378.

Stapleton A.E. 1999. Ultraviolet radiation and plants: burning questions. Plant Cell 4: 1353–1358.

Takeuchi Y., Ikeda S. and Kasahara H. 1993. Dependence on wavelength and temperature of growth inhibition induced by UV-B radiation. Plant Cell Physiol. 34: 913–917.

Tedrow J.C.F. 1977. Soils of the Polar Landscapes. The State University of New Jersey, Rutgers.

Taalas P., Kaurola J., Kylling A., Shindell D., Sausen R., Dameris M., Grewe V., Herman J., Damski J. and Steil B. 2000. The impact of greenhouse gases and halogenated species on future solar UV radiation doses. Geophys. Res. Lett. 27: 1127–1130.

Van Coppenolle B., McCouch S.R., Watanabe I., Huang N. and Van Hove C. 1995. Genetic diversity and phylogeny analysis of *Anabaena-Azolla* based on RFLPs detected in *Azolla-Anabaena azollae* DNS complexes using *nif* gene probes. Theor. Appl. Genet. 91: 589–597.

van de Staaij J., Rozema J. and Aerts R. 1999. The impacts on mutualistic plant/micro-organism interactions at the soil-root interface. In: Rozema J. (ed.), Stratospheric Ozone Depletion. The Effects of Enhanced UV-B Radiation on Terrestrial Ecosystems, Backhuys Publishers, Leiden, pp. 159–171.

Vincent W.F. 1988. Microbial ecosystems of Antarctica. Press Syndicate of the University of Cambridge, Cambridge.

Vincent W.F. 2000. Cyanobacterial dominance in the polar regions. In: Whitton B.A. and Potts M. (eds.), The Ecology of Cyanobacteria: Their Diversity in Time and Space, Kluwer Academics Publishers, Dordrecht London Boston, pp. 321–340.

Vincent W.F. and Neale P.J. 2000. Mechanisms of UV damage to aquatic organisms. In: de Mora S. , Demers S. and Vernet M. (eds.), The Effects of UV Radiation in the Marine Environment, Cambridge University Press, Cambridge, pp. 149–176.

Vitousek P.M., Walker L.R., Whittaker L.D., Mueller-Dombois D. and Matson P.A. 1987. Biological invasion by *Myrica faga* alters ecosystem development in Hawaii. Science 238: 802–804.

Zepp R.G., Callaghan T.V. and Erickson D.J. 1998. Effects of enhanced solar ultraviolet radiation on biogeochemical cycles. J. Photochem. Photobiol. B: Biol. 46: 69–82.

Zielke, M. 2004. Diversity and nitrogen fixation activity of cyanobacterial communities in terrestrial acrtic ecosystems. Doctoral thesis, Department of Biology. University of Tromsø, Tromsø.

Zielke M., Ekker A.S., Olsen R.A., Spjelkavik S. and Solheim B. 2002. The influence of abiotic factors on biological nitrogen fixation in different types of vegetation in the High Arctic, Svalbard. Arct. Antarct. Alp. Res. 34: 293–299.

Photo. Aerial view of Adventdalen. Adventdalen is a glaciervalley, with the Adventriver ending in Isfjorden at the background of the photograph. UV lamp plots of Tromso University and Vrije Universiteit Amsterdam are situated in the glaciervalley as well as at the slope of the glaciervalley at Isdammen (Photograph: J.Rozema).

Plant Ecology (2006) 182:121–135
DOI 10.1007/s11258-005-9035-x

Stratospheric ozone depletion: high arctic tundra plant growth on Svalbard is not affected by enhanced UV-B after 7 years of UV-B supplementation in the field

Jelte Rozema[1,*], Peter Boelen[1], Bjørn Solheim[2], Matthias Zielke[2], Alwin Buskens[1], Marieke Doorenbosch[1], Ruben Fijn[1], Jelger Herder[1], Terry Callaghan[3,4], Lars Olof Björn[5], Dylan Gwynn Jones[6], Rob Broekman[1], Peter Blokker[1] and Willem van de Poll[7]
*[1]Department of Systems Ecology, Institute of Ecological Science, Climate Center Vrije Universiteit, De Boelelaan 1087, 1081HV, Amsterdam, The Netherlands; [2]Department of Biology, University of Tromsø, N-9037, Tromsø, Norway; [3]Abisko Scientific Research Station, Royal Swedish Academy of Sciences, SE-981 07, Abisko, Sweden; [4]Sheffield Centre for Arctic Ecology, University of Sheffield, X Taptonville Road, Sheffield, S10 5BR, UK; [5]Department of Cell and Organism Biology, Lund University, Lund University, Box 117, SE-22362, Lund, Sweden; [6]Institute of Biological Sciences, University of Wales, Aberystwyth; [7]Department of Marine Biology, University of Groningen, Groningen, The Netherlands; *Author for correspondence (e-mail: jelte.rozema@ecology.falw.vu.nl)*

Received 1 September 2004; accepted in revised form 15 December 2004

Key words: Arctic, *Bistorta vivipara*, *Cassiope tetragona*, Cover, DNA-damage, *Dryas octopetala*, Gametophyte, Inflorescence, Leaf area, Ozone depletion, *Pedicularis hirsuta*, Plant density, *Polytrichum hyperboreum*, *Salix polaris*, *Sanionia uncinata*, Supplementation, Thickness, Tundra, Ultraviolet-b radiation

Abstract

The response of tundra plants to enhanced UV-B radiation simulating 15 and 30% ozone depletion was studied at two high arctic sites (Isdammen and Adventdalen, 78° N, Svalbard).The set-up of the UV-B supplementation systems is described, consisting of large and small UV lamp arrays, installed in 1996 and 2002. After 7 years of exposure to enhanced UV-B radiation, plant cover, density, morphological (leaf fresh and dry weight, leaf thickness, leaf area, reproductive and ecophysiological parameters leaf UV-B absorbance, leaf phenolic content, leaf water content) were not affected by enhanced UV-B radiation. DNA damage in the leaves was not increased with enhanced UV-B in *Salix polaris* and *Cassiope tetragona*. DNA damage in *Salix polaris* leaves was higher than in leaves of *C. tetragona*. The length of male gametophyte moss plants of *Polytrichum hyperboreum* was reduced with elevated UV-B as well as the number of *Pedicularis hirsuta* plants per plot, but the inflorescence length of *Bistorta vivipara* was not significantly affected. We discuss the possible causes of tolerance of tundra plants to UV-B (absence of response to enhanced UV-B) in terms of methodology (supplementation versus exclusion), ecophysiological adaptations to UV-B and the biogeographical history of polar plants

Introduction

While the depletion of stratospheric ozone as a consequence of the emission of CFC's has given rise to a reoccurring hole in the ozone layer over the Antarctic since 1974 (Molina and Rowland 1974; Farman et al.1985), severe ozone depletion during the Arctic spring since the early 1990's has also been observed (McPeters et al. 1996; Newman et al. 1997; http://www.esa.int). While stratospheric ozone at the Antarctic spring is reduced from 300 to 100 Dobson Units, i.e. a 60–70% decrease and the ozone depleted area is very extensive, spring arctic ozone levels are rarely less than 150–200 Dobson Units and comprise a smaller depleted area (http://www.esa.int./export/esaSA/SAHFROOSTCearth). As a result of stratospheric ozone breakdown, increased solar UV-B reaches the earth (McKenzie et al. 2003; WMO 2003).

In the present research project effects of enhanced UV-B radiation on plant species of a high arctic tundra ecosystem, Adventdalen (78°15′ N, 16°30′ E) on Svalbard are being studied. A UV-supplementation field experiment was started in 1996 on Svalbard (Björn 2002). The UV lamp supplementation experiment includes a long-term (7 years) and a short term (2 years) field experiment simulating 15 and 30% ozone depletion. A similar UV-B supplementation field experiment is running at the Antarctic Island of Signy (Boelen et al. 2005, this volume). The tundra ecosystems studied on Svalbard, form part of the arctic area with severe ozone depletion during springtime (Newman et al. 1997; http://www.esa.int). The UV-B lamp experiments on Svalbard are unique because they cover a longterm (7 years) assessment of effects of enhanced UV-B on tundra plants. The majority of the tundra plants is long-lived and some species (e.g. *Cassiope tetragona*, *Dryas octopetala*) have evergreen leaves and shoots, i.e. perennial aerial parts photosynthesizing more than one summer season and receiving high UV-B radiation for several years. Alternatively the deciduous polar willow is shedding its leaves at the end of the polar summer.

UV-B effects on cyanobacterial nitrogen fixation are discussed by Solheim et al. (2002); Bjerke et al. (2003); Solheim et al. (2006, this volume); (Zielke et al. 2003; Zielke 2004). In addition the study of effects of enhanced UV-B on the arctic tundra is relevant because the tundra plant species encounter severe climatic stress and may therefore be vulnerable to enhanced UV-B. Furthermore, cold and frost may exacerbate temperature dependent DNA damage caused by UV-B by preventing repair (Li et al. 2002).

It is therefore hypothesized that plant species of the high arctic tundra will be vulnerable to enhanced UV-B. We expect DNA damage to increase with enhanced UV-B with reduced growth and plant biomass as a result. In particular gametophyte moss plant length growth of *Polytrichum hyperboreum* and inflorescence length growth of *Bistorta vivipara* may be reduced. Changes of morphological and ecophysiological parameters may reflect adaptations to UV-B to prevent UV-B damage.

The purpose of the present paper is to describe and analyse effects of enhanced UV-B radiation on tundra species cover, plant density, morphological, ecophysiological and reproductive parameters.

Material and methods

Site descriptions

The field work was done in the summer period (June–September) of 1999, 2000, 2002 and 2003.

The Adventdalen site represents a flat valley floor on glaciofluvial and fluvial deposits with *Salix polaris* and the mosses *Polytrichum hyperboreum* and *Sanionia uncinata* as the dominant tundra plant species (Table 1). *P. hyperboreum* is frequently sporulating (see Figures 1 and 2), while *S. uncinata* bears no spore capsules. The Adventdalen site is ca 5 m above stream level surface and about 200 m away from the stream bed of the river in the center of Adventdalen. The site can be characterized as a moist open tundra vegetation on glaciofluvial and fluvial deposits, mainly sandur (Kristiansen and Sollid 1987). The Isdammen site is near the Isdammen water reservoir of Longyearbyen on a mountain slope, circa 30 m above sea level on marine deposits (Kristiansen and Sollid 1987). The Isdammen site is drier than the Adventdalen site, and the soil has a higher organic matter content.

Vegetation cover of the tundra plant species in the large UV plots irradiated from 1996–2002 was assessed (Tables 2, 3). The tundra plants, e.g. *Salix*

Table 1. List of tundra plant species present at the Isdammen and Adventdalen site.

Plant species	Isdammen	Adventdalen
Salix polaris	x	x
Cassiope tetragona	x	–
Dryas octopetala	x	–
Oxyria digina	x	–
Bistorta vivipara	x	x
(= Polygonum viviparum)		
Pedicularis hirsuta	x	x
Saxifraga hirculus	x	–
Luzula confusa	–	x
Festuca rubra	–	x
Polytrichum hyperboreum	x	x
Equisetum arvense	–	x
Peltigera aphtosa	–	x
Sanionia uncinata	x	x
Stellaria crassipes	x	–
Carex misandra	x	x
Saxifraga oppositifolia	x	–
Saxifraga hieracifolia	x	–
Alopecurus borealis	x	–

Nomenclature after Rønning (1996) and Elvebakk (1994).

polaris and *Cassiope tetragona* are small and have only rarely shoots longer than 10–15 cm (Johnstone and Henry 1997, see Figures 1 and 2), and considering the total number of tundra plant species at Isdammen and Adventdalen (about 15–20) (Table 1), the relief (flat valley floor Adventdalen site) and homogeneity of the tundra soil, as well as the obvious visual homogeneity of the tundra vegetation (Figure 1) at Adventdalen and Isdammen, it is reasonable to assume that the 135×270 cm^2 of the large UV plots cover representative parts of the tundra vegetation.

The spatial distribution of the small UV lamp plots was chosen randomly, but in more detail such that *Salix* and *Cassiope* (Isdammen) or *Salix* and *Polytrichum* (Adventdalen) were well represented. The total cover of the tundra vegetation was about 80% (Isdammen) and 65–70% (Adventdalen) in the 7 year irradiated plots. Total vegetation cover of the small UV lamp plots at Isdammen and Adventalen was 80% and 65–85% respectively (Table 2).

Experimental design, UV supplementation systems at Isdammen and Adventdalen

Both at Isdammen and Adventdalen eight sets of metal frames holding UV-B fluorescent tubes were installed over the tundra vegetation since 1996 (Björn 2002).

There are four control sets (ambient UV-B) and four sets simulating 15% ozone depletion (enhanced UV-B) as described by Johanson et al. (1995a, b) and Solheim et al. (2002).

In June 2002, 16 additional mini-UV lamp sets were installed over the tundra vegetation both at Isdammen and Adventdalen adjacent to the existing sets already installed in 1996 (Figure 1). At each site 4 of these mini UV lamp sets are controls (C) with wooden bars replacing the fluorescent UV-B tubes, 4 sets represent UV-A treatments, where Mylar foil blocked UV-B and UV-C, but transmitted UV-A radiation emitted by the lamps, and two sets of 4 lamps with cellulose acetate foil blocking UV-C radiation, transmitting UV-B (and UV-A) simulating 15 and 30% ozone depletion (referred to as UV-B1 and UV-B2), with a longer lamp burning period in the latter case. Radiation spectra of the four treatments applied and details of the UV-B dosimetry and electronics of the lamp switch control system are given by Boelen et al. 2005 (this volume). The sites for the small UV lamps were chosen in such a way that *Salix polaris* and *Polytrichum hyperboreum* were dominantly present (Adventdalen site) or *Salix polaris* and *Cassiope tetragona* (Isdammen site).Treatment and control plots were randomised.

In 2002 and 2003 both the large and small UV supplementation systems operated from mid June until late August-beginning of September. Then the large UV fluorescent lamps were removed, while frames with lamp-holders remained at the tundra. The small lamp systems were removed completely and stored during winter time.

Field sampling and measurements

Vegetation cover

Vegetation cover was assessed in the subplots of the large UV plots (47.5×57.5 cm^2, Adventdalen; 47.5×76.7 cm^2, Isdammen) at August 14, 2002 and August 13 in 2003 (Isdammen), July 27, 2002, and August 13, 2003 (Adventdalen) as well as in the small UV plots at August 14, 2002 and July 31, 2003 (Isdammen), July 27 in 2002 and at July 31 in 2003 (Adventdalen) by visual estimation. 1% vegetation cover relates to 5×5 cm^2 or 5×6 cm^2 in

Table 3. Vegetation cover of tundra plant species *S. polaris, P. hyperboreum, S. uncinata* and *Cassiope tetragona* exposed to enhanced UV-B for 1 (2002) and 2 (2003) years (small UV lamp plots), for the other species cover estimates of 2002, expressed as percentage ground cover, or number of plants per plot (*Oxyria digyna, Bistorta vivipara*).

Treatment		Control	UVA	UVB1	UVB2	*p* value
Adventdalen						
Salix polaris	2002	38.0(4.2)	42.5(4.8)	41.3(3.1)	55.0(20.4)	0.125
	2003	28.8(11.3)	43.6(6.3)	32.5(3.2)	37.5(6.0)	0.518
Polytrichum hyperboreum	2002	19.0(1.3)	32.5(6.6)	26.3(2.3)	18.0(5.2)	0.119
	2003	40.0(7.4)	38.8(10.7)	28.8(5.2)	28.8(7.5)	0.624
Sanionia uncinata	2002	7.7(1.4)	10.8(3.5)	10.8(5.7)	7.0(3.3)	0.837
	2003	13.3(3.6)	15.0(4.6)	16.3(6.3)	10.8(4.0)	0.892
Equisetum arvense	2002	7.8(3.5)	2.0(1.0)	9.8(9.0)	7.5(5.5)	0.792
Luzula confusa	2002	7.7(2.6)	4.3(5.3)	6.0(2.6)	2.3(1.9)	0.127
Festuca rubra	2002	0.0(0.0)	0.0(0.0)	3.7(2.3)	0.0(0.0)	0.113
Peltigera aphtosa	2002	0.75(0.75)	0.75(0.75)	0.50(0.28)	1.30(1.25)	0.932
Bistorta vivipara	2002	4.8(1.7)	4.5(0.6)	4.0(2.0)	2.5(1.5)	0.739
Crustose lichen	2002	35(5.4)	13.5(7.1)	17.5(6.0)	18.7(6.3)	0.129
Isdammen						
Salix polaris	2002	18.8(3.8)	22.0(7.3)	48.7(5.7)	33.1(2.7)	0.003
	2003	20.0(0.0	22.5(6.0)	15.0(2.0)	28.8(3.1)	0.058
Cassiope tetragona	2002	26.3(6.6)	28.8(7.5)	20.8(8.3)	4.5(3.9)	0.100
	2003	41.3(12.5)	33.8(13.0)	47.5(20.3)	7.0(6.0)	0.018
Polytrichum hyperboreum	2002	0.25(0.25)	0.0(0.0)	0.0(0.0)	0.0(0.0)	0.426
	2003	0.8(0.8)	0.8(0.8)	0.0(0.0)	9.8(0.8)	
Sanionia uncinata	2002	35.8(6.4)	41.3(2.9)	21.0(3.0)	26.8(4.4)	0.032
	2003	13.8(2.4)	18.8(1.3)	11.3(2.4)	21.3(8.8)	0.264
Dryas octopetala	2002	5.5(2.2)	5.0(1.2)	8.3(7.2)	7.8(2.4)	0.377
	2003	6.0(3.0)	8.3(1.8)	19.5(13.6)	22.5(12.5)	0.307
Stellaria crassipes	2002	0.0(0.0)	0.0(0.0)	2.5(2.5)	12.5(12.5)	0.483
Festuca rubra	2002	6.7(6.4)	0.2(0.2)	23.9(13.2)	0.6(0.1)	0.150
Oxyria digyna	2002	1.0(1.0)	1.5(0.9)	0.0(0.0)	0.0(0.0)	0.360
Peltigera aphtosa	2002	0.0(0.0)	0.3(0.3)	1.25(0.5)	0.7(0.4)	0.006
Bistorta vivipara	2002	17.5(4.3)	4.3(2.0)	4.0(2.3)	29.3(6.9)	0.004
Saxifraga hirculus	2002	0.25(0.25)	0.25(0.25)	0.0(0.0)	0.25(0.25)	0.802
Crustose lichen	2002	10.5(5.1)	0.5(0.5)	3.3(1.9)	5.0(1.7)	0.147

Average values and standard error of the mean (based on all subsamples). The anova's were carried out with the average values of the subplots of the four lamp units per treatment.

plants, bearing a sporophye or not, was measured with a ruler from the base to the tip of the longest leaf.

Ecophysiological parameters

UV-B absorbance acid methanol leaf extracts Salix polaris. About 5 mg dried leaf material of *Salix polaris* was ground with 0.5 ml methanol, then another 4.5 ml methanol was added and the suspension was transferred into a Pyrex tube filled with 4 ml of a methanol:H_2O:HCl (79:20:1) mixture, heated (90 °C) in a waterbath for 60 min, vortexed. The samples were centrifuged at 2500 rpm for 5 min. UV-B absorption (280–400 nm) was measured with a Shimadzu UV-160PC spectrophotometer.

Total phenolic content Salix polaris leaves

About 25 mg of dried *Salix polaris* leaves was ground with 0.5 ml 50% methanol, put into pyrex tubes with 2 ml 50% methanol. The tubes were shaken (1 h) and centrifuged at 2500 rpm (5 min). 50 μl of the extract was mixed with 3.95 ml distilled water. About 250 μl of the Folin Ciocalteu reagent (Merck) was added and after exactly 8 min 750 μl Na_2CO_3 (20 g/100 ml) was added. Absorbance was measured after 2 h at 760 nm with a Shimadzu UV-160PC spectrophotometer.

DNA damage

Leaves of tips of *Salix polaris* and *Cassiope tetragona* branches were sampled August 8, 2002, at the Adventdalen and Isdammen site, and stored at −85 °C within 1 h after leaf sampling. DNA was extracted from three leaves (*Salix*) or one or two smaller pedicels (*Cassiope*) following a protocol described in de Bakker (in prep.). In short: DNA damage was assessed as cyclobutane thymine dimers, and was quantified using the H3 antibody (Roza et al. 1988; Boelen 2002) applying an immunoblot procedure, see van de Poll (2003) for methodological details.

Reproductive parameters

Male and female catkins Salix polaris. Male and female catkins of *Salix polaris* were counted within squares of $20 \times 20\ cm^2$ in the small UV plots at Adventdalen. The apparent sex ratio was calculated using the ratio of maximum numbers of female and male catkins counted. It is based on visible above ground male or female flowers. It was not possible to determine an absolute sex ratio based on the number of male and female plants which would require destructive sampling of aboveground and belowground plant parts.

Inflorescence length Bistorta vivipara. Length of all inflorescences of *Bistorta vivipara* was measured at August 4 and 5, 2003, in the plots of the small UV lamps irradiated for 2 years at Adventdalen and at the site of the large UV lamps at Isdammen. The *Bistorta vivipara* plants were randomly distributed in the plots.

Statistical analysis. Effects of enhanced UV-B on vegetation and plant parameters were tested with a one way anova (with the Bonferroni test), with four replicates per treatment for the large and small UV lamp systems. In addition to analysis of UV-B effects after 6 or 7 years (large UV lamp systems) and 1 or 2 years of irradiance (small UV lamps) plant cover changes from 2002 to 2003 as affected by UV-B were statistically analysed. Normality was tested with Shapiro-Wilk or Kolmogorov-Smirnoff. Homogeneity of variance was tested with the Levene statistic. In cases where variances were not homogeneous, and transformations did not deliver homogeneity the non-parametric Mann–Whitney (2 independent samples) or Kruskal and Wallis (k independent samples) tests were applied following procedures described in SPSS 10.1 and background knowledge derived from Sokal and Rohlf (1995) and Quinn and Keough (2002).

Results

Plant cover, plant growth and plant density

Based on the vegetation cover estimated in the subplots of the large UV lamps there is a non significant decline in the cover of *Salix polaris* at Adventdalen in 2002, and at Isdammen in 2002 and 2003 (Table 2). Also, plant cover of *Cassiope tetragona, Sanionia uncinata* and *Polytrichum hyperboreum* and other tundra species was not significantly affected by enhanced UV-B after 6 and 7 years of irradiance (Table 2).

Vegetation cover of the tundra species in the small UV plots after 1 and 2 years of irradiation with UV-A, UV-B1 and UV-B2 radiation treatment is summarized in Table 3. Since the mini lamps were installed June 2002 the cover data for 2002 are regarded a test for homogeneity of the distribution of the cover of the tundra species per plot. Cover of species with *p*-values < 0.05 is not homogeneously distributed over the four treatments.

These plant cover data for separate plots may eventually allow year-to-year comparisons with elevated UV-B which could reveal UV-B effects. Since estimation of vegetation cover in 2002 and 2003 was differing two weeks for the plots with the small lamps, no proper year to year comparisons could be made as yet.

Overall, after 6 and 7 years of irradiance, UV-B has not significantly affected plant cover of any of the tundra plant species studied.

The length growth of male gametophyte moss plants of *Polytrichum hyperboreum*, measured from the last developed antheridium appeared to be reduced from 3.25 (s.e.m 0.06) mm to 2.20 (s.e.m 0.05) mm with elevated UV-B ($p = 0.015$), length of other moss plants in the plots was not affected by the UV-B treatments.

The number of *Pedicularis hirsuta* plants per plot did significantly decrease with enhanced UV-B both at Isdammen ($p = 0.047$) and at Adventdalen ($p = 0.027$), (Figure 4)

Morphological and ecophysiological parameters

Leaf area of *Salix polaris* (per leaf) was not significantly decreased with enhanced UV-B for Adventdalen plants (Table 4), but increased with enhanced UV-B for Isdammen plants.

Leaf thickness was not affected by UV-B, neither was fresh and dry weight per leaf, leaf UV-B absorbance, total leaf phenolic content, and water content (Table 4).

Table 4. Effects of enhanced UV-B radiation on leaf thickness, leaf area, UV-B absorbance, total phenolic content, fresh and dry weight per leaf, water content of arctic tundra plant species *Salix polaris* and *Cassiope tetragona* sampled July 2002 at Isdammen en Adventdalen after 1 and 6 years of UV-B radiation simulating 15% ozone depletion or 30% ozone depletion.

	Isdammen	Adventalen
Salix polaris		
Leaf thickness (mm)		
6 year radiation	$p=0.098$ n.s.	$p=0.192$ n.s.
1 year radiation	$p=0.001$ sign.	$p=0.092$ n.s
Leaf area (mm^2)		
6 year radiation	$p=0.001$ sign.	$p=0.651$ n.s.
1 year radiation	$p=0.121$ n.s.	$p=0.976$ n.s.
UV-B absorption leaves (absorbance area/mg dry weight)		
6 year radiation	$p=0.125$ n.s.	$p=0.912$ n.s.
1 year radiation	$p=0.626$ n.s.	$p=0.247$ n.s.
Total phenolic leaves (g tannic acid/g dry weight)		
6 year radiation	$p=0.968$ n.s.	$p=0.861$ n.s.
1 year radiation	$p=0.022$ sign.	$p=0.127$ n.s.
Fresh weight per leaf (g)		
6 year radiation	$p=0.182$ n.s.	$p=0.707$ n.s.
Dry weight per leaf (g)		
6 year radiation	$p=0.064$ n.s	$p=0.720$ n.s
Water content (% dry weight)		
6 year radiation	$p=0.202$ n.s.	$p=0.782$ n.s.
1 year radiation	$p=0.68$ n.s.	$p=0.444$ n.s
Cassiope tetragona		
Water content		
6 year radiation	$p=0.236$ n.s.	
1 year radiation	$p=0.065$ n.s.	
Fresh weight per leaf (g)		
6 year radiation	$p=0.101$ n.s	
Dry weight per leaf (g)		
6 year radiation	$p=0.172$ n.s.	

p-Values of one way anova.

DNA damage did not increase with exposure to enhanced UV-B in *Salix polaris* and *Cassiope tetragona* (Table 5). However DNA damage in *Salix* leaves was larger than in *C. tetragona* ($p=0.001$).

Reproductive parameters

The number of male and female catkins of *Salix polaris* per plot and the ratio of this under the small UV lamps was assessed in Adventdalen in 2002 and 2003 (Figure 3), there was no significant effect of enhanced UV-B on the apparent sex ratio (*n* female/*n* male catkins) in 2002 ($p=0.338$), and in 2003 ($p=0.434$), but the apparent sex ratio was male biased, i.e. significantly smaller than 1.0 ($p=0.05$ and $p=0.00$ in 2003). The number of female catkins in 2002 and 2003 did not differ significantly ($p=0.101$), neither did the number of male catkins ($p=0.150$). The UV-B treatment did not affect the number of female or male catkins in 2002 and 2003 ($p=0.760$; $p=0.212$; $p=0.063$; $p=0.499$, respectively).

The inflorescence length of *Bistorta vivipara* was measured 4th and 5th of August 2003 both under the large (31.5–31.9 mm) and the small UV lamps (26.3–37.7 mm).There was no significant effect of the UV-B treatment on the inflorescence length ($p=0.877$ and $p=0.855$, respectively).

Discussion

Plant cover, plant growth, plant density and plant morphology

There were no significant UV-B effects on plant cover and plant parameters of most of the tundra plants e.g. *Salix polaris, Cassiope tetragona, Sanionia uncinata* and *Polytrichum hyperboreum* after 7 and 2 years of UV radiation (Tables 3, 4).The evergreen ericaceous dwarf shrub *Cassiope tetragona* is considered to be rich in secondary compounds such as the flavonoids myricetin and quercetin (Björn et al. 1997) with a characteristic, apparent smell and is not grazed by the reindeer and appears to be tolerant to enhanced UV-B (Callaghan et al. 1989).

The compound responsible for the characteristic smell of *Cassiope* leaves has recently been identified as eudesmol (Blokker et al. 2005). Probably the smell of eudesmol acts as a deterrent and prevents reindeer grazing.

Table 5. DNA damage expressed as CPD's per megabase DNA.

Treatment *Salix polaris*	Control	UVA	UVB1	UVB2	*p*-value
Isdammen 2 years	5.4(3.5)	6.9(6.3)	3.1(1.1)	8.7(3.5)	0.31
Isdammen 6 years	7.7(3.1)	–	3.0(2.3)	–	0.05
Cassiope tetragona 2 years	1.1(0.8)	1.8(2.1)	2.6(3.4)	2.3(0.8)	0.27
Cassiope tetragona 6 years	0.7(0.6)	–	0.2(0.3)	–	0.28
Salix polaris					
Adventdalen 2 years	8.8(3.5)	7.0(6.0)	7.2(6.1)	7.0(5.3)	0.96
Adventdalen 6 years	7.5(5.7)	–	5.0(4.1)	–	0.69

Average values of 4 replicate samples and standard deviation. P values of 1 way anova's.

Figure 3. Male (a) and female (b) catkin of *Salix polaris*. Photograph Jelger Herder.

Cover of the moss *S. uncinata* was not affected by enhanced UV-B, which agrees with results of Lud et al. (2002), who regard the moss *S. uncinata* a UV-B tolerant Arctic and Antarctic moss species. We have not observed spore capsules of *S. uncinata* at Isdammen and Adventdalen and obviously this moss species is reproducing vegetatively at the Svalbard tundra. By contrast, *P. hyperboreum* is frequently sporulating in the arctic summers (Table 2). Measured plant parameters (Table 4) as well as quantification of DNA damage of *Salix polaris* cannot easily be related to other plant parameters. Leaf area is not significantly reduced, the leaf weight of high UV-B plants is not reduced, neither is leaf thickness significantly affected by enhanced UV-B.

Individual *Salix* leaves develop, grow and photosynthesize during one tundra summer season (June-end of August) and then senesce, and for the growth, morphological and ecophysiolocal parameters studied here effects of enhanced UV-B would be similar for the 6 year and 1 year irradiated plots, assuming such effects to occur each summer in developing leaves. Leaf buds of *Salix polaris* for the forthcoming summer season are already developed at the end of the preceding growing season, and have been exposed to enhanced UV-B in that season.

Plant ecophysiological parameters

There were only few significant effects of enhanced UV-B on leaf water content, leaf UV-B absorption and leaf phenolic content measured in *S. polaris* and *C. tetragona* (Table 4).

Leaf phenolic content increased with enhanced UV-B at the small lamps at Isdammen, but not at Adventalen and leaf UV-B absorbance was unaffected for *S. polaris* at the large and the small UV lamp systems.

DNA damage in *Salix* leaves did not increase with enhanced UV-B (Table 5), neither was there a

DNA damage increase in *Cassiope* leaves with enhanced UV-B. However leaf DNA damage in *Salix* was significantly greater than in *Cassiope.* Lud et al. (2002) found UV-B induced DNA damage in *S. uncinata*, combined with efficient photorepair and the circumbipolar moss was judged UV-B tolerant (cf Rousseaux et al. 1999; Boelen et al. 2005).

Even with more plant growth, morphological or physiological evidence, it may remain difficult to predict or explain unambiguously a change in plant cover. Often the effect of enhanced UV-B cannot be characterized as inhibition or damage, but rather as a photomorphogenetic effect. Shoot length growth of *Deschampsia antarctica* was decreased, but the number of tillers increased with enhanced UV-B (Rozema et al. 2001a) and the RGR remained unaffected. Also within one species parameters may respond differentially to enhanced UV-B. Male plants of *P. hyperboreum* growing after having developed antheridia, showed reduced length growth with enhanced UV-B, but no such effect was seen in *P. hyperboreum* moss plants bearing no sporophytes nor antheridia. We have no further evidence to support the assumption that the measured length growth of the last developed antheridium, started at the same time. Tagging other moss plants of *P. hyperboreum* with woollen threads indicated length growth of moss plants to be about 1 mm per summer season. This may imply that the measured length growth of male moss gametophyte plants may be the result of at least three summer seasons. This would indicate that antheridium formation in moss plants in the tundra will not take place every year, as occurs with *Polytrichum commune* in Dutch peat lands (Rozema et al. unpublished). It is unknown if sporophyte formation in the tundra occurs every year or less frequently.

Reproductive parameters

Salix polaris is a dioecious deciduous perennial woody tundra plant (Rønning 1996; Elvebakk 1994), with male and female catkins on separate plants and leaves rapidly developing when frost and snow disappear in June (Figure 3). Shining and hairy leaves of *Salix polaris* are in a more or less horizontal position, less than a few centimeters above the tundra soil, remain green until mid August and then senesce. Individual male and female plants vary in size, but generally length of stems and rhizomes close to the tundra soil surface is less than 10–20 cm (Callaghan et al. 1989; Havstrom et al. 1995).

We found no effects of enhanced UV-B on the number of male and the number of female catkins of *Salix polaris*, neither there was a UV-B effect on the apparent sex ratio. More male catkins than female catkins were found. This male biased sex ratio is in contrast with female biased sex ratio's for arctic dwarf willows including *Salix polaris* reported by Crawford and Balfour (1983, 1990). Preferential herbivory by Svalbard reindeer has been suggested to explain female biased sex ratio's in *Salix polaris* (more female flowers than male flowers) (Dormann and Skarpe (2002). Exclusion of reindeer possibly preferring to graze on male *Salix* flowers led to an increase in male flowers. A fence surrounds the large UV lamp plots and there is no or very limited grazing by reindeer in our large (installed 1996) and small UV (installed 2002) lamp plots at Isdammen and Adventdalen. The small UV lamps are close to the large UV lamps and reindeer grazing has not been observed in 2002 and 2003. A male biased sex ratio in five wind pollinated gymnosperms and angiosperms has been be attributed to dry conditions (Freeman et al. 1976). Male plants preferred dry sites to maximize pollen dispersal and female plants preferred moist sites maximizing seed set.

With the aim to determine UV-B absorbing compounds in pollen (Rozema et al. 2001b, c) male catkins have been collected in all UV plots. Collecting male flowers does not affect the sex ratio of *Salix polaris* since the number of male catkins per plot did not differ in 2002 and 2003. At the end of the summer season dried and senesced male flowers drop off the parental plant naturally.

Of the tundra species studied, the number of *Pedicularis hirsuta* plants per plot, both at Isdammen en Adventdalen reduced (Figure 4). *Pedicularis hirsuta* is a perennial hemiparasite with a taproot and root xylem bridges to neighbouring tundra plants, providing the hemiparasite with xylem water and nutrients. So far, clear ecophysiological causes of reduced numbers of *Pedicularis* plants in the UV-B plots remain unclear. Small seedlings of *Pedicularis* occur within the plots (personal observation), and if no root contact with host tundra plants occurs, these seedlings die.

Chlorophyll development is obviously impaired in this hemiparasite and one may speculate that a limited capacity of synthesizing protective UV-B absorbing compounds may cause *Pedicularis* to be sensitive to enhanced UV-B. Arctic hemiparasites often have nutrient-rich leaves with little phenolic or tannin content and therefore generally decompose faster than surrounding species (Quested et al. 2003).

Inflorescences of *Bistorta vivipara* develop within one summer season, and will be exposed during this length growth to UV-B treatments. However unlike the density of *P. hirsuta* plants, the inflorescence length of *Bistorta* was not affected by UV-B.

Enhanced UV-B does not affect tundra plant growth

The absence of UV-B effects in many plant growth, morphological and ecophysiological parameters for these tundra plants, as well as the lack of reduction of plant cover or numbers per plot (except for plant density *Pedicularis hirsuta* in Isdammen and Adventdalen, and the length growth of male gametophytes of *P. hyperboreum*) indicates that many tundra species are tolerant to enhanced UV-B. This would indicate that many plants species of tundra ecosystems do no differ in their response to enhanced UV-B from species of other terrestrial ecosystems of other climate zones (Rozema et al. 1997; Caldwell et al. 1998; Sullivan and Rozema 1999; Searles et al. 2001, 2002; Paul 2001; Aphalo 2003; Paul and Gwynn Jones 2003). However other causes may also explain absence of UV-B effects, which we discuss below.

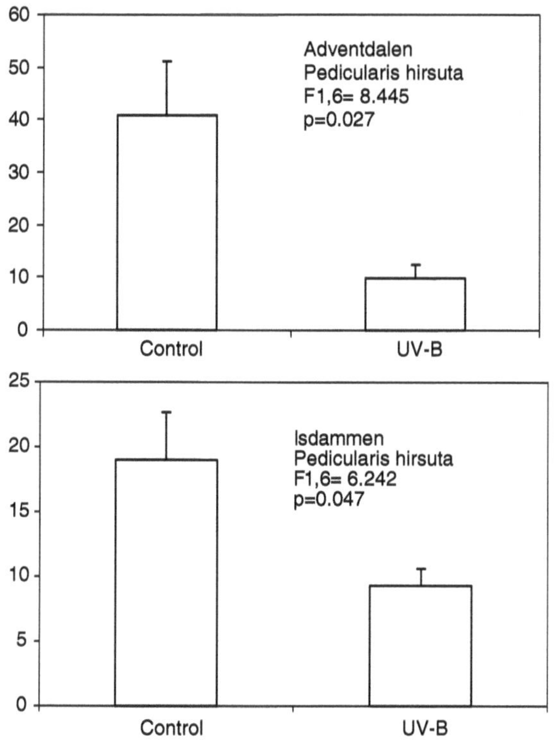

Figure 4. Effect of 6 year exposure (1996–2002) to enhanced UV-B, simulating 15% ozone depletion, on the number of *Pedicularis hirsuta* plants per plot. Mean values with standard error of the mean.

Spatial variation and statistical power

Obvious spatial variation in nitrogen fixation by terrestrial cyanobacteria homogeneously irradiated by the UV lamps at Adventdalen and Isdammen has also been recognized by Solheim et al. 2006 (this volume). Similar spatial variation in the distribution of tundra plants may prevent detection of UV-B treatment effects on for example percentage cover of *Salix polaris* (Table 2).

An experimental design with a larger number of replicated UV plots than the current four replications per treatment would improve statistical power, though logistically it would be more complicated and expensive. Limited power of the current experimental design may obstruct detection of (small) UV-B effects.

UV-B supplementation and UV-B exclusion experiments in polar regions

Only few UV-B enhancement field experiments or UV-B field manipulations experiments in high arctic or antarctic regions are known to us (Arctic Svalbard: Björn 2002; Solheim et al. 2002; Bjerke et al. 2003; Zielke et al. 2003; Zielke 2004; Rozema et al. 2005; Antarctic: Day et al. 1999, 2001; Ruhland and Day 2000; Rozema 1999; Rozema et al. 2001a; Lud et al. 2002; Newsham 2003). There are some more similar studies in the subarctic (e.g. Johanson et al. 1995a, b; Björn et al. 1997; Gehrke 1998, 1999; Phoenix et al. 2001, 2003; Sonesson et al. 2002; Semerdjieva et al.

2003) and subantarctic (Rousseaux et al. 1999; Ballaré et al. 2001; Searles et al. 2001, 2002; Robson et al. 2003a,b,c) which are partially comparable to the polar studies. Most of these reports indicate small, subtle or no UV-B effects (see Rozema et al. 2005). UV-B exclusion experiments with UV-B absorbing foils in the field indicate length growth reduction of *Deschampsia antarctica* with UV-B levels varied from below ambient UV-B to near ambient UV-B (Day et al. 1999; Ruhland and Day 2000). Such UV-B exclusion manipulations may be relevant from an ecological point of view, but they do not demonstrate that growth of Antarctic flowering plants is reduced with enhanced UV-B as may be shown with UV-B supplementation in the field (Figure 5). With UV-B supplementation, UV-B is varied from ambient levels to above ambient levels, e.g. from 2.5 kJ UV-B_{be} m^{-2} day^{-1} to 5.0 kJ UV-B_{be} m^{-2} day^{-1}. Because of different UV-B levels involved the outcome of UV-B exclusion field studies cannot be compared with those of UV-B supplementation field studies. In addition, the UV-B plant growth response curve (Figure 5) shows that growth reduction related with UV-B varied from ambient to above ambient is much less (18%) than from below ambient to near ambient UV-B (78%).

Based on the above, growth reduction with enhanced UV-B as in the Isdammen and Adventdalen UV lamp supplementation experiment is expected to be limited and not easy to detect.

In case absence of UV-B effects reflects real tolerance of tundra plants to enhanced UV-B, adaptations to UV-B may be expected such as effective UV-B absorbing pigments.

In a recent meta-analysis of about 100 reports of field studies of enhanced UV-B effects Searles et al. (2001) indicate a 10% increase of UV-B absorbing compounds with 15% ozone depletion in plant species of terrestrial ecosystems from various climate zones.

UV-B absorbing compounds may be induced by UV-B, e.g. the flavonoid quercitin however constitutively high levels of other UV-B pigments of tundra plants will also prevent UV-B damage, and may explain absence of UV-B induced UV-B damage.

Tolerance of polar plants to enhanced UV-B also implies that there is no apparent vulnerability of polar plants to frost, cold, short summer season

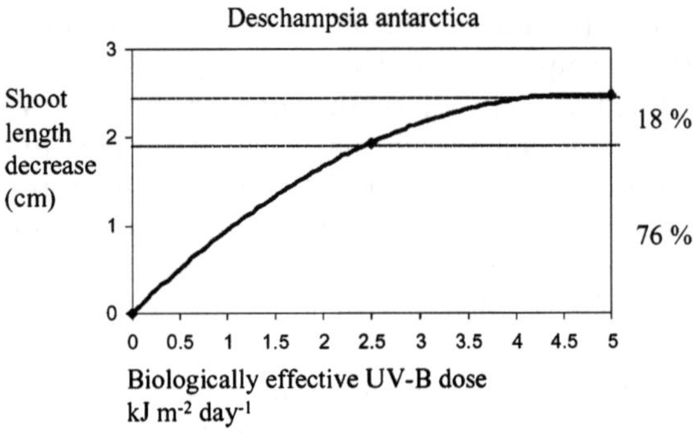

Figure 5. Shoot length decrease (cm) of *Deschampsia antarctica* relative to shoot length at 0 kJ m^{-2} day^{-1} of plants exposed to 0, 2.5 or 5.0 kJ m^{-2} day^{-1} biologically effective UV-B for 78 days. Plants were grown in a climate room (4 °C, PAR 150 μmol m^{-2} s^{-1}, 75% Relative Humidity). Recalculated after Rozema et al. (2001a).

and that low temperatures do not markedly inhibit repair of DNA damage.

From a historical and biogeographical point of view, the tundra biome and ecosystems at Svalbard are relatively young and recent tundra vegetation of the northern hemisphere will have developed since the last glacial period, i.e. the Younger Dryas some 11,000 years ago (Isarin 1997) from northward migrating plants originating from lower than polar latitudes with higher natural solar UV-B fluxes and associated tolerance to UV-B.

Perspective

Although the use of CFC's has been phased out, recovery of stratospheric ozone may last until 2050–2060 (World Meteorological Organization 2002) and enhanced UV-B levels may occur several decades.

Ozone depletion is most severe in the Antarctic and Arctic spring and UV-B levels at ground level in polar regions are increased accordingly (Rozema et al. 2005; McKenzie et al. 2003; Rex et al. 2004). While terrestrial Antarctic ecosystems have only two species of higher plants: *Deschampsia antarctica* and *Colobanthus quitensis* (Smith 1994; Convey et al. 2005), more than 160 higher plant species occur in the high arctic tundra ecosystem of Svalbard (Elvebakk 1994; Rønning 1996), allowing more species interactions and feedbacks and perhaps providing a more general, representative ecosystem response to enhanced UV-B than the more simple two-species Antarctic ecosystem (Rozema et al. 2005).

The world-wide unique longterm (1996–2004) UV-B supplementation field experiment at the Adventdalen and Isdammen high arctic tundra ecosystem should therefore be continued.

Acknowledgements

The UV-B supplementation system at Isdammen and Adventdalen was started in 1996 under the European Commission contract EV5V-CT910031. Dr. B. Solheim, University of Tromsø, is greatly acknowledged for continuation and maintenance of the UV lamp facilities after 1998. Field work at Svalbard of J.R. in 2000 was funded by EC contract UVAQTER number ENV-CT97-0580. The installation of UV minilamps in 2002 is financially supported by NWO-ALW-NAAP grant number 851.20.010 (UVANTARTIC). We acknowledge the permission for the field work by Sysselmannen, Longyearbyen and the cooperation and support of UNIS. The field work of A.B., M.D (2002), J.H. and R.F.(2003) at Svalbard forms part of their MSc research projects, jointly supervised by J.R. P.B., Dr B. Solheim and Dr M. Zielke, which has been funded partially by the Vrije Universiteit, (Dr K. Kits) which is greatly appreciated. The support by and cooperation with Dr S. Spjelkavik (UNIS), Prof Dr I. Jonsdottir (UNIS) and Dr Rene van der Wal (Centre for Ecology & Hydrology – Banchory, Scotland) is appreciated. Prof Dr A.S. Blix is greatly acknowledged for permission for storage of UV equipment and for mains power at the Adventsdalen site.

References

Aphalo P. 2003. Do current levels of UV-B affect vegetation? The importance of long-term experiments. New Phytol. 160: 273–280.

Ballaré C.L., Rousseaux M.C., Searles P.S., Zaller J.G., Giordano C.V., Robson T.M., Caldwell M.M., Sala O.E. and Scopel A.L. 2001. Impacts of solar ultraviolet-B radiation on terrestrial ecosystems of Tierra del Fuego (southern Argentina) – An overview of recent progress. J. Photochem. Photobiol. B-Biol. 62: 67–77.

Bjerke J.W., Zielke M. and Solheim B. 2003. Long-term impacts of simulated climatic change on secondary metabolism, thallus structure and nitrogen fixation activity in two cyanolichens from the Arctic. New Phytol. 159: 361–367.

Björn L.O., Callaghan T.V., Gehrke C., Gunnarsson T., Holmgren B., Johanson U., Snogerup S., Sonesson M., Steiner O. and Yu S.G. 1997. Effects on subarctic vegetation of enhanced UV-B radiation. In: Lumsden P.J. (ed.), Plants and UV-B. Cambridge University Press, pp. 233–246.

Björn L.O. 2002. Effects of UV-B radiation on terrestrial organisms and ecosystems with special reference to the arctic. In: Hessen D.O. (ed.), UV Radiation and Arctic Ecosystems Ecological Studies 153, Springer, Berlin, pp. 93–121.

Blokker P., Boelen P. Broekman R. and Rozema J. 2005. The potential of *p*-coumaric and ferulic acid as UV proxies: occurrence in pollen and spores, preservation and pyrolytic analysis. Plant Ecology (this volume).

Boelen P. 2002. Sundazed to the Core. Accumulation of DNA damage in (sub)tropical picoplankton assemblages exposed to solar ultraviolet radiation. Ph D Thesis. Groningen University. Stichting Drukkerij C. Regenboog, Groningen.

Boelen P., de Boer M. K., de Bakker N., Blokker P. and Rozema J. 2006. Field studies on the effects of solar UV-B on polar bryophytes: overview and methodology. Plant Ecology (this volume).

Caldwell M.M., Björn L.O., Bornman J.F., Flint S.D., Kulandaivelu G., Teramura A.H. and Tevini M. 1998. Effects of increased solar ultraviolet radiation on terrestrial ecosystems. J. Photochem. Photobiol. B-Biol. 46: 40–52.

Callaghan T.V., Carlsson B.A. and Tyler N.J.C. 1989. Historical records of climate-related growth in Cassiope tetragona from the arctic. J. Ecol. 77: 823–837.

Convey, P. and Smith, R.I.L., 2006. Responses of terrestrial Antarctic ecosystems to climate change. Plant Ecology. (this volume).

Crawford R.M.M. and Balfour J. 1983. Female predominant sex ratios and physiological differentiation in arctic willows. J. Ecol. 71: 149–160.

Crawford R.M.M. and Balfour J. 1990. Female biased sex ratios and differential growth in arctic willows. Flora 184: 291–302.

Day T.A, Ruhland C.T, Grobe C.W and Xiong F. 1999. Growth and reproduction of Antarctic vascular plants in response to warming and UV radiation reductions in the field. Oecologia 119: 24–35.

Day T.A., Ruhland C.T. and Xiong F. 2001. Influence of solar UV-B radiation an Antarctic terrestrial plants: result from a 4-year field study. J. Photochem. Photobiol. B-Biol. 62: 78–87.

Dorman C., F. and Skarpe C. 2002. Flowering, growth and defence in the two sexes: consequences of herbivore exclusion for *Salix polaris*. Funct. Ecol. 16: 4–17.

Elvebakk A. 1994. A survey of plant associations and alliances from Svalbard. J. Veg. Sci. 5: 791–802.

Farman J.C, Gardiner B.G and Shanklin J.D. 1985. Large losses of total ozone in Antarctica reveal seasonal CLOx/NOx interaction. Nature 315: 207–210.

Freeman D.C., Klikoff L.G. and Harper K.T. 1976. Differential resource utilization by the sexes of dioecious plants. Science 193: 597–599.

Gehrke C. 1999. Impacts of enhanced ultraviolet-B radiation on mosses in a subarctic heath ecosystem. Ecology 80: 1844–1851.

Gehrke C. 1998. Effects of enhanced UV-B radiation on production-related properties of a *Sphagnum fuscum* dominated subarctic bog. Funct. Ecol. 12: 940–947.

Havstrom M., Callaghan T.V., Jonasson S. and Svoboda J. 1995. Little ice-age temperature estimated by growth and flowering differences between subfossil and extant shoots of *Cassiope tetragona*, an arctic heather. Funct. Ecol. 9: 650–654.

Isarin R. 1997. The climate in north-western Europe during the younger Dyas. A comparison of multi-proxy climate reconstructions with simulation experiments. Ph D thesis Vrije Universiteit.

Johanson U., Gehrke C., Björn L.O. and Callaghan T.V. 1995a. The effects of enhanced UV-B radiation on the growth of dwarf shrubs in a subarctic heathland. Funct. Ecol. 9: 713–719.

Johanson U., Gehrke C., Björn L.O., Callaghan T.V. and Sonesson M. 1995b. The effects of enhanced UV-B radiation on a subarctic heath ecosystem. Ambio 24: 106–111.

Johnstone J.F. and Henry G.H.R. 1997. Retrospective analysis of growth and reproduction in *Cassiopa tetragona* and relations to climate in the Canadian high arctic. Arctic Alpine Res. 29: 459–469.

Kristiansen K.J. and Sollid J.L. 1987. Svalbard. Map of surficial material. Norsk Polar Institutt.

Li S., Paulsson M. and Björn L.O. 2002. Temperature-dependent formation and photorepair of DNA damage induced by UV-B radiation in suspension-cultured tobacco cells. J. Photochem. Photobiol. B: Biol. 66: 67–72.

Lud D., Moerdijk T.C.W., van de Poll W.H., Buma A.G.J. and Huiskes A.H.L. 2002. DNA damage and photosynthesis in Antarctic and Arctic *Sanionia uncinata* (Hedw.) Loeske under ambient and enhanced levels of UV-B radiation. Plant Cell Environ. 25: 1579–1589.

McKenzie R.L., Björn L.O., Bais A. and Ilyasd M. 2003. Changes in biologically active ultraviolet radiation reaching the Earth's surface. Photochem. Photobiol. Sci. 2: 5–15.

McPeters R., Hollandsworth S., Flynn L., Herman J. and Seftor C. 1996. Long-term ozone trends derived from the 16-year combined 7/meteor 3 TOMS version 7 record. J. Geophys. Res. 23: 3699–3702.

Molina M.J. and Rowland F.S. 1974. Stratospheric sink for chlorofluoromethanes: chlorine atom-catalysed destruction of ozone. Nature 249: 810–812.

Newman P, Gleason J, McPeters R and Stolarski R. 1997. Anomalously low ozone over the arctic. Geophys. Res. Lett. 24: 2689–2692.

Newsham K. 2003. UV-B radiation arising from stratospheric ozone depletion influences the pigmentation of the moss *Andraea regularis*. Oecologia 135: 327–331.

Paul N.D. 2001. Plant responses to UV-B: time to look beyond stratospheric ozone depletion?. New Phytol. 150: 5–8.

Paul N.D. and Gwynn-Jones D. 2003. Ecological roles of solar UV radiation: towards an integrated approach. Trends Ecol. Evol. 18: 48–55.

Phoenix G.K, Gwynn-Jones D, Callaghan T.V and Sleep D. 2001. Effects of global change on a sub-Arctic heath: effects of enhanced UV-B radiation and increased summer precipitation. J. Ecol. 89: 256–267.

Phoenix G.K., Gwynn-Jones D., Lee J.A. and Callaghan T.V. 2003. Ecological importance of ambient solar ultraviolet radiation to a sub-arctic heath community. Plant Ecol. 165: 263–273.

Quinn G.P. and Keough M.J. (2002) Experimental Design and Data Analysis for Biologists. Cambridge University Press.

Poll W.H., vande Bischof K., Buma A.G.J and Breeman A.M. 2003. Habitat related variation in UV tolerance of tropical marine red macrophytes is not temperature dependent. Physiol. Plantarum 118: 74–83.

Quested H.M, Cornelissen J.H.C., Press M.C., Callaghan T.V., Aerts R., Trosien F., Riemann P., Gwynn-Jones D., Kondratchuk A. and Jonasson S.E. 2003. Decomposition of sub-arctic plants with differing nitrogen economies: a functional role for hemiparasites. Ecology 84: 3209–3221.

Rex M., Rex M., Salawitch R.J., von der Gathen P., Harris N.R.P., Chipperfield M.P. and Naujokat B. 2004. Arctic ozone loss and climate change. Geophys. Res. Lett. 31: Art. No. L04116.

Robinson S.A., Wasley J. and Tobin A.K. 2003a. Living on the edge - plants and global change in continental and maritime Antarctica. Global Change Biol. 9: 1681–1717.

Robson T.M., Pancotto V.A., Flint S.D., Ballare C.L., Sala O.E., Scopel A.L. and Caldwell M.M. 2003b. Six years of solar UV-B manipulations affect growth of *Sphagnum* and vascular plants in a Tierra del Fuego peatland. New Phytol. 160: 379–389.

Robson T.M., Pancotto V.A., Flint S.D., Ballare C.L., Sala O.E., Scopel A.L. and Caldwell M.M. 2003c. Six years of solar UV-B manipulations affect growth of Sphagnum and vascular plants in a Tierra del Fuego peatland. New Phytol. 160: 379–389.

Rønning O.I. 1996. Svalbards flora. Norsk Polar Institutt.

Roza L., vander Wulp K.J.M., Mc Farlane S.J., Lohman, P.H.M., and Baan R.A. 1988. Detection of cyclobutane thymine dimers in DNA of human cells with monoclonal antibodies raised against a thymine dimer containing tetranucleotide. Photochem. Photobiol. 48: 627–633.

Rozema J., Broekman R., Lud D., Huiskes L., Moerdijk A.H.T., de Bakker N., Meijkamp B. and van Beem A. 2001a. Consequences of depletion on stratospheric ozone for terrestrial Antarctic ecosystems: the response of *Deschampsia antarctica* to enhanced UV-B radiation in a controlled environment. Plant Ecol. 154: 101–115.

Rozema J., Noordijk A.J., Broekman R.A., van Beem A., Meijkamp B.M., Bakker N.V., de Stroetenga M., Bohncke S.J.P., Konert M., Kars S.M., Peat H.E., Smit R.I.L. and Convey P. 2001b. (Poly)phenolic compounds in pollen and spores of plants as indicators of solar UV-B: a new proxy for the reconstruction of past solar UV-B. Plant Ecol. 154: 49–58.

Rozema J., Broekman R., Blokker P., Meijkamp B.M., Bakker N.V., de Staaij J., van, de Beem A., Ariese F. and Kars S.M. 2001c. UV-B absorbance and UV-B Absorbing Compounds (*para-coumaric acid*) in pollen and sporopollenin: the perspective to track historic UV-B. J. Photochem. Photobiol. B-Biol. 6: 108–117.

Rozema J., Boelen P, Doorenbosch M, Boekel C, Solheim B., Bohncke S.J.P., Konert M., Fijn R.C., Herder J.E. and Kars S.M. 2006. Vegetation and climate reconstruction based on palynological analyses of high arctic tundra peat cores. (5000 years B.P) from Svalbard. Plant Ecol. (this volume).

Rozema J. 1999. Stratospheric Ozone Depletion the effects of Enhanced UV-B Radiation on Terrestrial Ecosystems. Backhuys Publishers, Leiden, p. 355.

Rozema J. Boelen P. and Blokker P. 2005. Depletion of stratospheric ozone over the Antarctic and Arctic: responses of plants of polar terrestrial ecosystems to enhanced UV-B, an overview. Environ. Poll. 137: 428–4.

Rousseaux M.C., Ballaré C.L., Giordano C.V., Scopel A.L., Zima A.M., Szwarcberg-Bracchitta M., Searles P.S., Caldwell M.M. and Diaz S.B. 1999. Ozone depletion and UV-B radiation: Impact on plant DNA damage in southern South America. Proc. Nat. Acad. Sci. USA 96: 15310–15315.

Ruhland C.T. and Day T.A. 2000. Effects of ultraviolet-B radiation on leaf elongation, production and phenylpropanoid concentrations of *Deschampsia antarctica* and *Colobanthus quitensis* in Antarctica. Physiol. Plantarum 109: 244–251.

Searles P.S., Flint S.D. and Caldwell M.M. 2001. A meta-analysis of plant field studies simulating stratospheric ozone depletion. Oecologia 127: 1–10.

Searles P.S., Flint S.D., Diaz S.B., Rousseaux M.C., Ballare C.L. and Caldwell M.M. 2002. Plant response to solar ultraviolet-B radiation in a southern South American Sphagnum peatland. J. Ecol. 90: 704–713.

Semerdjieva S.I., Sheffield E., Phoenix G.K., Gwynn-Jones D., Callaghan T.V. and Johnson G. 2003. Contrasting strategies for UV-B screening in sub-Arctic dwarf shrubs. Plant Cell Environ. 26: 957–964.

Smith R.I.L. 1994. Vascular plants as bioindicators of regional warming in Antarctica. Oecologia 99: 322–328.

Sokal R.R. and Rohlf F.J. 1995. Biometry. W.H. Freeman and Company.3rd edn. pp.887.

Solheim B., Johanson U., Callaghan T.V., Lee J.A., Gwynn-Jones D. and Björn L.O. 2002. The nitrogen fixation potential of arctic cryptogram species is influenced by enhanced UV-B radiation. Oecologia 133: 90–93.

Solheim B., Bjerke J.W., Zielke M. and Rozema J. 2006. Effects of enhanced UV-B radiation on nitrogen fixation in arctic ecosystems. Plant Ecol. (this volume).

Sonesson M., Carlsson B.A., Callaghan T.V., Halling S., Björn L.O., Bertgren M. and Johanson U. 2002. Growth of two peat-forming mosses in subarctic mires: species interactions and effects of simulated climate change. Oikos 99: 151–160.

Sullivan J. and Rozema J. 1999. UV-B effects on terrestrial plant growth and photosynthesis. In: Rozema J. (ed.), Stratospheric Ozone Depletion, the Effects of Enhanced UV-B Radiation on Terrestrial Ecosystems, Backhuys, Leiden, pp. 39–57.

World Meteorological Organization 2003. Scientific assessment of ozone depletion 2002. Global ozone research and monitoring project. Report no 47. WMO.

Zielke M., Ekker A.S., Olsen R.A., Spjelkavik S. and Solheim B. 2003. The influence of abiotic factors on biological nitrogen fixation in different types of vegetation in the High Arctic, Svalbard.. Arct. Antarct. Alp. Res. 35: 538–538.

Zielke M. 2004. Diversity and nitrogen fixation activity of cyanobacterial communities in terrestrial arctic ecosystems. Doctoral thesis, Department of Biology. University of Tromsø, Tromsø.

Photo. Mini UV-B supplementation systems installed in the Antarctic (Signy, South Orkney Islands) (Photograph by P. Boelen).

Plant Ecology (2006) 182:137–152
DOI 10.1007/s11258-005-9023-1

Outdoor studies on the effects of solar UV-B on bryophytes: overview and methodology

Peter Boelen[1,]*, M. Karin de Boer[2], Nancy V. J. de Bakker[1] and Jelte Rozema[1]
[1]*Institute of Ecological Science, Department of Systems Ecology, Faculty of Earth and Life Sciences, Vrije Universiteit, De Boelelaan 1087, 1081 HV, Amsterdam, The Netherlands;* [2]*Department of Marine Biology, University of Groningen, 14, 9750 AA, Haren, The Netherlands; *Author for correspondence (e-mail: peter.boelen@ecology.falw.vu.nl; phone: +31 20 5987048; fax: +31 20 5987123)*

Received 1 September 2004; accepted in revised form 15 December 2004

Key words: Antarctic, Arctic, Bryophytes, *Chorisodontium aciplyllum*, Cyclobutane Pyrimidine dimers, DNA damage, mosses, Ozone depletion, *Polytrichum strictum*, *Sanionia uncinata*, Terrestrial polar ecosystems, Ultraviolet-B radiation, UV-B exposure systems, UVBR, *Warnstorfia sarmentosa*

Abstract

In this review all recent field studies on the effects of UV-B radiation on bryophytes are discussed. In most of the studies fluorescent UV-B tubes are used to expose the vegetation to enhanced levels of UV-B radiation to simulate stratospheric ozone depletion. Other studies use screens to filter the UV-B part of the solar spectrum, thereby comparing ambient levels of UV-B with reduced UV-B levels, or analyse effects of natural variations in UV-B arising from stratospheric ozone depletion. Nearly all studies show that mosses are well adapted to ambient levels of UV-B radiation since UV-B hardly affects growth parameters. In contrast with outdoor studies on higher plants, soluble UV-B absorbing compounds in bryophytes are typically not induced by enhanced levels of UV-B radiation. A few studies have demonstrated that UV-B radiation can influence plant morphology, photosynthetic capacity, photosynthetic pigments or levels of DNA damage. However, there is only a limited number of outdoor studies presented in the literature. More additional, especially long-term, experiments are needed to provide better data for statistical meta-analyses. A mini UV-B supplementation system is described, especially designed to study effects of UV-B radiation at remote field locations under harsh conditions, and which is therefore suited to perform long-term studies in the Arctic or Antarctic. The first results are presented from a long-term UV-B supplementation experiment at Signy Island in the Maritime Antarctic.

Introduction

Stratospheric ozone depletion occurs mainly over polar regions during spring. A relatively small decrease in stratospheric ozone concentration, owing to anthropogenic emissions of chlorofluorocarbons (CFCs), will strongly affect the intensity of ultraviolet-B radiation (UV-B radiation: 280–315 nm) at ground level. Since the 1970s springtime stratospheric ozone over the Antarctic has decreased up to 50%, leading to increases in biologically effective UV-B radiation of more than 130% (Madronich et al. 1998; WMO 2003).

Many studies on the effects of enhanced UV-B radiation have focused on higher plants. However, Antarctic terrestrial ecosystems are dominated by cryptogams, with mosses predominating in the moister, more sheltered habitats and lichens in the more arid and windswept situations (Fenton 1982). There are approximately 75 moss species but relatively few achieve dominance (Fowbert 1996) and higher plants are poorly represented in the Antarctic.

Until several years ago relatively little was known about the effects of UV-B radiation on mosses. This is remarkable because beforehand it was assumed that the majority of bryophytes were particularly susceptible to UV-B radiation and that they were more sensitive to UV-B radiation than higher plants (Gehrke 1999). Firstly, most bryophytes have (undifferentiated) leaves of only one cell layer thickness and no protective cuticle and epidermis to attenuate UV-B. Secondly, since leaf water content in bryophytes depends on air humidity, high levels of solar radiation can lead to physiological inactivity, which may hinder processes such as DNA photorepair. Finally, since most bryophytes are rootless, mosses are not able to compensate for aboveground stresses, such as UV-B radiation, through the belowground storage of assimilates.

Studies on the effects of UV-B radiation on mosses are not only important to predict the effects of future ozone depletion scenarios but are also essential for studies that attempt to reconstruct historical ozone levels. Rozema et al. (2001b) suggested that polyphenolic compounds in pollen and spores that have been incorporated into moss banks may be applied as a new proxy for the reconstruction of historic variation in solar UV-B levels. In the Antarctic, e.g. at Signy Island, two species of moss, *Chorisodontium aciphyllum* and *Polytrichum strictum*, form characteristic peat banks which may be over 2 m deep, may cover more than 2500 m^2 in area and are as much as 5000 years old (Fenton 1980). The moss peat banks consist mainly of the relatively uncompacted remains of the two moss species. The mosses become incorporated into the permafrost at a depth of 20 to 30 cm. Since most parts of Antarctic moss peat banks are permanently frozen, these banks may represent a unique archive of historical UV-B levels (Rozema et al. 2001c; 2002). Besides the analysis of UV-B induced polyphenolic compounds pollen and spores that are incorporated into these moss banks, the preservation of frozen moss material may allow DNA extraction and quantification of cyclobutane pyrimidine dimers (CPDs) caused by exposure to solar UV-B radiation (Caldwell et al. 1998).

In this paper we give an overview of recent field studies on the effects of altered levels of UV-B radiation on bryophytes. The experimental methods that have been used to manipulate levels of UV-B radiation in the field are discussed. Furthermore, a mini UV-B supplementation system, especially designed to study the effects of enhanced UV-B radiation on terrestrial vegetation at remote field locations, is described. This UV-B supplementation system has already been used for two consecutive years at Signy Island in the Maritime Antarctic where natural vegetation (four moss species and the higher plant *Deschampsia antarctica*) is exposed to enhanced UV-B radiation in a long-term experiment.

Outdoor studies on the effects of solar UV-B on bryophytes: an overview

UV-B radiation can have many direct and indirect effects on organisms. UV-B radiation induces damage to essential molecules like proteins, pigments and DNA. This damage can influence important cellular processes such as nutrient uptake, DNA transcription and replication or photosynthesis, which eventually can lead to decreased growth rate or impaired reproduction (see Rozema et al. 1997a; Caldwell et al. 1998; Searles et al. 2001; Robinson et al. 2003). There are several strategies to minimise the deleterious effects of UV-B radiation. In higher plants UV-B damage to essential molecules is avoided by producing UV-B absorbing compounds (see Meijkamp et al. 1999; Searles et al. 2001) or by increasing leaf thickness (Bornman and Vogelmann 1991). Another strategy is to utilise repair mechanisms to ameliorate UV-B induced damage.

Studies on the ecophysiological responses of higher plants to UV-B are mainly based on greenhouse or climate room experiments. In a meta-analysis of field based studies on higher plants, Searles et al. (2001) concluded that the accumulation of UV-B absorbing compounds in leaves was the most apparent effect of enhanced UV-B radiation while most morphological and photosynthetic parameters were not affected. In contrast to higher plants, most studies on bryophyte responses to UV-B are conducted under field conditions. These outdoor studies investigating the impact (on growth, morphology, photosynthetic parameters, UV-B screening pigments and DNA damage) of UV-B radiation on bryophytes are summarised in Table 1. All field experiments considered here were carried out in polar or sub-polar regions. There is a substantial variation in the duration of the studies, with some performed for 10 h and others for up to 6 years. Long-term studies are more difficult to maintain but have the advantage that slow processes, such as growth, can be monitored. The majority of the studies have employed UV-B lamps to simulate a decrease in stratospheric ozone concentration. In other studies screens were used to reduce solar UV-B below ambient levels. These studies can be particularly valuable at Antarctic sites with severely depleted stratospheric ozone concentrations, but comparisons between UV-B supplementation studies and exclusion experiments should be done with care. Covering plants with screens can have artefactual effects on temperature and water availability.

Growth, morphology and reproduction

In higher plants many effects of UV-B radiation involve morphogenetic changes. Plant morphogenetic parameters that may change are plant height, leaf area, branching and plant phenology (Rozema et al. 1997a).

Only a few field studies relating growth and morphology to UV-B exposure and, to our knowledge, no studies relating reproduction have been reported for bryophytes. Long-term field studies of *Sphagnum* sp. showed that UV-B radiation can influence shoot biomass and length but that biomass production per area is not affected (Gehrke 1998; Searles et al. 2002; Robson et al. 2003). Lud et al. (2002) showed that, while shoot biomass and length were unaffected, branching of the Antarctic moss *Sanionia uncinata* was reduced by reduction of ambient UV-B levels. In a greenhouse experiment where the moss *Hylocomium splendens* was grown under enhanced UV-B levels the phenological development of the moss was accelerated by up to 2 weeks (Johanson et al. 1995).

Photosynthetic activity

UV-B radiation can directly or indirectly affect many photosynthetic processes in higher plants such as performance of Photosystem II, the integrity of the thylakoid membrane and enzymatic processes in the Calvin Cycle (Bornman 1989; Teramura and Sullivan 1994; Jansen et al. 1998). However, most of these studies have been carried out under unrealistic irradiance conditions with low levels of UV-A radiation and/or PAR (Photosynthetic Active Radiation; 400–700 nm). Field studies, with plants exposed to more realistic enhanced levels of UV-B radiation, reveal that in most cases photosynthetic performance is not affected (Allen et al. 1998; Searles et al. 2001; Dormann and Woodin 2002).

Although Montiel et al. (1999) observed a decreased photochemical yield in the moss *Sanionia uncinata*, in most field studies on terrestrial bryophytes (i.e. *Sanionia uncinata*, *Bryum argenteum*, *Sphagnum fuscum*) photosynthetic activity is not affected by reduced or enhanced UV-B radiation (Gehrke 1998; Green et al. 2000; Huiskes et al. 2001; Lud et al. 2002, 2003; Newsham et al. 2002). Also in the majority of the bryophytes (including three *Sphagnum* sp.) that have been investigated, concentrations of chlorophylls are not influenced by UV-B exposure (Gehrke 1999; Searles et al. 1999, 2002; Niemi et al. 2002a, b, Newsham et al. 2002; Lud et al. 2003; Newsham 2003). However in *Sphagnum fuscum*, chlorophyll *a* concentrations decreased, while chlorophyll levels in *Sphagnum balticum* increased after exposure to supplemental UV-B radiation (Gehrke 1998; Niemi et al. 2002a).

Oxidative damage

Although ultraviolet-A radiation (UV-A: 315–400 nm) is considered to be the major generator of intracellular oxidative stress (Pourzand and Tyrrel 1999), several studies have shown that UV-B radiation exposure can lead to increases in reactive oxygen species (ROS) in plants. ROS can react with lipids, pigments, proteins and nucleic acids, leading to inactivation of enzymes, membrane damage or oxidative DNA damage (see Day 2001). Plants cells usually contain enzymatic and non-enzymatic defences against oxygen toxicity. These

Table 1. Summary of outdoor studies on the effects of UV-B radiation on terrestrial bryophytes.

Species	Location (Latitude /Longitude)	Duration and type of study	Growth morphology and reproduction	Photosynthetic activity	Photosynthetic and photoprotective pigments	Membranes	UV-B absorbing compounds (UACs)	DNA damage	References
Andreaea regularis	Antarctic Adelaide Island (67°34′ S, 68°07′ W)	1 month *in situ*			Increased carotenoids under naturally elevated UV-BR		Increased UACs under naturally elevated UV-BR		Newsham (2003)
Cephaloziella varians	Antarctic Adelaide Island (67°34′ S, 68°07′ W)	1 month *in situ*		No effect of naturally elevated UV-BR on Fv/Fm (maximum quantum yield of PSII)	Increased carotenoids under naturally elevated UV-BR		Increased UACs under naturally elevated UV-BR		Newsham et al. (2002)
Chorisodontium aciphyllum	Maritime Antarctic Signy Island (60°42′ S, 45°35′ W)	3–5 weeks of enhanced UV-BR simulating 15 and 30% ozone depletion					No effect of enhanced UV-BR on UACs	No effect of enhanced UV-BR on CPD accumulation	This paper
Dicranum elongatum	Northern Sweden Abisko (68°21′ N, 18°49′ E)	2 years enhanced UV-BR simulating 20% ozone reduction	No effect of UV-BR on growth		No effect of UV-BR on chlorophyll		No effect of UV-BR on flavonoids		Sonesson et al. (2002)
Hylocomium splendens	Northern Sweden Abisko (68°35′ N, 18°82′ E)	3–6 years enhanced UV-BR simulating 15% ozone depletion	Growth reduced after 3 years but enhanced after 6 years, shoot morphology altered after 3 years		No effect on chlorophyll and carotenoid concentrations after 3 years of enhanced UV-BR		No effect on UACs after 3 years		Gehrke (1999); Phoenix et al. (2001)
Polytrichum commune	Northern Sweden Abisko (68°35′ N, 18°82′ E)	3 years enhanced UV-BR simulating 15% ozone depletion	Annual length increment reduced, no effect on biomass, shoot morphology altered		Concentration of chlorophyll *a* and carotenoids decreased, no effect on chlorophyll *b* concentration		Less UACs after 3 years of enhanced UV-BR		Gehrke (1999)
Polytrichum strictum	Maritime Antarctic Signy Island (60°42′ S, 45°35′ W)	3–5 weeks of enhanced UV-BR simulating 15 and 30% ozone depletion					No effect of enhanced UV-BR on UACs	No effect of enhanced UV-BR on CPD accumulation	This paper
Sanionia uncinata	Maritime Antarctic Signy Island (60°42′ S, 45°35′ W)	3–5 weeks of enhanced UV-BR simulating 15 and 30% ozone depletion					No effect of enhanced UV-BR on UACs	No effect of enhanced UV-BR on CPD accumulation	This paper

Antarctic Léonie Island (67°35′ S, 68°20′ W)	3–4 weeks screening study		No effect on PSII quantum yield					Huiskes et al. (2001)
Antarctic Léonie Island (67°35′ S, 68°20′ W)	2 years screening study	No effect on biomass or length growth, less branching under reduced UV-BR						Lud et al. (2002)
Antarctic Léonie Island (67°35′ S, 68°20′ W)	2 days of enhanced UV-BR (8.4 kJ m^{-2} day^{-1}, weighted with Setlow, normalised at 300 nm)		No effect on actual and maximum quantum yield of PSII, no effect on photosynthetic gas exchange	No effect of UV-BR on chlorophyll			DNA damage increased under enhanced UV-BR	Lud et al. (2002)
Arctic Ny-Alesund, Svalbard (78°55′ N, 11°56′ E)	2 days of enhanced UV-BR (8.4 kJ m^{-2} day^{-1}, weighted with Setlow, normalised at 300 nm)		No effect on actual and maximum quantum yield of PSII	No effect of UV-BR on chlorophyll			DNA damage increased under enhanced UV-BR	Lud et al. (2002)
Antarctic Léonie Island (67°35′ S, 68°20′ W)	3 short-term screening studies		No effect on actual and maximum quantum yield of PSII, no effect on photosynthetic gas exchange	No effect on chlorophyll concentrations				Lud et al. (2003)
Antarctic Léonie Island (67°35′ S, 68°20′ W)	10 h of enhanced UV-BR (in total 8.7 kJ m^{-2}, weighted with Setlow, normalised at 300 nm)			No effect on chlorophyll and carotenoid concentrations		No effect of enhanced UV-BR on UACs	DNA damage increased under enhanced UV-BR	Lud et al. (2003)
Antarctic Adelaide Island (67°34′ S, 68°07′ W)	7 days of enhanced UV-BR (30% of background solar levels, no UV-A control) and screening study		PSII quantum yield reduced under enhanced UV-BR compared to ambient and reduced UV-BR					Montiel et al. (1999)

Table 1. Continued.

Species	Location (Latitude /Longitude)	Duration and type of study	Growth morphology and reproduction	Photosynthetic activity	Photosynthetic and photoprotective pigments	Membranes	UV-B absorbing compounds (UACs)	DNA damage	References
	Antarctic Adelaide Island (67°34′ S, 68°07′ W)	1 month *in situ*		No effect of naturally elevated UV-BR on Fv/Fm (maximum quantum yield of PSII)	Increased carotenoids under naturally elevated UV-BR		Increased UACs under naturally elevated UV-BR		Newsham et al. (2002)
Sphagnum angustifolium	Central Finland Kuopio (62°13′ N, 27°35′ E)	14 weeks of enhanced UV-BR simulating 30% ozone depletion	No effect of UV-BR on capitulum or stem dry mass		No effect on chlorophyll and carotenoid concentrations after 14 weeks of enhanced UV-BR	Membrane leakage of Mg and Ca after 6 weeks exposure to enhanced UV-BR, No effect of UV-BR on membrane permeability after 12 weeks exposure	No effect of enhanced UV-BR on UACs		Niemi et al. (2002b)
Sphagnum balticum	Central Finland Kuopio (62°13′ N, 27°35′ E)	14 weeks of enhanced UV-BR simulating 30% ozone depletion, no UV-A control	No effect of UV-BR on capitulum or stem dry mass		Increase of chlorophyll *a*, chlorophyll *b* and carotenoids after exposure to enhanced UV-BR	Increase of membrane conductivity	No effect of enhanced UV-BR on UACs		Niemi et al. (2002a)
Sphagnum fuscum	Northern Sweden Abisko (68°35′ N, 18°82′ E)	2 years of enhanced UV-BR simulating 15% ozone depletion	Decrease in annual length increment, dry mass per unit length increased, no effect on production of biomass per area	Maximum net photosynthesis (NPmax) increased under UV-BR when calculated per unit chlorophyll, no effect on NPmax on a dry-mass basis, dark respiration decreased	Chlorophyll *a* and carotenoids decreased under enhanced UV-BR, no effect on chlorophyll *b*		No effect of enhanced UV-BR on UACs		Gehrke (1998)
Sphagnum fuscum	Northern Sweden Abisko (68°21′ N, 18°49′ E)	2 years enhanced UV-BR	No effect of UV-B R on growth		No effect of UV-BR on chlorophyll		No effect of UV-BR on flavonoids		Sonesson et al. (2002)

Sphagnum magellanicum	Central Finland Kuopio (62°13′ N, 27°35′ E)	14 weeks of enhanced UV-BR simulating 30% ozone depletion			No effect on chlorophyll and carotenoid concentrations after 14 weeks of enhanced UV-BR	Membrane leakage of Mg after 12 weeks exposure to enhanced UV-BR			Niemi et al. (2002b)
	Southern Argentina Tierra del Fuego (54°51′ S, 68°36′ W)	3–6 years screening study (near ambient and reduced UV-BR)	Height growth decreased but capitulum density increased under near-ambient conditions, no effect on production of biomass per area		No effect on chlorophyll and carotenoids		No effect on UACs		Searles et al. (1999, 2002); Robson et al. (2003)
Sphagnum papillosum	Central Finland Kuopio (62°13′ N, 27°35′ E)	14 weeks of enhanced UV-BR simulating 30% ozone depletion, no UV-A control			No effect on chlorophyll and carotenoid concentrations after 14 weeks of enhanced UV-BR	Increase of membrane conductivity, leakage of Mg and Ca	No effect of enhanced UV-BR on UACs		Niemi et al. (2002a)
	Central Finland Kuopio (62°13′ N, 27°35′ E)	14 weeks of enhanced UV-BR simulating 30% ozone depletion			No effect on chlorophyll and carotenoid concentrations after 14 weeks of enhanced UV-BR	No effect of UV-BR on membrane permeability after 12 weeks exposure	No effect of enhanced UV-BR on UACs		Niemi et al. (2002b)
Warnstorfia sarmentosa	Maritime Antarctic Signy Island (60°42′ S, 45°35′ W)	3–5 weeks of enhanced UV-BR simulating 15 and 30% ozone depletion					No effect of enhanced UV-BR on UACs	No effect of enhanced UV-BR on CPD accumulation	This paper

so-called antioxidants can remove, neutralise or scavenge ROS in cells, thereby protecting cellular components from oxidative damage. Enzymatic antioxidants are superoxide dismutase (SOD), catalase (CAT), ascorbate peroxidase (AP) dehydroascorbate reductase and glutathione reductase. Non-enzymatic defences are carotenoids, flavonoids, ascorbate and glutathione (Dai et al. 1997; Jansen et al. 1998).

Very little is known about UV-induced oxidative stress in bryophytes. Niemi et al. (2002a, b) showed that enhanced UV-B levels could lead to increased membrane permeability in *Sphagnum* spp. Although the authors did not mention this, an increase in membrane permeability could be the result of oxidative stress (Dai et al. 1997).

In a growth chamber experiment, Markham et al. (1998) studied the effects of varying UV-B levels on the liverwort *Marchantia polymorpha*. They found a strong correlation between increasing UV-B levels and the ratio of two individual flavonoids. Since the total UV-B screening effectiveness of the flavonoids was not altered, they suggested that an increase of one particular flavonoid (a luteolin glycoside) improved the level of antioxidant defence.

UV-B-absorbing compounds

Many plants can adapt to high UV-B fluxes by the production of secondary metabolites which absorb UV-B radiation. In a meta-analysis of 62 outdoor irradiation studies (no studies of moss species included), Searles et al. (2001) concluded that the accumulation of UV-B absorbing compounds in leaves was the most apparent effect of enhanced UV-B radiation. UV-B radiation stimulates the production of phenylalanine ammonium lysase (PAL) and other key enzymes of the phenylpropanoid pathway (Rozema et al. 1997a; Meijkamp et al. 1999). PAL catalyses the formation of phenylalanine to *trans*-cinnamic acid, which leads to the formation of complex phenolic compounds such as flavonoids, tannins and lignin. Although flavonoids can be found in many (but not all) moss species (Markham 1990), most studies show that UV-B absorbing compounds in bryophytes were not enhanced by supplemental levels of UV-B radiation applied from fluorescent UV lamps (Barsig et al. 1998; Gehrke 1998; Markham et al. 1998; Gehrke 1999; Niemi et al. 2002b; Searles et al. 2002; Lud et al. 2003). In an outdoor UV-B supplementation experiment (see below) at Signy Island (Maritime Antarctic) soluble UV-B absorbing compounds in four moss species (*Sanionia uncinata*, *Chorisodontium aciphyllum*, *Warnstorfia sarmentosa* and *Polytrichum strictum*) were not induced by enhanced levels of UV-B radiation (Figure 1). However, studies by Newsham et al. (2002) and Newsham (2003) suggested that in Antarctic bryophytes UV-B absorbing pigments may be induced within 24 h under natural elevated irradiances of UV-B arising from ozone depletion. In addition, Markham et al. (1990) found that flavonoid concentrations in herbarium specimens of *Bryum argenteum* were correlated with historical ozone levels for the period 1960–1990.

UV-B induced DNA damage and repair

UV-B radiation can cause dimerization of DNA bases, leading to the formation of photoproducts like cyclobutane pyrimidine dimers (CPDs) and pyrimidine (6–4) pyrimidone photoproducts. Cells need to repair this damage before they can proceed with DNA transcription or replication. There are two main pathways to repair dimerised pyrimidines: excision (dark) repair and photodependent repair (Sancar and Sancar 1988).

Lud et al. (2002, 2003) studied the effects of ambient and enhanced levels of UV-B radiation on the bipolar bryophyte *Sanionia uncinata*. They concluded that ambient summer levels of UV-B radiation did not induce significant levels of DNA damage. However, DNA damage, measured as CPDs, clearly increased as a result of artificially enhanced UV-B radiation. DNA damaged during the day was repaired during the (non dark) night. In 2003 an outdoor UV-B supplementation experiment (see below) was started at Signy Island in the Maritime Antarctic. The effects of UV-B radiation on four moss species (*Sanionia uncinata*, *Chorisodontium aciphyllum*, *Warnstorfia sarmentosa* and *Polytrichum strictum*) was investigated (Figure 2). Levels of DNA damage were low but comparable with concentrations found by Lud et al. (2002, 2003). A significant difference between mosses exposed to enhanced levels of

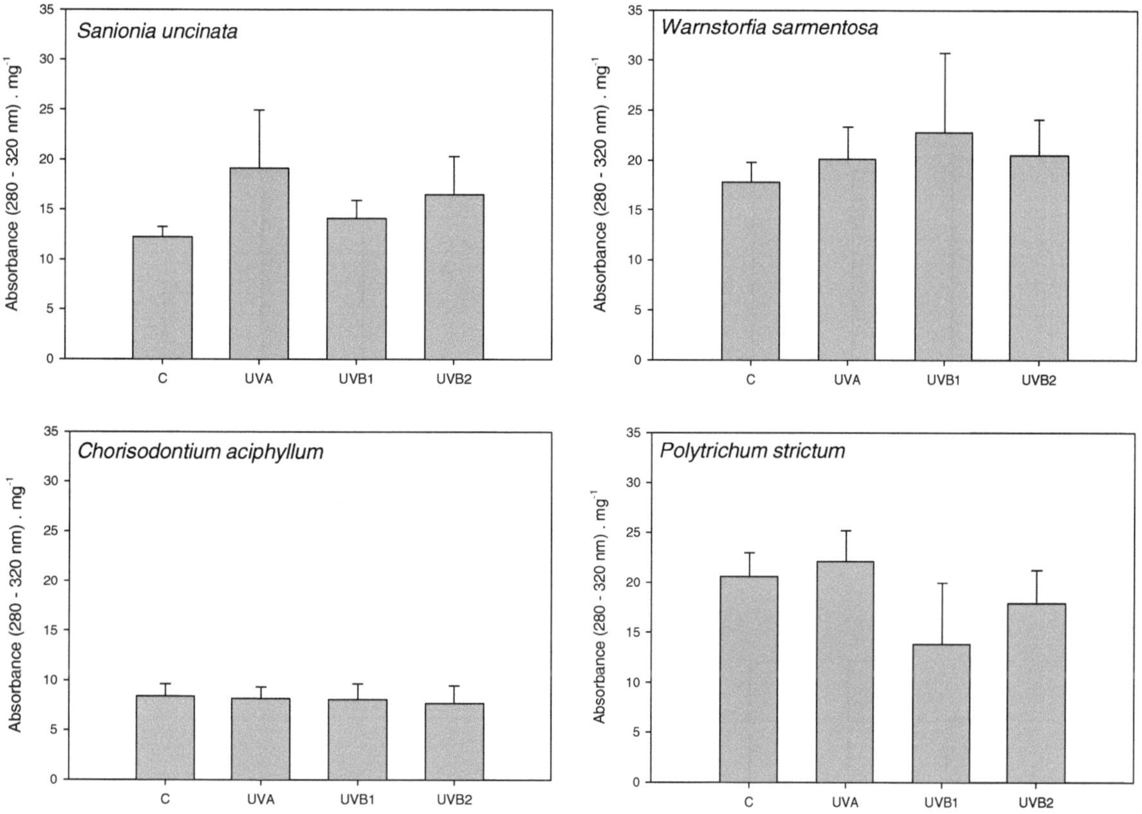

Figure 1. Soluble UV-B absorbing compounds (acidified methanol extracts, see Rozema et al. 2001b) in four moss species from outdoor experiments at Signy Island (60°42′ S, 45°35′ W) after 5 weeks (2 February until 6 March 2003) of exposure to UV-B radiation simulating 15 or 30% ozone depletion (C = ambient control, UVA = UV-A control, UVB1 = 15% ozone depletion, UVB2, 30% ozone depletion, error bars represent standard deviation, $n = 4$). Around noon five randomly selected shoots per species per plot (four plots per treatment) were collected . The top sections (circa 0.5 cm) of air dried shoots were pooled and used for methanol extraction.

UV-B and the controls (ambient UV-B levels) could not be demonstrated. This suggests that these mosses are well adapted to ambient levels of UV-B radiation.

Outdoor studies on the effects of solar UV-B on bryophytes: methodology

Outdoor experiments are important tools for studying changes to vegetation that might occur as a result of stratospheric ozone depletion. There are several methods to manipulate UV-B radiation in the field. In many field studies screens are used to filter the UV-B portion of the solar spectrum (Rozema et al. 1997a; Ballare et al. 2001; Lud et al. 2003). In this way ambient solar UV-B levels can be compared with reduced levels of solar UV-B radiation. Filtration of the UV-B band of solar radiation leaves the natural relations of the solar spectrum intact and avoids the UV-A radiation produced by fluorescent tubes which is thought to perhaps influence plant growth (Newsham et al. 1996). These studies do not simulate ozone depletion scenarios but are valuable and have substantially increased our knowledge of the mechanisms behind UV-B effects on plants. However, screens alter several other abiotic factors other than incident radiation, notably temperature and water availability (Kennedy 1995). Recently, Krizek and Mirecki (2004) showed that cellulose diacetate films, used in some UV-exclusion studies (ambient control), release phytotoxic compounds.

Newsham et al. (2002) and Newsham (2003) performed non-manipulative studies at Antarctic sites experiencing natural substantial stratospheric ozone depletion. Multiple regression analyses

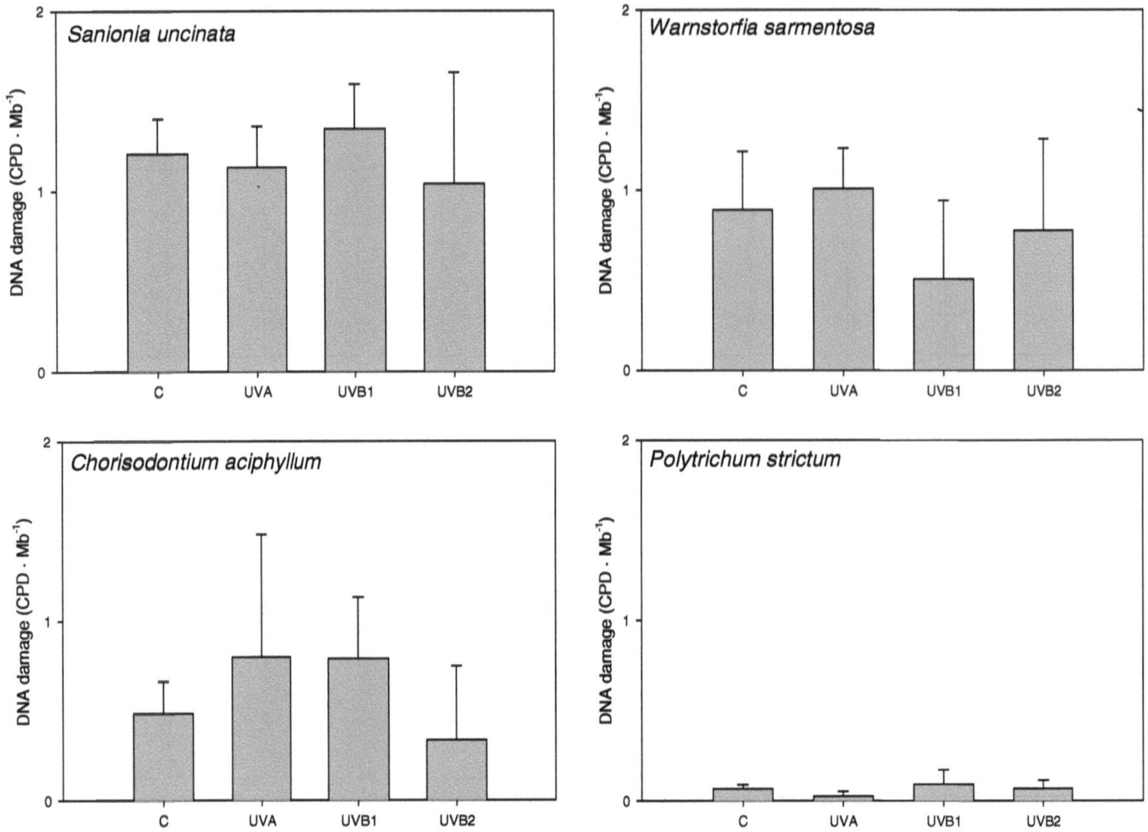

Figure 2. DNA damage, expressed as cyclobutane pyrimidine dimers (CPDs) per million nucleotides (Boelen et al. 2002) in four moss species from outdoor experiments at Signy Island after 3 weeks (2 February until 22 February 2003) of exposure to UV-B radiation simulating 15 or 30% ozone depletion (C = ambient control, UVA = UV-A control, UVB1 = 15% ozone depletion, UVB2, 30% ozone depletion, error bars represent standard deviation, $n = 4$). Around noon five randomly selected shoots per species per plot (four plots per treatment) were collected and immediately frozen at −80 °C. The top sections (circa 0.5 cm) of the shoots were pooled and used for DNA extraction.

indicated that the ratio of UV-B radiation to PAR irradiance was the best predictor for concentrations of UV-B screening pigments and carotenoids in foliage of Antarctic bryophytes. Although this kind of study gives insight into the effects of enhanced levels of UV-B radiation arising from ozone depletion, levels of UV-B depend on local conditions and cannot be modified to simulate future ozone depletion scenarios.

In UV-B supplementation studies, the vegetation is exposed to UV-B radiation emitted by fluorescent lamps, thus simulating a decrease in stratospheric ozone concentration. In modulated outdoor supplementation systems ambient UV-B levels are monitored and the output of the UV-B lamps is varied to supply a UV-B dose proportional to ambient levels (Caldwell et al. 1983, Sullivan et al. 1994, Niemi et al. 2002b). The advantage of this method is that local changes in irradiance, e.g. caused by cloud cover, are accommodated for by the system. However, modulated experiments need complex and expensive lamp dimming systems and sensors to monitor ambient and treatment UV-B levels. Most of the lamp field experiments that have taken place have used simpler and more robust systems in which lamps have been switched on for a specific period of time centred around solar noon (e.g. Johanson et al. 1995; Gehrke 1998). In these so-called "square wave" systems, exposure times are calculated with mathematical models (e.g. Green 1983) and expressed in terms of a certain level of ozone depletion. Switching times are adjusted every day or week to follow seasonal patterns of solar radiation. In some experiments the diurnal pattern of exposure has been improved by switching on alternate lamps in several stages ("step wave" systems; e.g. Gehrke 1999). Since both systems do not compensate for overcast days, ratios of UV-B:UV-A:PAR can differ between days.

Outdoor UV-B supplementation systems need an electricity supply, which is sometimes not available in the field. Besides this, the frames that are necessary to support the lamps are usually large, which can create problems on steep slopes or places exposed to strong winds. Gehrke et al. (1996), Gehrke (1998) developed a small UV-B supplementation system to study effects of enhanced UV-B radiation on an ombrotrophic peatland ecosystem close to the Abisko Scientific Research Station in northern Swedish Lapland.

A mini UV-B supplementation system

In 2002 the Vrije Universiteit Amsterdam started several long-term experiments to study the effects of enhanced UV-B radiation on natural vegetation. A mini UV-B supplementation system was developed to expose vegetation in the Antarctic, the Arctic and the Netherlands. A similar system has been used for short-term experiments in the Antarctic (Rozema et al. 1997b; Huiskes et al. 2001; Rozema et al. 2001a; Lud et al. 2002, 2003).

The mini UV-B supplementation system uses mini-fluorescent UV-tubes (Vilber-Lourmat T-15M, 12 V, 15 W) that can be powered by batteries when a regular electricity supply is not available. Each lamp frame contains one UV-tube (44 cm in length , 30 mm diameter, preburned for 70 h). Since these lamps are small, and cheap and easy to transport, many individual lamps can be used to increase the number of statistically independent replicates. Our experimental design incorporates two UV-B treatments (UV-B1 and UV-B2) where short-wave UV-C radiation (< 280 nm) is excluded by cellulose acetate (CA) filters placed on UV-B transmitting plexiglass underneath each tube, one UV-A control treatment (UV-A treatment, Mylar foil, same exposure time as UV-B2 treatment) and a further control where the fluorescent tubes are replaced by wooden bars (control C). Each treatment is replicated four times. The tubes are controlled by timers (Grässlin, Germany) which automatically adjust the exposure time every week to follow seasonal changes in natural UV-B radiation (see below).

Lamp spectra at sample height (48 cm) were measured using an Optronics 752 spectroradiometer, calibrated against a NIST-traceable OL752-10E standard lamp (Figure 3). The spatial distribution of UV-B radiation emitted by the UV-B lamps was measured using a UVX radiometer (UVP Inc, Upland, USA) with a UVX-31 sensor. A 25–30% reduction of radiation levels was recorded at the edges of each 30×35 m plot (Figure 4a). Since UV-B levels increase strongly as the distance between the vegetation and the UV-B lamp decreases (Figure 4b), the lamp distance or exposure time have to be adjusted as plant height increases.

Biologically effective irradiances (BEIs) were calculated from lamp spectra using Caldwell's (1971) generalised plant action spectrum, the DNA action spectrum from, Setlow (1974) and the new biological spectral weighting function (BSWF) for plant growth from Flint and Caldwell (2003a), all normalised at 300 nm. Lamp output and terminal voltage of the batteries, that are used to power the lamps when a regular power supply is not available, are temperature dependent. To estimate the influence of temperature and lamp voltage on lamp output, these parameters were investigated in a temperature controlled climate room. Lamp output was dependent on temperature and tube voltage (Figure 5). In general lamp output was lower at lower temperatures and voltages.

Since CA foils degrade in time (Steeneken et al. 1995) the BEI under the lamp was recorded over 40 h. From these data the average lamp output (average irradiance at sample height during 20 h: Caldwell weighted: 0.079 W m^{-2}, Setlow weighted: 0.076 W m^{-2}, weighted with new BSWF from Flint & Caldwell: 0.071 W m^{-2}) was calculated. In the field foils were replaced after 20 h exposure time. The extra UV-B doses required to simulate a 15 or 30% reduction in ozone were calculated using the irradiance model of Green (1983) and computer encoded by Björn and Murphy (1985). The outcome of this simulation depends on the biological weighting function that is used (Flint and Caldwell 2003a, b). We have used the generalised plant action spectrum from Caldwell (1971) since this is the most common

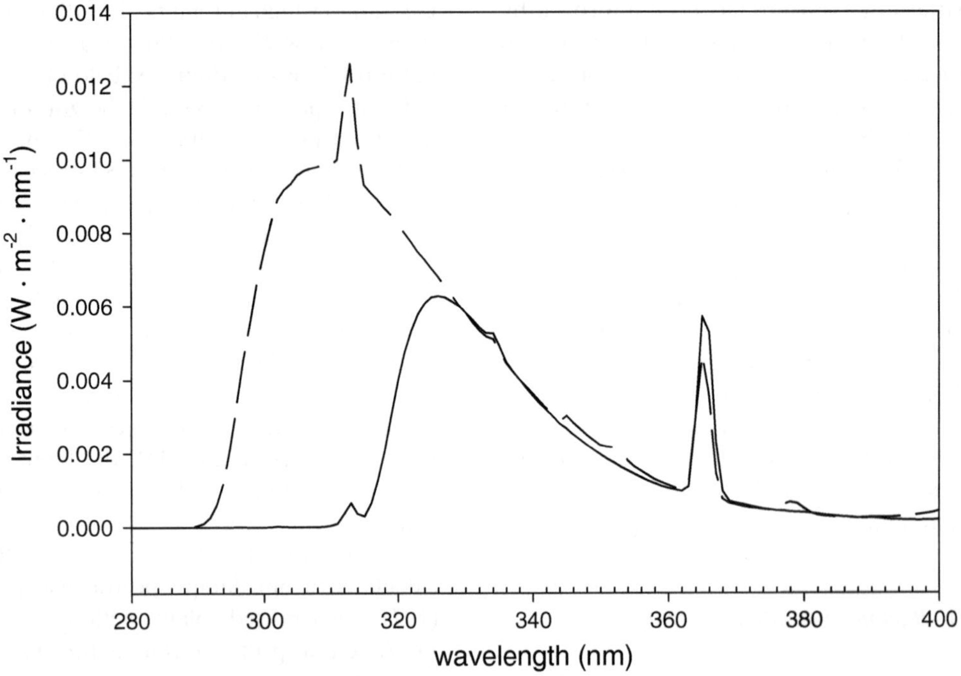

Figure 3. Lamp spectra from UV-B minilamps at sample height (48 cm) after 9 h of exposure; dashed line: UV-C radiation excluded with cellulose acetate (CA) foil (UV-B treatment), solid line: UV-C and UV-B radiation excluded with Mylar foil (UV-A treatment).

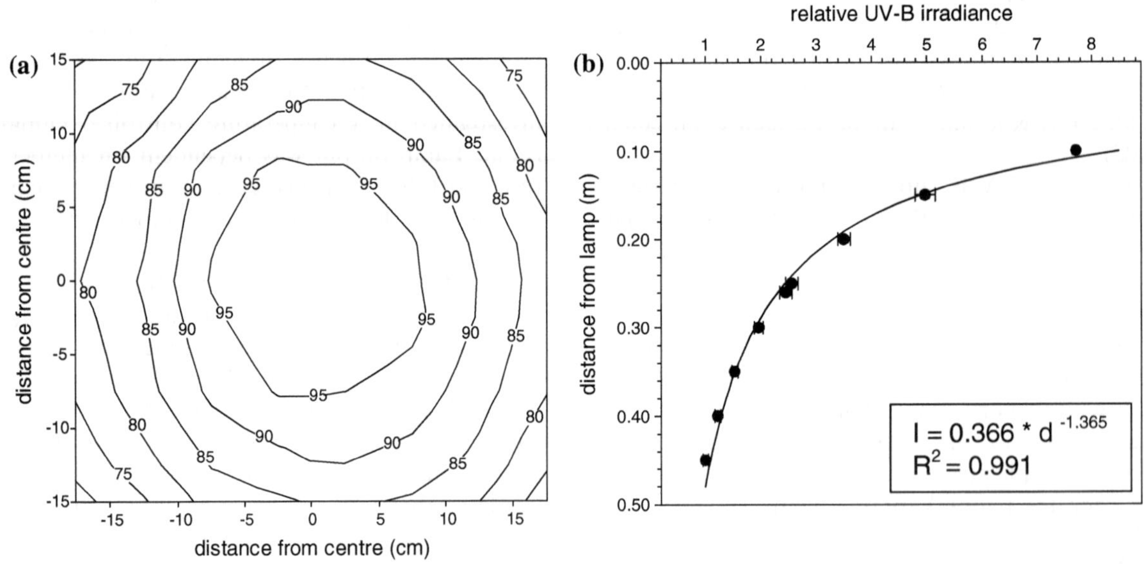

Figure 4. Spatial distribution of UV-B radiation emitted by the mini UV-B lamp system. (a) Contour plot of relative UV-B irradiance (centre = 100%) at sample height (48 cm). (b) Relative irradiance in the centre plotted against lamp distance.

used BSWF and in this way is possible to compare our experiments with other (previous) field experiments. Average lamp output and calculated doses were used to determine the duration of exposure time. The daily irradiation was centred around solar noon.

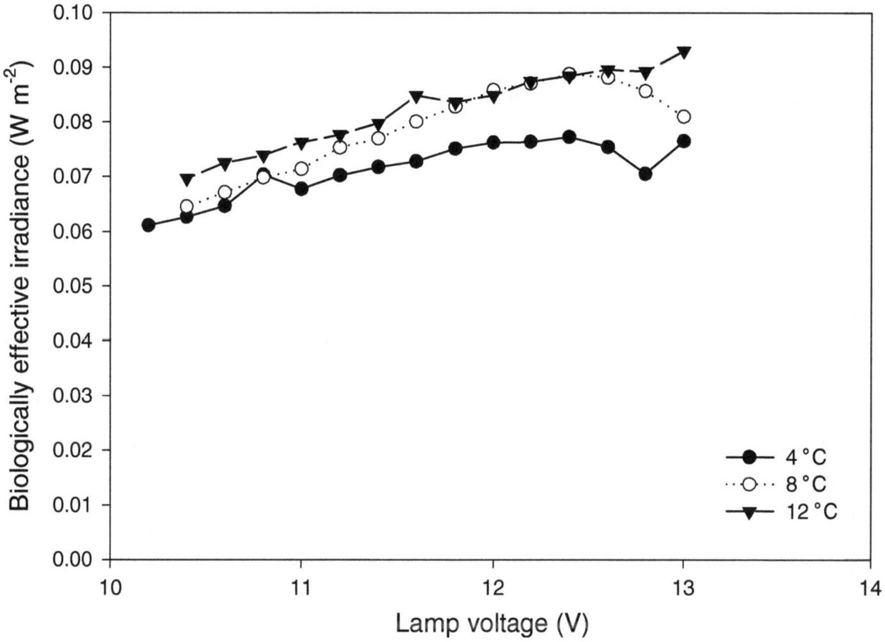

Figure 5. Lamp output (Caldwell weighted biologically effective irradiance at sample height, 48 cm) and terminal voltage of the batteries, that are used to power the lamps when a regular power supply is not available, are temperature dependent. To estimate the influence of temperature and lamp voltage on lamp output these parameters were investigated in a temperature controlled climate room. Lamp output was slightly dependent on temperature and tube voltage. In general lamp output is lower at lower temperatures and at lower voltages.

Concluding remarks

Bryophytes are a significant species group of terrestrial plants, especially in polar and subpolar regions where the relative increase in UV-B radiation due to stratospheric ozone depletion is high. In this review an attempt has been made to summarise all outdoor studies on the effect of UV-B radiation on bryophytes. Since there are only a limited number of field studies on bryophytes presented in the literature, a statistical meta-analysis of these data has not been made so far. However, nearly all studies demonstrated that mosses are well adapted to ambient levels of UV-B radiation since UV-B hardly affects growth parameters. In all screening studies, where reduced UV-B levels were compared with ambient levels, no effect of UV-B reduction or exclusion on photosynthesis, accumulation of DNA damage and levels of UV-absorbing compounds was found. DNA damage can accumulate under enhanced levels of UV-B radiation, but is quickly repaired when lamps are switched off. A few studies have demonstrated that supplemental UV-B radiation can influence photosynthetic activity and photosynthetic pigments.

In a meta-analysis of outdoor studies on higher plants, Searles et al. (2001) concluded that the accumulation of UV-B absorbing compounds in leaves was the most apparent effect of enhanced UV-B radiation. This is probably not true for bryophytes. In all studies where UV-B was artificially enhanced no increase of UACs was found. However, two *in situ* studies from Newsham et al. (2002) and Newsham (2003) suggested that UACs were induced within 24 h under naturally elevated UV-B radiation. Furthermore, in all studies described above, soluble UACs have been measured. Meijkamp et al. (in preparation) showed that non-soluble (cell-wall bound) compounds contribute largely to the constitutive UV-screen in *Vicia faba* leaves. Therefore in future studies the responses of insoluble flavonoids to UV-B in bryophyte tissues should be investigated.

Long-term studies have shown that even though biomass production per area is not altered, shoot morphology can be influenced by UV-B radiation. Changes in plant morphology might affect ecosystem structure and function. For example, more branching or a decrease in annual length increment can lead to reduced plant size, which can shift the competitive balance by altering interspecific competition for visible light. To study these subtle interactions it is essential to continue these long-term experiments.

The mini UV-B supplementation system that we have described here is especially designed to study effects of UV-B radiation at remote field locations under harsh conditions and is therefore highly suited to perform long-term studies in the Arctic or Antarctic. Polar ecosystems are characterised by extreme aridity and low temperatures and within this context it would be interesting to consider potential interactions between UV-B stress and other environmental factors, such as water availability and temperature.

Acknowledgements

This work was funded by the Dutch council for Scientific Research (NWO-ALW grant UVANTARCTIC, No. 851.20.010). We would like to thank Rod Strachan, Judith Dickson and Pete Convey from the British Antarctic Survey and Stef Bokhorst from the Netherlands Institute of Ecology for their support at the long-term field experiment at Signy Island. Logistic support from the British Antarctic Survey is also gratefully acknowledged. We would like to thank Eveline Schut, Mariska Wagenaar and Rob Broekman for technical assistance. We also thank Martien van Vilsteren, Daan van Marum and Flip de Kriek (Electronics department and Mechanical department from de Vrije Universiteit Amsterdam) for their skilful support in designing and constructing the mini UV-B supplementation system. Anita G. J. Buma and Willem van de Poll, University of Groningen, are thanked for their help with DNA damage analysis. Further we wish to thank K. Newsham and L.O. Björn for constructive suggestions on an earlier version of this manuscript.

References

Allen D.J., Nogues S. and Baker N.R. 1998. Ozone depletion and increased UV-B radiation: is there a real threat to photosynthesis? J. Exp. Bot. 49: 1775–1788.

Ballaré C.L., Rousseaux M.C., Searles P.S., Zaller J.G., Giordano C.V., Robson T.M., Caldwell M.M., Sala O.E. and Scopel A.L. 2001. Impacts of solar ultraviolet-B radiation on terrestrial ecosystems of Tierra del Fuego (southern Argentina) – An overview of recent progress. J. Photochem. Photobiol. B-Biol. 62: 67–77.

Barsig M., Schneider K. and Gehrke C. 1998. Effects of UV-B radiation on fine structure, carbohydrates, and pigments in *Polytrichum commune*. Bryologist 101: 357–365.

Boelen P., Post A.F., Veldhuis M.J. and Buma A.G.J. 2002. Diel patterns of UVBR-induced DNA damage in picoplankton size fractions from the Gulf of Aqaba. Red Sea. Microb. Ecol. 44: 164–174.

Björn L.O. and Murphy T.M. 1985. Computer calculation of solar ultraviolet-radiation at ground-level. Physiol. Veg. 23: 555–561.

Bornman J.F. 1989. Target sites of UV-B radiation in photosynthesis of higher plants. J. Photochem. Photobiol. B-Biol. 4: 145–158.

Bornman J.F. and Vogelmann T.C. 1991. Effect of UV-B radiation on leaf optical-properties measured with fiber optics. J. Exp. Bot. 42: 547–554.

Caldwell M.M. 1971. Solar UV irradiation and the growth and development of higher plants. In: Giece A.C. (ed.), Photophysiology, Academic Press, New York, pp. 131–177.

Caldwell M.M., Gold W.G., Harris G. and Ashurst C.W. 1983. A modulated lamp system for solar UV-B (280–320 Nm) – supplementation studies in the field. Photochem. Photobiol. 37: 479–485.

Caldwell M.M., Bjorn L.O., Bornman J.F., Flint S.D., Kulandaivelu G., Teramura A.H. and Tevini M. 1998. Effects of increased solar ultraviolet radiation on terrestrial ecosystems. J. Photochem. Photobiol. B-Biol. 46: 40–52.

Dai Q.J., Yan B., Huang S.B., Liu X.Z., Peng S.B., Miranda M.L.L., Chavez A.Q., Vergara B.S. and Olszyk D.M. 1997. Response of oxidative stress defense systems in rice (*Oryza sativa*) leaves with supplemental UV-B radiation. Physiol. Plant. 101: 301–308.

Day T.A. 2001. Ultraviolet radiation and plant ecosystems. In: Cockell C.S. and Blaustein A.R. (eds.), Ecosystems, evolution and ultraviolet radiation, Springer-Verlag, NY, pp. 80–117.

Dormann C.F. and Woodin S.J. 2002. Climate change in the Arctic: using plant functional types in a meta-analysis of field experiments. Funct. Ecol. 16: 4–17.

Fenton J.H.C. 1980. The rate of peat accumulation in Antarctic moss banks. J. Ecol. 68: 211–228.

Fenton J.H.C. 1982. The formation of vertical edges on Antarctic moss peat banks. Arctic Alpine Res. 14: 21–26.

Flint S.D. and Caldwell M.M. 2003a. A biological spectral weighting function for ozone depletion research with higher plants. Physiol. Plant. 117: 137–144.

Flint S.D. and Caldwell M.M. 2003b. Field testing of UV biological spectral weighting functions for higher plants. Physiol. Plant. 117: 145–153.

Fowbert J.A. 1996. An experimental study of growth in relation to morphology and shoot water content in maritime Antarctic mosses. New Phytol. 133: 363–373.

Gehrke C., Johanson U., Gwynn-Jones D., Bjorn L.O., Callaghan T.V. and Lee J.A. 1996. Effects of enhanced ultraviolet-B radiation on terrestrial subarctic ecosystems and implications for interactions with increased CO_2. Ecol. Bull. 45: 192–203.

Gehrke C. 1998. Effects of enhanced UV-B radiation on production-related properties of a *Sphagnum fuscum* dominated subarctic bog. Funct. Ecol. 12: 940–947.

Gehrke C. 1999. Impacts of enhanced ultraviolet-B radiation on mosses in a subarctic heath ecosystem. Ecology 80: 1844–1851.

Green A.E.S. 1983. The penetration of ultraviolet-radiation to the ground. Physiol. Plant. 58: 351–359.

Green T.G.A., Schroeter B. and Seppelt R. 2000. Effects of temperature, light and ambient UV on the photosynthesis of the moss *Bryum argenteum* Hedw. in continental Antarctica. In: Davison W., Howard-Williams C. and Broady P. (eds.), Antarctic Ecosystems: Models for Wider Understanding, Caxton Press, Christchurch, pp. 165–170.

Huiskes A.H.L., Lud D. and Moerdijk-Poortvliet T.C.W. 2001. Field research on the effects of UV-B filters on terrestrial Antarctic vegetation. Plant Ecol. 154: 77–86.

Jansen M.A.K., Gaba V. and Greenberg B.M. 1998. Higher plants and UV-B radiation: balancing damage, repair and acclimation. Trends Plant Sci. 3: 131–135.

Johanson U., Gehrke C., Bjorn L.O., Callaghan T.V. and Sonesson M. 1995. The Effects of enhanced UV-B Radiation on a sub-Arctic heath ecosystem. Ambio 24: 106–112.

Kennedy A.D. 1995. Simulated climate-change – are passive greenhouses a valid microcosm for testing the biological effects of environmental perturbations. Global Change Biol. 1: 29–42.

Krizek D.T. and Mirecki R.M. 2004. Evidence for phytotoxic effects of cellulose acetate in UV exclusion studies. Environ. Exper. Bot. 51: 33–43.

Lud D., Moerdijk T.C.W., van de Poll W.H., Buma A.G.J. and Huiskes A.H.L. 2002. DNA damage and photosynthesis in Antarctic and Arctic *Sanionia uncinata* (Hedw.) Loeske under ambient and enhanced levels of UV-B radiation. Plant Cell Environ. 25: 1579–1589.

Lud D., Schlensog M., Schroeter B. and Huiskes A.H.L. 2003. The influence of UV-B radiation on light-dependent photosynthetic performance in *Sanionia uncinata* (Hedw.) Loeske in Antarctica. Polar Biol. 26: 225–232.

Madronich S., McKenzie R.L., Bjorn L.O. and Caldwell M.M. 1998. Changes in biologically active ultraviolet radiation reaching the Earth's surface. J. Photochem. Photobiol. B-Biol. 46: 5–19.

Markham K.R. 1990. Bryophyte flavonoids, their structures distribution, and evolutionary significance. In: Zinsmeister H.D. and Mues R. (eds.), Bryophytes: Their Chemistry and Chemical Taxonomy, Oxford University Press, New York, pp. 143–159.

Markham K.R., Franke A., Given D.R. and Brownsey P. 1990. Historical Antarctic ozone level trens from herbarium specimen flavonoids. Bulletin de Liaison Groupe de Polyphenols 15: 230–235.

Markham K.R., Ryan K.G., Bloor S.J. and Mitchell K.A. 1998. An increase in the luteolin: apigenin ratio in *Marchantia polymorpha* on UV-B enhancement. Phytochemistry 48: 791–794.

Meijkamp B., Aerts R., van de Staaij J., Tosserams M., Ernst W.H.O. and Rozema J. 1999. Effects of UV-B on secondary metabolites in plants. In: Rozema J. (ed.), Stratospheric Ozone Depletion: The Effects of Enhanced UV-B Radiation on Terrestrial Ecosystems, Backhuys Publishers, Leiden, pp. 71–99.

Montiel P., Smith A. and Keiller D. 1999. Photosynthetic responses of selected Antarctic plants to solar radiation in the southern maritime Antarctic. Polar Res. 18: 229–235.

Newsham K.K., Mcleod A.R., Greenslade P.D. and Emmett B.A. 1996. Appropriate controls in outdoor UV-B supplementation experiments. Global Change Biol. 2: 319–324.

Newsham K.K., Hodgson D.A., Murray A.W.A., Peat H.J. and Smith R.I.L. 2002. Response of two Antarctic bryophytes to stratospheric ozone depletion. Global Change Biol. 8: 972–983.

Newsham K.K. 2003. UV-B radiation arising from stratospheric ozone depletion influences the pigmentation of the Antarctic moss *Andreaea regularis*. Oecologia 135: 327–331.

Niemi R., Martikainen P.J., Silvola J., Sonninen E., Wulff A. and Holopainen T. 2002a. Responses of two Sphagnum moss species and *Eriophorum vaginatum* to enhanced UV-B in a summer of low UV intensity. New Phytol. 156: 509–515.

Niemi R., Martikainen P.J., Silvola J., Wulff A., Turtola S. and Holopainen A. 2002b. Elevated UV-B radiation alters fluxes of methane and carbon dioxide in peatland microcosms. Global Change Biol. 8: 361–371.

Phoenix G.K., Gwynn-Jones D., Callaghan T.V., Sleep D. and Lee J.A. 2001. Effects of global change on a sub-Arctic heath: effects of enhanced UV-B radiation and increased summer precipitation. J. Ecol. 89: 256–267.

Pourzand C. and Tyrrell R.M. 1999. Apoptosis, the role of oxidative stress and the example of solar UV radiation. Photochem. Photobiol. 70: 380–390.

Robinson S.A., Wasley J. and Tobin A.K. 2003. Living on the edge - plants and global change in continental and maritime Antarctica. Global Change Biol. 9: 1681–1717.

Robson T.M., Pancotto V.A., Flint S.D., Ballare C.L., Sala O.E., Scopel A.L. and Caldwell M.M. 2003. Six years of solar UV-B manipulations affect growth of Sphagnum and vascular plants in a Tierra del Fuego peatland. New Phytol. 160: 379–389.

Rozema J., van de Staaij J., Bjorn L.O. and Caldwell M. 1997a. UV-B as an environmental factor in plant life: stress and regulation. Trends Ecol. Evol. 12: 22–28.

Rozema J., Van de Staaij J., Meijkamp B., Lud D., Moerdijk T. and Huiskes A. 1997b. Stratosperic ozone depletion and effects of enhanced UV-B radiation on terrestrial antarctic plants: the methodology of a mini-UV-B supplementation system. Circumpolar J. 12: 43–48.

Rozema J., Broekman R., Lud D., Huiskes A.H.J., Moerdijk T., de Bakker N., Meijkamp B. and van Beem A. 2001a. Consequences of depletion of stratospheric ozone for terrestrial Antarctic ecosystems: the response of *Deschampsia antarctica* to enhanced UV-B radiation in a controlled environment. Plant Ecol. 154: 103–115.

Rozema J., Broekman R.A., Blokker P., Meijkamp B.B., de Bakker N., van de Staaij J., van Beem A., Ariese F. and Kars S.M. 2001b. UV-B absorbance and UV-B absorbing compounds (para-coumaric acid) in pollen and sporopollenin: the perspective to track historic UV-B levels. J. Photochem. Photobiol. B-Biol. 62: 108–117.

Rozema J., Noordijk A.J., Broekman R.A., van Beem A., Meijkamp B.M., de Bakker N.V.J., van de Staaij J.W.M., Stroetenga M., Bohncke S.J.P., Konert M., Kars S., Peat H., Smith R.I.L. and Convey P. 2001c. (Poly)phenolic compounds in pollen and spores of Antarctic plants as indicators of solar UV-B – A new proxy for the reconstruction of past solar UV-B? Plant Ecol. 154: 11–26.

Rozema J., van Geel B., Bjorn L.O., Lean J. and Madronich S. 2002. Paleoclimate: toward solving the UV puzzle. Science 296: 1621–1622.

Sancar A. and Sancar G.B. 1988. DNA-repair enzymes. Annu. Rev. Biochem. 57: 29–67.

Searles P.S., Flint S.D., Diaz S.B., Rousseaux M.C., Ballare C.L. and Caldwell M.M. 1999. Solar ultraviolet-B radiation influence on Sphagnum bog and Carex fen ecosystems: first field season findings in Tierra del Fuego, Argentina. Global Change Biol. 5: 225–234.

Searles P.S., Flint S.D. and Caldwell M.M. 2001. A meta-analysis of plant field studies simulating stratospheric ozone depletion. Oecologia 127: 1–10.

Searles P.S., Flint S.D., Diaz S.B., Rousseaux M.C., Ballare C.L. and Caldwell M.M. 2002. Plant response to solar ultraviolet-B radiation in a southern South American Sphagnum peatland. J. Ecol. 90: 704–713.

Setlow R.B. 1974. The wavelengths in sunlight effective in producing skin cancer: a theoretical analysis. Proc. Natl. Acad. Sci. USA 71: 3363–3366.

Sonesson M., Carlsson B.A., Callaghan T.V., Halling S., Björn L.O., Bertgren M. and Johanson U. 2002. Growth of two peat-forming mosses in subarctic mires: species interactions and effects of simulated climate change. Oikos 99: 151–160.

Steeneken S.F., Buma A.G.J. and Gieskes W.W.C. 1995. Changes in transmission characteristics of polymethylmethacrylate and cellulose-(III) acetate during exposure to ultraviolet-light. Photochem. Photobiol. 61: 276–280.

Sullivan J.H., Teramura A.H., Adamse P., Kramer G.F., Upadhyaya A., Britz S.J., Krizek D.T. and Mirecki R.M. 1994. Comparison of the response of soybean to supplemental UV-B radiation supplired by either square-wave or modulated irradiation systems. In: Biggs R.H. and Joyner M.E.B. (eds.), Stratospheric ozone depletion/UV-B radiation n the biosphere, Springer-Verlag, Berlin, pp. 211–220.

Teramura A.H. and Sullivan J.H. 1994. Effects of UV-B Radiation on photosynthesis and growth of terrestrial plants. Photosynth. Res. 39: 463–473.

WMO (World Meteorological Organization). 2003. Scientific Assessment of Ozone Depletion: 2002. Global Ozone Research and Monitoring Project—Report No. 47. Geneva, pp. 498.

Photo. Arctic tundra peat in front of Stuphallet, Brøggerhalvøya, not far from Ny Alesund.Nutrient enrichment by nesting seabirds at the steep rocks of Stuphallet stimulates growth of many arctic plants. Because of the permafrost at 15–20 cm below ground level, and soil melt-water in the summer period, the wet tundra soil is waterlogged. A peat profile was sampled with a depth of 105 cm using a motor-driven soil corer with a saw-tooth end. Photograph by J. Rozema.

Plant Ecology (2006) 182:155–173
DOI 10.1007/s11258-005-9024-0

A vegetation, climate and environment reconstruction based on palynological analyses of high arctic tundra peat cores (5000–6000 years BP) from Svalbard

J. Rozema[1,*], P. Boelen[1], M. Doorenbosch[1], S. Bohncke[2], P. Blokker[1], C. Boekel[1], R. A. Broekman[1] and M. Konert[1]
*[1]Department of Systems Ecology, Institute of Ecological Science, Climate Center, Vrije Universiteit, 1081 HV, Amsterdam, The Netherlands; [2]Department of Quaternary Geology and Geomorphology, Vrije Universiteit, De Boelelaan 1085, 1081 HV, Amsterdam, The Netherlands; *Author for correspondence (e-mail: jelte.rozema@ecology.falw.vu.nl)*

Received 1 September 2004; accepted in revised form 15 December 2004

Key words: Arctic, Brøggerhalvøya, *Cassiope tetragona*, Climate change, Little Ice Age, Nutrient-enrichment, Peat, Permafrost, Pollen record, *Salix polaris*, *Saxifraga oppositifolia*, Svalbard, Tundra vegetation composition

Abstract

As a reference for ongoing studies reconstructing past vegetation, climate and environment, pollen spectra in tundra peat profiles from Svalbard, were investigated. The base of tundra peat cores collected from Ny Ålesund, Stuphallet, Blomstrand and Isdammen has been ^{14}C dated to 350–490 BP, 5710 BP, 4670 BP and 700–900 BP, respectively. The Stuphallet and Blomstrand (Brøggerhalvøya) peat profiles were composed of a peat developed in a nutrient enriched and wet tundra environment of steep birdcliffs. Pollen concentrations were low, *Brassicaceae* pollen dominated the whole profile. In contrast, the Ny Ålesund and Isdammen profiles contained high pollen concentrations and suggest a nutrient-poor, dry tundra environment. Pollen of the polar willow, *Salix polaris*, occurred commonly throughout all four peat profiles. In the relatively high resolution (10 years per peat core sample) analysis of the Ny Ålesund core, starting before or at the beginning of the Little Ice Age (LIA, 16th-mid 19th century), dominance of *Saxifraga oppositifolia* indicates a cold and dry climate, followed by a decline of *Saxifraga oppositifolia* and gradual increase of *Salix polaris* after the LIA, which indicates a moist and milder climate.

Introduction

The research presented here forms part of a project aimed at assessing and analysing the interactions between climate change and the plant species of arctic terrestrial ecosystems. This approach includes field experimental study of the response of terrestrial plant species to global warming with Open Top Chambers (e.g. Aerts et al. 2005; Blokker et al. in prep.; Rozema et al. in prep.), and to enhanced UV-B as a result of changes of stratospheric ozone concentrations (Boelen et al. 2005; Blokker et al. 2005; Rozema et al. 2005). Antarctic and arctic terrestrial ecosystems are pre-eminently suitable to this research purpose since both global warming and ozone depletion are most pronounced in polar regions (Farman et al. 1985; Mc Peters et al. 1996; Newman et al. 1997; Hassol, ACIA 2004).

While the average global warming is estimated broadly at 2.5 °C for the period 2000–2100 (IPCC 2001) winter temperatures in arctic areas (60–90° North) tended to increase by 2–4 °C during the past 50 years (Hassol 2004) and the projected increase of arctic winter temperature is 4–7 °C for 2000–2100 (IPCC 2001).

In addition to various abiotic proxies (e.g. isotope ratio's $^{18}O/^{16}O$, Isaaksson et al. 2003; periglacial features, Isarin 1997), reconstruction of past temperature and UV regimes can be based on palynological analyses and plant-climate indicator relationships (Iversen 1954; Zagwijn 1994; Isarin 1997; Isarin and Bohncke 1999), extended with experimentally obtained plant temperature and plant UV-B dose response or transfer relationships. The latter comprises climate induced chemical changes in plants, e.g. flavonoids, para-coumaric acid and ferulic acid in pollen and macrofossil plant remains (Blokker et al. 2005).

The Spitsbergen archipelago (74–84° N, 10–35° W) forms part of the Arctic, far north of the polar circle. Owing to the warm Gulfstream water along the west coast of Spitsbergen, the climate is less extreme than at more northern and eastern parts of Svalbard. The mean annual temperature for 1975–1996 for Longyearbyen was −5 °C and the mean July temperature for this 22 year period was +6.6 °C (Hisdal 1985, 1998; Rønning 1965, 1996). Overall on Svalbard, average winter temperature varies from −15 to −20 °C and mean summer temperature from +5 to +8 °C.

Precipitation is 400 mm per year in the west, where a major part falls as rain or moist fog, and 200–300 mm further inland (Steffensen 1982). The vegetative summer period is relatively short, i.e. between 40 and 70 days. Vegetation-climate regions have been recognized consisting of (a) a barren zone, (b) the White arctic bell-heather zone, (c) the Mountain Avens zone and (d) the inner fjord zone, corresponding to some extent with dominance of *Cassiope tetragona* (b), *Dryas octopetala* (c) and *Salix polaris* (d) (Elvebakk 1994, 1997; Rønning 1996). Vegetation at the foot of steep cliffs used by nesting birds, enriched with droppings (guano) is characterized by *Brassicaceae*, *Polygonaceae* and (some) grass species. According to Elvebakk (1997), the mean temperature of the warmest month (MTWM) of polar deserts with *Saxifraga oppositifolia* is 3 °C , the MTWM of the northern arctic tundra with *Salix polaris* is 3–5 °C and the MTWM of the middle arctic-tundra zone (inner fjord area) with *Cassiope tetragona* and *Dryas octopetala* is 5–7 °C.

Based on this and the literature, vegetation, environment and climate relationships are used and described in Table 1. This data set has subsequently been used to infer climate and environmental changes from the pollen diagrams.

More generally, the practice of inferring past climate change with Climate Indicator Species has been described by Zagwijn (1994), Isarin (1997), Isarin and Bohncke (1999). For this purpose, the relationship between plant parameters and the mean annual air temperature (MAAT) is used, or the minimum (or maximum) mean temperature of the warmest (or coldest) month. Often these climate parameters represent threshold values, i.e.

Table 1. Climate indicator species used for climate–plant relationships on Svalbard used for reconstruction of climate and environmental changes inferred from the pollen diagrams.

Species	Indication	Reference
Caryophyllaceae: *Silene acaulis, S. uralensis, Cerastium*	Cold	Elvebakk 1985, 1997; Rønning 1965, 1996
Brassicaceae Cruciferae (Draba, Cardamine, Cochlearia, Braya)	Nutrient rich	v/d Knaap 1988a
Polygonaceae: (Oxyria, Bistorta)	Nutrient rich	Elvebakk 1985, 1997; Rønning 1965, 1996
Ranunculaceae	Wet, moist	Elvebakk 1985, 1997; Rønning, 1965, 1996
Cassiope tetragona	Warm, dry, nutrient poor	Elvebakk 1985, 1997
Poaceae: (Gramineae)	Nutrient rich	v/d Knaap 1988a, b
Salix polaris	Mild, wet	v/d Knaap 1989a
Saxifraga oppositifolia	Cold, dry, nutrient poor	v/d Knaap 1985, 1988a, b; Birks 1991, 1996; Aiken et al. 1999
Saxifraga stellaris	Wet and cold	Elvebakk 1985, 1997; Rønning 1965, 1996
Rosaceae (Dryas octopetala)	Dry, calcareous	Elvebakk 1985, 1997; Rønning, 1965, 1996

temperatures below which plants die, or above which plants flower and reproduce (Iversen 1954). It should be noted that no such quantitative transfer functions are available for high arctic tundra plants. Therefore the climate and environmental changes at the sites in this study can only be described qualitatively. Moreover, the climate and environmental factor relationships used here may be correlated and compound (e.g. cold and moist), obstructing reconstruction of single climate parameters such as temperature.

Antarctic moss peat banks at Signy Island, comparable in origin and age to the Svalbard peat cores of this paper, did not contain substantial pollen concentrations (Boelen et al. 2005). The Antarctic moss peat banks mainly consist of remains of mosses and their spores. This absence of pollen is not remarkable since only two flowering plant species occur in antarctic terrestrial environments. In addition, terrestrial Antarctic vegetation is often sparse and patchy. In contrast, the arctic flora of Svalbard includes 173 vascular plant species and high arctic tundra vegetation covers large areas on Svalbard (Barkman 1987; Summerhayes and Elton 1923; Elvebakk 1994; Elven and Elvebakk 1996; Elvebakk and Prestud 1996; Rønning 1996). Arctic peat cores may therefore contain more pollen than Antarctic peat deposits.

Peat formation in polar regions is much slower than at lower latitudes, since tundra plant growth is limited to a short and cold arctic summer. However melt and rainwater in the soil above the omnipresent permafrost layer at the depth of 10–100 cm (Rønning 1996) helps to preserve dead plant material. Once dead plant material has left the active layer and forms part of the permafrost, microbial and chemical decay is very slow or absent, and preservation of organic plant matter including pollen and spores is optimal.

The aim of this paper is (1) to reconstruct the past vegetation of Svalbard during the last few thousands of years by palynological analyses of peat cores and (2) to compare these pollen records with climate and environment indicator species and their present-day plant climate and environment relationships to reconstruct past climatic and environmental conditions of the arctic tundra.

Dependent on the age of the peat cores more detailed research aims are to describe vegetational and climatical changes during periods of marked or abrupt climatic change such as the Little Ice Age, as derived from earlier palynolological and abiotic proxy reconstructions (e.g. $^{18}O/^{16}O$ Isaakson et al. 2003).

Materials and methods

Core sampling sites and soil core description

Limited information is available on peat formation in the high arctic and locations on Svalbard and palynological analysis of terrestrial pollen deposits (cf van der Knaap 1985, 1987, 1988a, b; 1989a, b; 1991). The Ny Ålesund and Blomstrand cores were collected after reconnaissance trips around Kongsfjorden and based on van de Knaap (1988a) (Figures 1a, b and 4a–c). The Isfjorden peat core site near Longyearbyen was chosen in the vicinity of a UV-B supplementation field experiment and ITEX open top roofs, where present day pollen and vegetation analyses have been conducted (Rozema et al. 2005)

Ny Ålesund

In August 2000 a core was collected near Ny Ålesund (78° N, 11° E) (Figures 2, 4b). The average temperature for Ny Ålesund in February, the coldest month, is c. −15.0 °C. In July, which commonly is the warmest month, the average temperature is c. +5.0 °C. The core (Figures 2, 4, 6) taken with a stainless steel corer, had a diameter of 40 mm and was 9.7 cm deep. The lithology is as follows: 0–3.8 cm: fresh moss peat, 3.8–6.0 cm: strongly humified peat, 6.0–9.7 cm: silt.

Peat material for ^{14}C dating was taken at 5.8–5.7 cm below the top of the core and was dated 280 ± 45 BP, indicating an age of 490 BP (1460 AD), (BP Before Present is standardized at AD 1950) , for the lower part (9.7 cm) of the core. The core was collected on a sun exposed dry arctic heather site dominated by *Cassiope tetragona* with a *Salix polaris* understorey (Figure 4b).

The composition of the present-day vegetation is: *Cassiope tetragona* (10–15%), *Salix polaris* (5–10%), the moss *Sanionia uncinata* (10–20%), *Oxyria digina*, *Bistorta vivipara*, *Saxifraga oppositifolia*, *S. cernua*, *S. hieracifolia*, *S. hirculus*, *Silene acaulis*, *Dryas octopetala*, *Cerastium arcticum*, *Alopecuris borealis*, *Poa alpina*, *Carex misandra*, *Luzula confusa*, *Stereocaulon alpinum* and white crustose lichens.

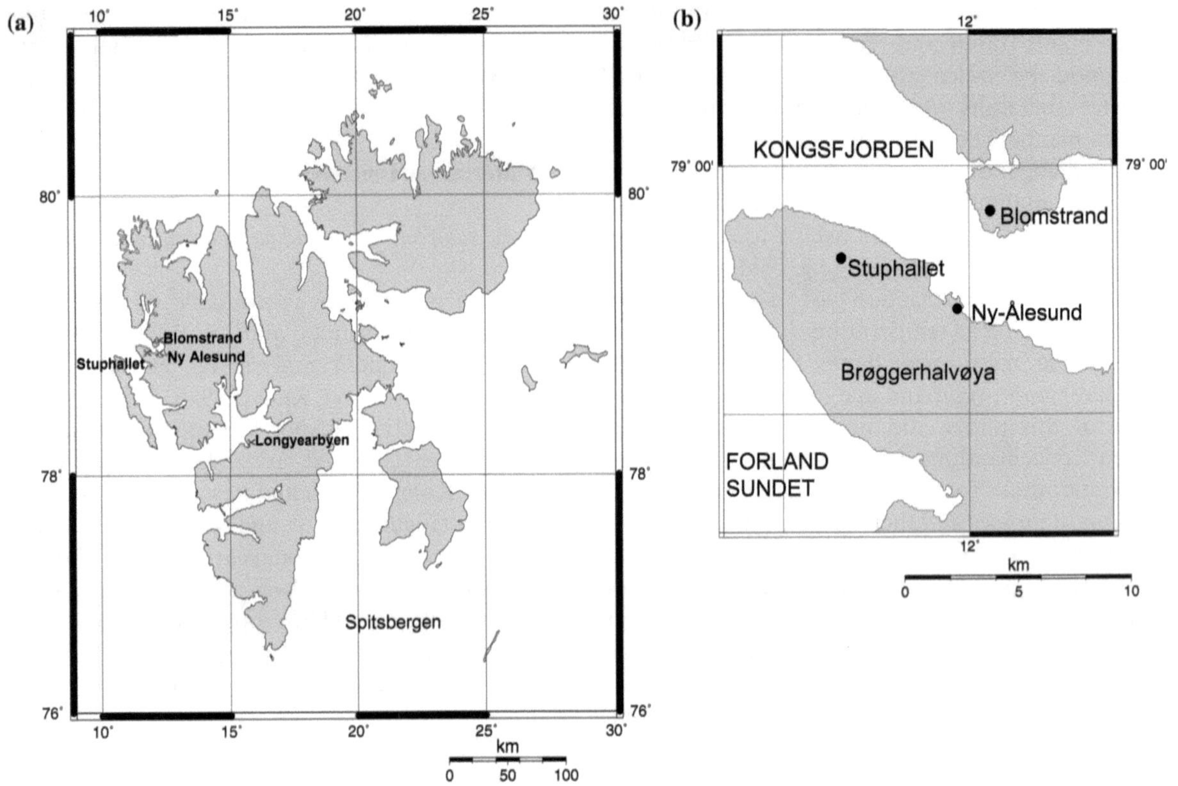

Figure 1. (a) Map of the Svalbard archipelago with the location of the peat core sites. 1. Stuphallet, 2. Blomstrand, 3. Ny Ålesund, 4. Isdammen. Peat cores 1 and 2 were collected on July 2002; peat cores 3 and 4 on August 2000. (b) Location of the peat core sites on Brøggerhalvøya, in the neighourhood of Ny Ålesund.Stuphallet and Blomstrand are peat areas adjacent to steep birdcliffs.

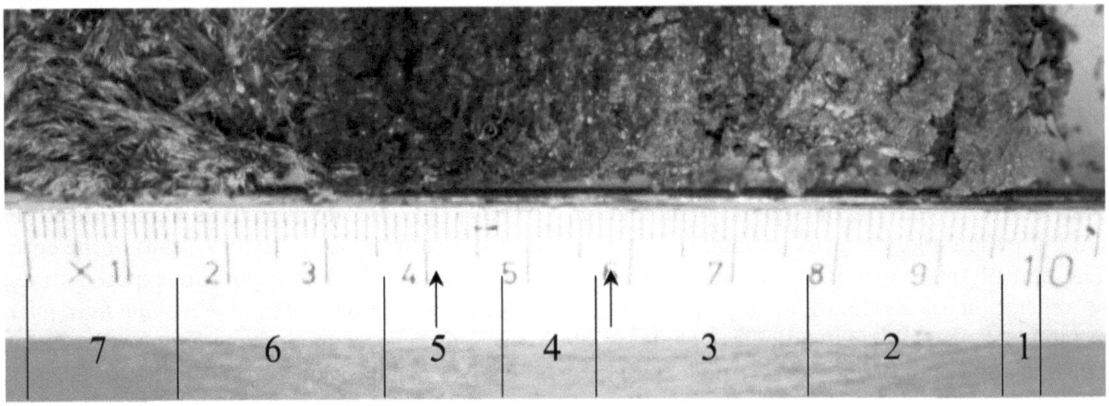

Figure 2. The Ny Ålesund peat core 9.7 cm long, collected on August 2000. The arrows indicate sampling for ^{14}C dating.

Blomstrand

In July 2002, a core was collected at the edge of a peat area in front of the bird cliffs at Blomstrand on Blomstrandøya (Figures 1, 7), which is a sun-exposed location, with the permafrost layer deeper than 70 cm at the edge of the peat area. The whole core was collected intact using a spade and a long knife. The peat core (Figure 3) was 67.4 cm

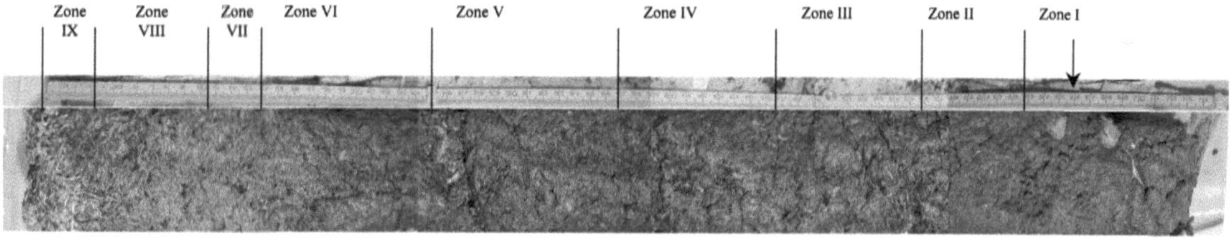

Figure 3. The Blomstrand peat core, 67.4 cm long, sampled July 2002.

deep. The lithology is as follows: Top layer 0–6 cm: living mosses, well preserved, yellow, 6–22 cm: moderately decomposed, dark brown, 22–32 cm: well preserved, dark brown, 32–62 cm: well decomposed, dark brown, 62–67.4 cm: moderately well preserved, dark/light brown. The core length was 67.4 cm with a diameter of 10 cm. Peat material for ^{14}C dating was taken at 65.2–67.4 cm depth, without reaching the permafrost layer. The composition of the present-day vegetation is: *Alopecurus borealis, Bistorta vivipara, Cassiope tetragona, Cerastium arcticum, Dryas octopetala, Oxyria digyna, Carex misandra, Luzula confusa, Draba* species, *Cochlearia groenlandica, Ranunculus hyperboreus, Salix polaris, Saxifraga cernua, Saxifraga hieracifolia, Saxifraga hirculus, Saxifraga oppositifolia, Silene acaulis, Stereocaulon alpestre* (lichen), *Sanionia uncinata, Polytrichum hyperboreum* (mosses), *Huperzia selago* (club moss).

Stuphallet

The permafrost layer was at 15–20 cm below ground level. In July 2002, samples were taken from the frozen peat using a motor-driven soil corer with a saw-tooth end with a diameter of 30 mm. The depth of the profile was 105 cm. Eighty sub samples, about 1.3 cm thick, were cut from the peat core. Forty subsamples were analysed at even intervals. Peat material for ^{14}C dating was taken at 103.7–105 cm depth. The species composition of the present-day vegetation is: the moss *Calliergon* spec. (20%), *Salix polaris* (5–15%), *Equisetum arvense* (horsetail), *Draba alpina, Draba* spec, *Braya purpurascens, Oxyria digina, Bistorta vivipara* (= *Polygonum viviparum*), *Dryas octopetala, Cerastium arcticum, Ranunculus pygmaeus, R. hyperboreus, Saxifraga hirculus, S. oppositifolia, S. cernua, S. hieracifolia, Sanionia uncinata* (5–10%), *Cardamine nymanii, Cochlearia groenlandica, Alopecurus borealis, Poa alpina.*

Isdammen

In August 2000, a core was collected near Longyearbyen, next to the drinking water reservoir Isdammen. The vegetation is dry and dominated by *Cassiope tetragona* and *Salix polaris*. The core was drilled with a stainless steel soil corer. The permafrost layer was not reached. Bedrock did not allow further coring. The core had a diameter of 3 cm and was 19 cm long. The lithology is at follows: 0–3.2 cm: humified peat with plant remains, 3.2–19 cm: humified peat. The species composition of the present-day vegetation, with percentage cover, is: *Salix polaris* (20–30%), *Cassiope tetragona* (40–50%), *Dryas octopetala* (5–10%), *Sanionia uncinata* (10–20%), *Polytrichum hyperboreum, Oxyria digyna, Bistorta vivipara* (= *Polygonum viviparum*), *Pedicularis hirsuta, Stellaria crassipes, Saxifraga oppositifolia, Saxifraga hieracifolia, Saxifraga hirculus, Alopecurus borealis, Poa alpina, Luzula confusa, Carex misandra, Festuca rubra, Equisetum arvense, Peltigera aphtosa* (lichen).

Collected peat cores were stored in open PVC tubes, (Ny Ålesund, Isdammen, Stuphallet) kept at 0 °C on Svalbard and after air transport and delivery, lasting less than 24 h, the peat cores were kept frozen at −20 °C. Subsampling for pollen analysis was performed on the defrosted peat core.

Core age and time resolution

Samples of organic material were sampled from the peat profile and C-14 dated by the Centrum voor Isotopen Onderzoek (CIO), Rijks Universiteit Groningen. Obtained radio-carbon age data (years BP) were transferred to Calender years BC using the cal25 programme (van der Plicht 1993):

Peat profile site	Radio carbon age Years BP (1950)	Calendar years BC (1 σ (sigma) confidence level)
GrN-28890 Blomstrand Collected July 2002	4670 ± 60 BP	3517–3482 cal BC 3479–3369 cal BC
GrN-28891 Stuphallet Collected July 2002 collected	5710 ± 150 BP	4718–4443 cal BC 4421–4395 cal BC 4387–4371 cal BC
GrA-20057 Ny Ålesund Collected August 2000	280 ± 45 BP	1521–1579 AD 1583–1594 AD 1619–1558 AD
GrA-20058 Ny Ålesund Collected August 2000	101 ± 6 BP	1689–1731 AD 1808–1826 AD 1829–1892 AD 1908–1924 AD

Figure 4. (a) Landscape and vegetation of (a) the Blomstrandhalvøya with the edge of the peat area bordering Kongsfjorden, (b), the Ny Ålesund arctic bell heather site, with Kongsfjorden and the Tre Kronar at the background, (c) the arctic peat in front of the Stuphallet birdcliffs, (d) *Cassiope tetragona* dominated Adventdalen glacier valley slope at Isdammen. Photographs J.Rozema.

Ny Ålesund

For a core taken 20 cm next to this core, which was about 1.5 cm shorter, the ^{14}C age has been determined at 101 ± 6 BP at 4.2 cm below the top and 280 ± 45 years BP at a depth of 5.7 cm below the soil surface. Based on these measurements the age of the lower part of the core of Ny Ålesund (9.7 cm) is estimated between 350 and 490 BP. With subsamples of 3 mm thick for acetolyis treatment and a core length of 9.7 cm the time resolution of the palynological analyis is estimated at 10–15 years for this core.

Blomstrand

A core taken at Stuphallet by van der Knaap (1988a) was dated at 45 cm depth at about 1900 years BP. Present-day Blomstrand is comparable to Stuphallet, a nutrient enriched environment at the foot of steep bird clifs. The age of the Blomstrand core has been ^{14}C estimated at 4670 ± 60 BP at a depth at 65.2–67.4 cm.

With 0.8–2.2 cm thick subsamples for acetolysis treatment and a core length of 67.4 cm the time resolution of the palynological analysis is much larger than for the Ny Ålesund core.The volume of the subsamples for the acetolysis treatment varied from 2.2–4.4 cm^3. Forty seven subsamples were taken and analysed from the core.

Stuphallet

The age of the Stuphallet core has been ^{14}C estimated at 5710 ± 150 BP at a depth of 103.7–105 cm. This value agrees reasonably well with ^{14}C dating of a similar core of precisely the same location by van der Knaap et al. 1988a: a Stuphallet 2 core was dated at 4300 ± 100 BP at 90 cm depth, estimated to have an age of 5020 BP at 105 cm after extrapolation. Subsamples for acetolysis were 1.3 cm thick.

Isdammen

This core still has not been ^{14}C dated. Since this core (19 cm) is longer than the core from Ny Ålesund (10 cm) and shorter than the core from Blomstrand, the core age is estimated to be between 700–980 years BP, under the assumption of accumulation rates similar to the Ny Alesund site.

Acetolysis, pollen identification

Acetolysis treatment of the peat profile subsamples was done according to Moore et al. (1991). Pollen were identified using Faegri and Iversen (1989) and Moore et al. (1991) as well as by comparison with light microscopy and scanning electron microscopy images of modern pollen as described in Rozema et al. (2001a, b) (Figure 5). Collection of present-day pollen of arctic tundra species was conducted during the summers of 2000–2004 at Adventdalen and Isdammen.

For calculation and graphic presentation of the pollen data Tilia and Tilia Graph (Grimm 1992) software was used.

Lycopodium marker solution

500 tablets were put into an 800 ml beaker and HCl 10% was added, as much as necessary and a solution of 1 tablet/ml was obtained. After dissolving it was topped up with water and left to settle overnight. The next day the clear liquid was decanted. The procedure was repeated one more time. The solution was transferred quantitatively into a 500 ml flask. It was topped up to 500 ml exactly and mixed thoroughly. The freshly mixed solution was transferred into an 800 ml beaker. It was stirred with a magneto stirrer while the required volume was added to the sample beakers.

Results and interpretation of the pollen record

For interpretation of the pollendiagrams obtained we use plant- climate and environment relationships as specified in Table 1.

Ny Ålesund (Figure 6)

Zone NyA-1 (9.7–9.55 cm) and NyÅ – 2 (9.7–7.75 cm)

These zones contain a herb dominated vegetation in which tussocks of *Saxifraga nivalis* and *Saxifraga oppositifolia*, pioneers of bare ground are prominent. Among the other pioneer taxa are *Cassiope tetragona*, Cyperaceae and Caryophyllaceae. The combination of *Saxifraga oppositifolia* and Caryophyllaceae indicates cold and dry conditions with poor soils (van der Knaap 1985,

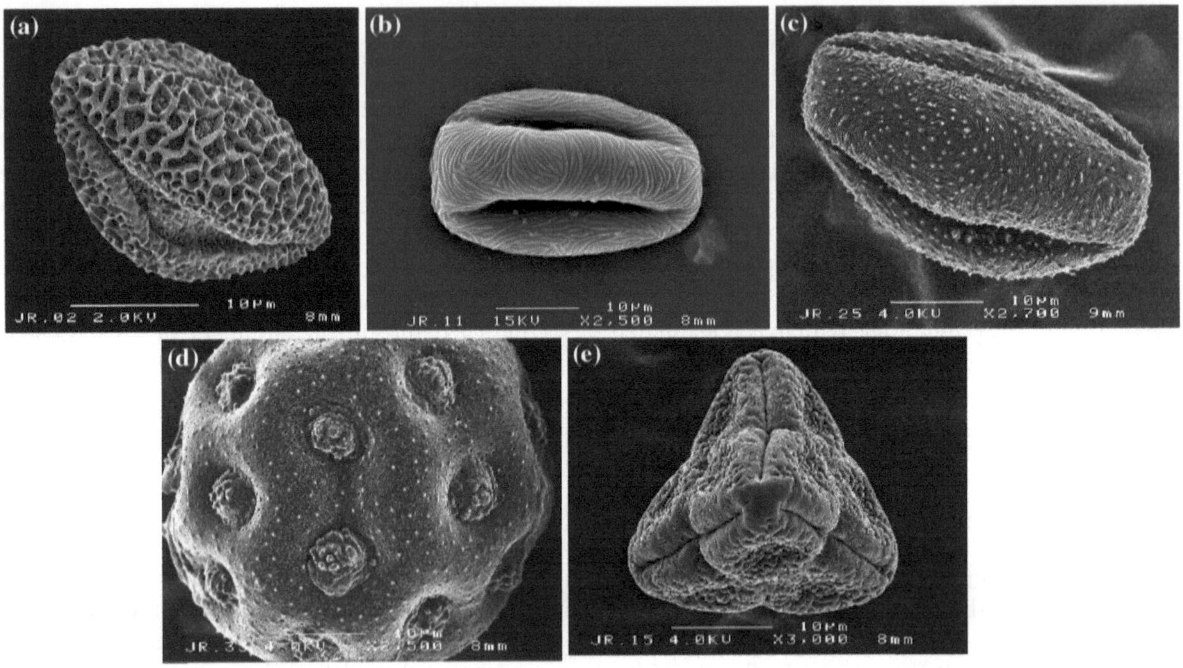

Figure 5. Scanning Electron Microscopy pictures of recent pollen grains. (a) *Salix polaris*, (b) *Saxifraga oppositifolia*, (c) *Saxifraga hirculus*, (d) *Cerastium arcticum*, (e) *Cassiope tetragona*.

1988a, b; Birks 1991; Elvebakk 1997; Rønning 1996). The presence of silt in the lower part of the profile indicates a pioneer vegetation colonising a (partly) barren landscape. Towards the top of zone NyÅ-2, *Cassiope tetragona* shows a clear increase indicating milder summer temperatures. Simultaneously the colonisation of the landscape proceeds with the spread of *Salix polaris*.

Zone NyÅ – 3 (7.75–5.65 cm)
A further colonisation by *Salix polaris* (60–75%) at the expense of the Caryophyllaceae, *Saxifraga oppositifolia* and the Cyperaceae is observed. Towards the top of this zone *Cassiope* rapidly declines and *Salix* starts to become dominant in the vegetation suggesting a rise in temperature. At the start of this zone, *Saxifraga hirculus* and Lamiaceae (Labiatae) occur temporarily.

Zones NyÅ 1–3
The increase of *Salix polaris* and the decrease of the tundra herbs might indicate a succession from tundra herbs to *Salix polaris*, possibly caused by warming of the climate or it might indicate a recolonization of the landscape after disappearance of an ice-sheet.

Zone NyÅ – 4 (5.65–4.75 cm)
The stable high values of *Salix polaris* coincide with the appearance of moss peat in the lithology. The transition to moss peat occurs at 5.8 cm and was dated to 280 ± 45 BP.

In this zone also the continuous curve of *Pinus* starts. *Pinus* does (and did not) not occur on Svalbard and this pollen must therefore originate from long distance dispersal.

The tundra herbs dominating below this have a low proportion, only Cyperaceae pollen increases slightly. The appearance of *Oxyria digyna* suggests nutrient-rich soils (van der Knaap 1988a).

Zone NyA-5 (4.75–3.55 cm)
The curves of Cyperaceae, *Casssiope tetragona* and Caryophyllaceae show a small peak, when *Salix polaris* decreases. *Oxyria* and *Saxifraga oppositifolia* are low or absent.

Zone NyA-6 (3.55–1.45 cm)
Salix polaris is still the predominant species. Cyperaceae, *Casssiope tetragona* and Caryophyllaceae are present in very low amounts. *Oxyria* and *Saxifraga oppositifolia* even disappear, although

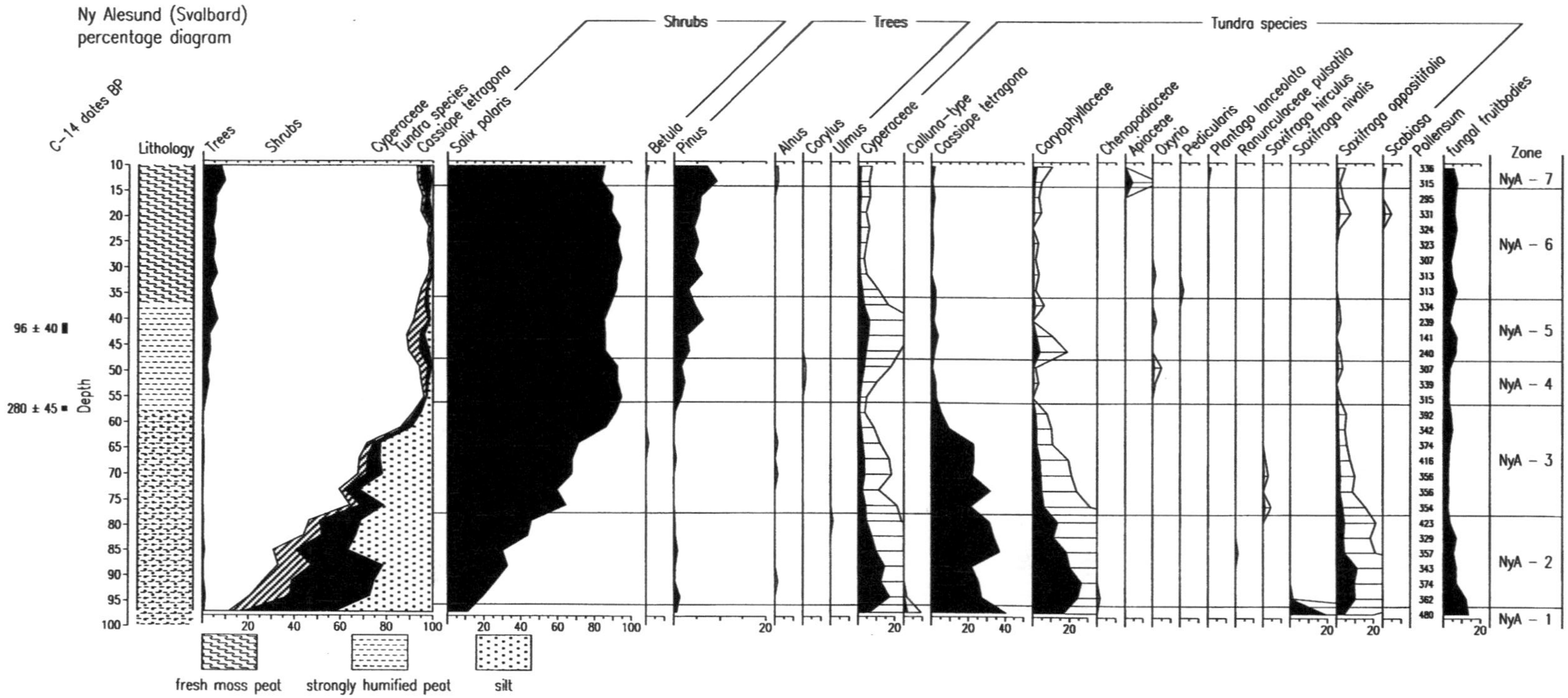

Figure 6. Percentage pollendiagram of the Ny Ålesund peat core.

Saxifraga oppositifolia appears again. This suggests relatively warm and wet conditions.

Zone NyA-7 (1.45 cm-top)
Salix polaris shows a small decrease. Apiaceae (Umbelliferae) and *Plantago lanceolata* pollen appear in this zone. The presence of *Plantago lanceolata* may be the result of human influence, which also could explain a decrease in the amount of *Salix polaris* pollen. However, *Plantago lanceolata* can also be considered a long-distance transported species (van der Knaap 1991).

An overview of the results indicates a succession from tundra herbs starting with cold adapted tundra species followed by a warmer period and an increase of the nutrient supply. Once *Salix polaris* had reached a constant high level and dominance, locally varying conditions resulted only in small variations in the assemblage.

Blomstrand (Figure 7)

Pollen numbers in some samples of the peat core have been very low and variable, and these results can therefore only be used to a limited extent as an indication of vegetation, environmental and climate changes.

Zone I (67.4–62.8 cm)
This zone dated at 4670 BP is characterized by a decrease in *Salix polaris* pollen (50–22.5%), which may indicate drier and colder conditions (van der Knaap 1989a). This might also explain the increase in *Saxifraga oppositifolia* (van der Knaap 1985; van der Knaap 1988a; b; Birks 1991). The relative increase in *Brassicaceae* suggests nutrient enrichment. Furthermore there is a decrease in the percentage and pollen number of *Saxifraga stellaris.*

Zone II (62.8–55.75 cm)
This zone is characterized by a maximum in the percentage of *Salix polaris* pollen, suggesting wetter conditions. This maximum is accompanied by a decrease in all tundra herb species. At the end of the zone *Salix polaris* decreases again and tundra herbs increase.

Zone III (55.75–46.4 cm)
In this zone *Salix polaris* pollen increase and *Brassicaceae* decrease, indicating wetter and less nutrient-rich conditions. The percentages of Cyperaceae and Gramineae pollen seem to be stable. *Saxifraga oppositifolia* starts higher than in zone II, but shows a decrease, which may be caused by warmer and wetter conditions. *Saxifraga stellaris*, a species which nowadays occurs at wet and cool sites (site 3), seems to increase, and *Koenigia islandica* appears.

Zone IV (46.4–36.8 cm)
This zone is characterized by a high and stable percentage of *Salix polaris* pollen (ca. 50%), suggesting relatively wet conditions. The pollen concentration diagram shows a decrease of the number of *Salix polaris* pollen at the end of the zone. However, the other species show a decrease as well, which results in a decrease in total pollen concentration. The number of *Poaceae (Gramineae)* pollen increases and *Cassiope tetragona* appears.

Zone V (36.8–25.4 cm)
This zone shows fluctuations in both *Salix polaris* and the tundra herb pollen. The percentage of *Salix polaris* is lower than in the previous zone. In the pollen concentration diagram it can be seen that the concentration of *Salix polaris* decreases slightly. Poaceae (Gramineae) decrease as well, and *Caryophyllaceae* disappear in this zone.

Zone VI (25.4–13.95 cm)
This zone is characterised by a low percentage of *Salix polaris* pollen and a high percentage of *Brassicaceae* pollen, suggesting nutrient rich conditions. However, the total pollen concentration is decreasing and the concentration of *Brassicaceae* starts very high, but decreases towards the end. *Cyperaceae* decrease, peaks of *Saxifraga oppositifolia* and *Saxifraga stellaris* may be the result of a cooling in temperature. *Gramineae* have disappeared; only a small peak can be noted. The pollen numbers in the last three peat subsamples are very low.

Zone VII (13.95–10.5 cm)
In this zone there is a transition from *Salix polaris* and *Saxifraga granulata* to *Brassicaceae* and *Cyperaceae*, indicators of a higher nutrient supply, to *Saxifraga stellaris* and *Papaver dahlianum*, which may be caused by a lowering in temperature.

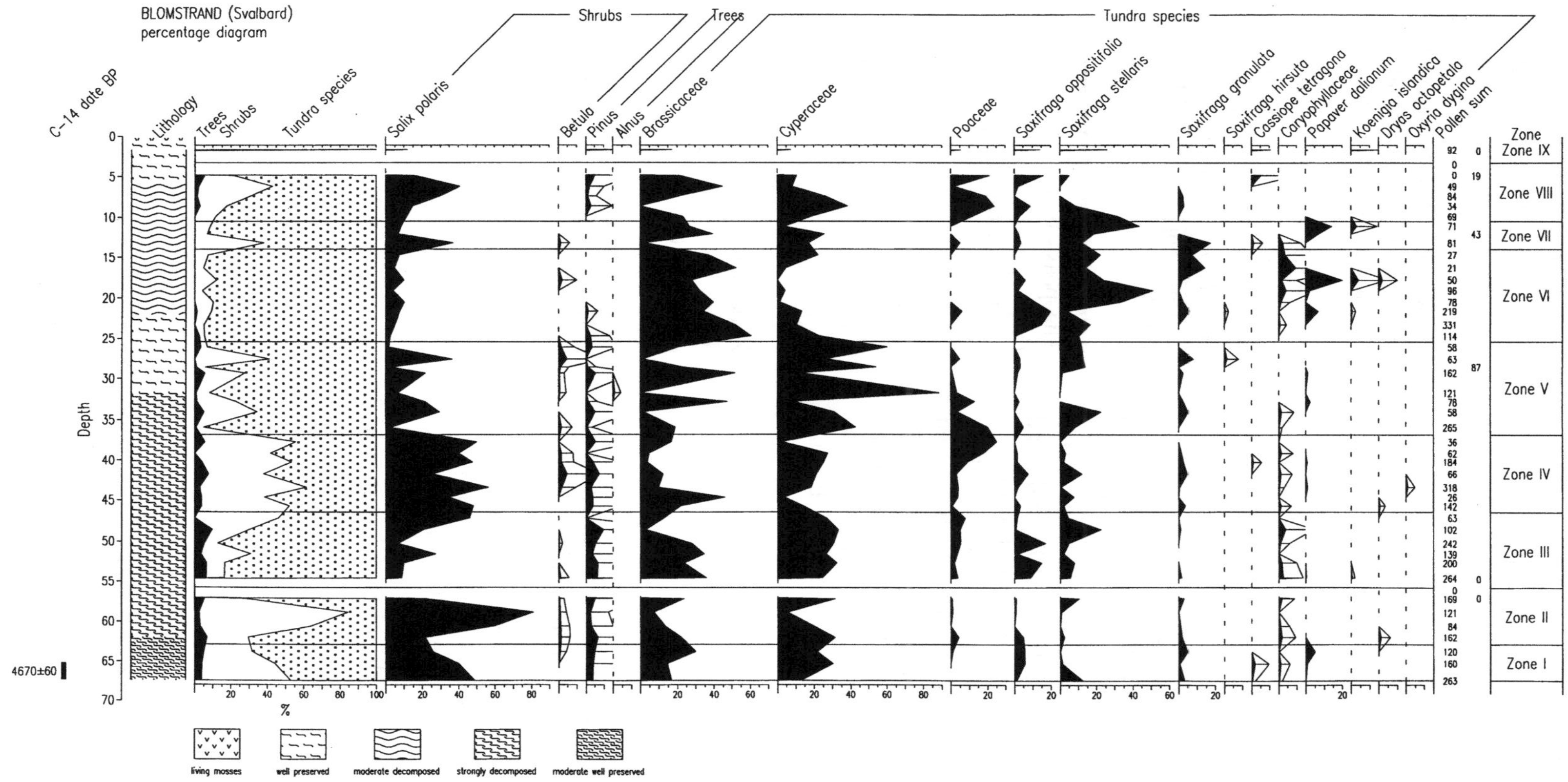

Figure 7. Percentage pollendiagram of the Blomstrand peat core.

Zone VIII (10.5–3.3 cm)
This zone is marked by a decrease in total pollen concentration. The proportion of Poaceae (Gramineae) pollen is higher than in the previous zone.

Zone IX (3.3cm-top)
This zone includes only one peat subsample per sample, i.e. the top layer of the core. The pollen numbers of the tundra species are higher than in the previous zones.

Pinus pollen are present in zone I–V
They disappear in zone VI. *Pinus* appears again in zone VIII again. The percentage is about 5% in these zones. *Pinus* is a long-distance transported pollen type (van der Knaap 1987, 1989b) and could be present because of afforestation in Europe.

In the Blomstrand peat profile, *Salix polaris*, *Brassicaceae* and *Cyperaceae* appear to be present in relatively high pollen numbers throughout the entire period. Conditions might have been wet and nutrient rich; local environmental variations may have resulted in variations in the diagram. Because of the low numbers of pollen grains in some intervals interpretation should be restrained.

Stuphallet (Figure 8)

Total pollen sum was very low and varied markedly. Only half of the 80 peat sampling intervals have been analysed. At a depth of 95–100 cm, there was an ice layer in the core. Brassicaceae pollen dominate the pollen diagram for the entire period and appear to increase the last thousands of years, reflecting the continuous input of nutrients by bird droppings from the nests on the nearby bird cliffs. Cyperaceae also occur the entire profile, in addition to *Saxifraga granulata* and *Saxifraga stellaris* and *Salix polaris*.

Isdammen (Figure 9)

Zone ISD-1 (19–13.2 cm)
In this zone *Salix polaris* dominates, indicating wetter conditions (van der Knaap 1989a). Proportions of the tundra herbs are more or less constant.

Zone ISD-2 (13.2–9.2 cm)
A decrease of *Salix polaris* and an increase of *Cyperaceae*, *Gramineae*, *Cassiope tetragona* and *Oxyria* can be seen. *Cassiope tetragona* suggests a milder climate (Rønning 1996; Elvebakk 1997), the increase of *Gramineae* and *Oxyria* also suggest richer soils (van der Knaap 1988a, b).

Zone ISD-3a (9.2–7.4 cm)
At the beginning of this period there is a decrease of *Cyperaceae*, *Gramineae* and *Cassiope tetragona*, and *Papaver dahlianum* and *Dryas octopetala* appear, suggesting colder conditions (Elvebakk 1997).

Zone ISD-3b (7.4–5.2 cm)
The peaks of *Caryophyllaceae*, *Oxyria* and *Saxifraga oppositifolia* may indicate colder conditions combined with a small decrease of *Salix polaris*.

Zone ISD-4 (5.2–3.2 cm)
Proportions of *Salix polaris* are relatively high and *Caryophyllaceae*, *Oxyria* and *Saxifraga oppositifolia* are low. *Salix polaris* dominates, which suggests higher temperatures and wetter conditions.

Zone ISD-5 (2.7 cm-top)
There is a small decrease of *Salix polaris* and a small increase of *Cassiope tetragona*. Proportions of *Saxifraga oppositifolia* are very low, indicating a warmer period.

In summary *Salix polaris*, indicating rather wet and warm conditions, dominated the environment. Some relatively brief periods of cooling and drying may have caused *Saxifraga oppositifolia* and *Caryophyllaceae* to increase, while during warmer periods *Cassiope tetragona* increased.

Discussion

Peat formation in the high Arctic

Peat formation under high Arctic or Antarctic climate conditions with very limited plant growth, is not common. Generally, optimal conditions for peat formation are annual mean temperatures between 5–10 °C (Clymo 1998). The mean annual temperature on the West-coast of Svalbard is currently about −6 °C. Another factor supporting peat formation is deficient drainage. Because of

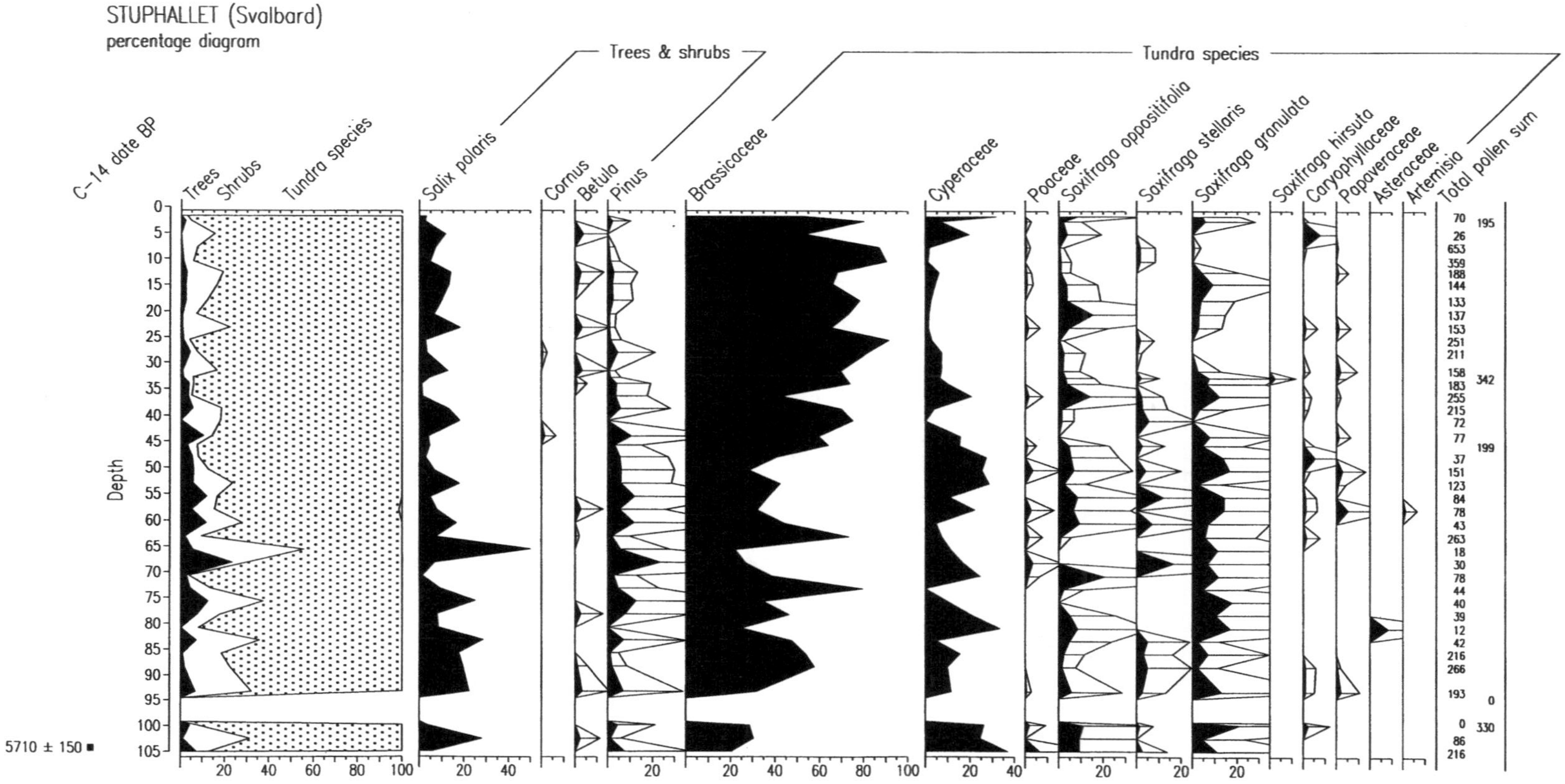

Figure 8. Percentage pollendiagram of the Stuphalllet peat core.

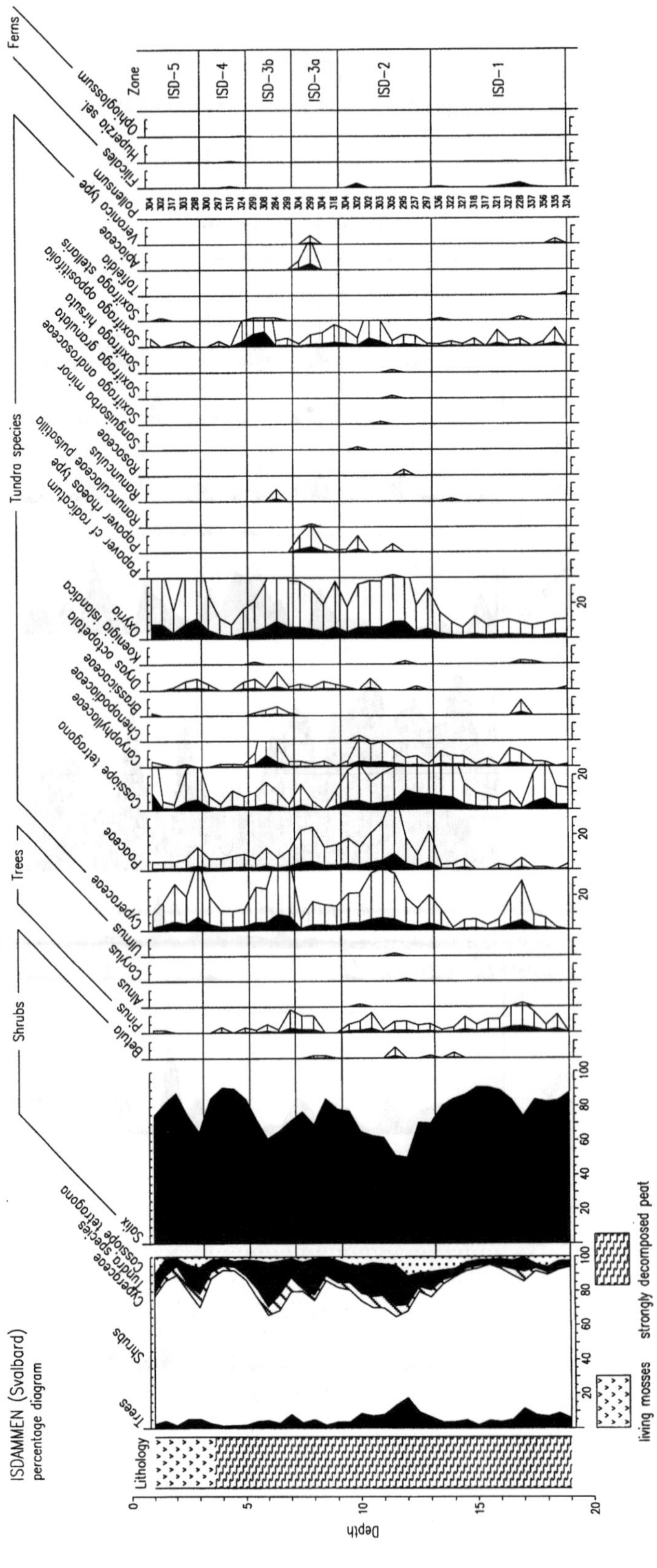

Figure 9. Percentage pollendiagram of the Isdammen peat core.

the permafrost present all over Svalbard, and soil melt-water in the summer period, the wet tundra soil is waterlogged. This leads to oxygen deficiency, which, together with low arctic temperatures strongly delays the decomposition of organic material (Bliss and Wiegolaski 1973; Stolbovoi 2002).

Peat formation on Svalbard seems to be increased in the environment of the bird cliffs (van der Knaap 1988b). Due to massive guano deposits, the nutrient concentration in the soil moisture increases supporting the enhanced development of peat forming vegetation. *Brassicaceae*, like *Cochlearia* and *Draba* are predominant taxa in the Blomstrand profile, indicating high nutrient availability (van der Knaap 1988a). This is comparable to the pollen records reported by van der Knaap (1985) at Stuphallet, a location under a bird cliff close to Blomstrand. The tundra vegetation at Stuphallet also reflects nutrient enriched soil conditions.

Permafrost soil conditions support good preservation of the peat layers. Discovering and coring peat core sections sufficiently long to represent most of the Holocene (last 10,000 years) is difficult. Many sites in the Arctic contain only peat core sections of approximately 2500 years of age or less (Boulton et al., 1976; Short and Andrews 1980; Ovenden 1988).

Pollen numbers in arctic peat

A general problem with Arctic palynology is the low concentration of pollen in the samples and consequently low pollen counts in the slides (Birks 1991).

For this reason paleo-reconstruction of arctic climate and environment is more often derived from lake deposits in limnological studies, where pollen, spores and macrofossil plant remains may accumulate (Birks et al. 2004).

Pollen concentrations were obviously low in the Blomstrand core (Figure 7) van der Knaap (1985) found similarly low pollen number in the Stuphallet peat core, in an environment, like Blomstrand, with continuous nutrient supply (guano) from adjacent birdcliffs. There is a ridge of smaller and larger stones between the foot of the bird cliffs and the Kongsfjorden. The area is relatively flat, the active layer in the summer was about 20 cm deep and both the melt water and rain caused waterlogged conditions. The nutrient supply and moist conditions allow abundant plant growth of some arctic plants, but with only few flowering plant species.

Pollen numbers were relatively high in the Ny Ålesund and Isdammen peat core. Flowering and the production of tundra plant species may be limiting factors, in addition to the low cover of tundra species.

Vegetational, environmental and climate history of Arctic Svalbard 5000–6000 years BP

The soil cores taken at Ny Ålesund, Blomstrand, Stuphallet and Isdammen are all estimated to be younger than 5760 years, i.e. representing the later part of the Holocene, which started about 10,000 years ago. The Holocene has been a relatively warm period, probably an interglacial between two ice-ages (Svendsen and Mangerud 1992, 1997). However, there have been both warmer and colder periods than present on Svalbard. Paleoenvironmental records from Svalbard show that in the period from ca. 9500 to 4000 years BP, summer air temperatures may have been higher and glaciers may have been smaller than in the present time (Birks 1991; Svendsen and Mangerud 1997). Circa 2500 years BP temperatures had fallen to near or below-present values and generally remained lower until the present century (Birks 1991). The relatively cold period in the interglacial Holocene lasting from about 16th to mid 19th century (Grove1988; Williams et al. 1998) is called the Little Ice Age (LIA). During the LIA glaciers expanded considerably on Svalbard (Svendsen and Mangerud 1997; Grove 2001). The uninterrupted palynological records of the peat cores presented in this paper indicate that none of the sites studied were glaciated during the period covered.

Isaakson et al. (2003) reconstructed past temperatures based on $^{18}O/^{16}O$ values of ice cores drilled in northern and north eastern parts of Svalbard for the period 1400–2000 AD. There is no evidence for an LIA decrease of Svalbard temperature at these ice core sites (far north of the sites where the peat profiles were collected), but reconstructed temperature decreased from about 1750–1850 AD, and increased from 1900–2000 AD. Since the LIA most glaciers on Svalbard have retreated because of gradual summer (and winter) warming (Svendsen and Mangerud 1997).

Despite significant temperatures changes in the Arctic regions during the past thousand years, it remains uncertain if and how these temperature changes affected plant growth and vegetation composition on Svalbard. Frost during the Little Ice Age may have strongly affected plant and vegetation in temperate climate regions such as Atlantic Europe, but may have left Arctic tundra plant growth unchanged (Van Geel et al. 1996, 1998; Mauquoy et al. 2002).

During the last 100 years a global warming has been marked (Hassol 2004). According to Karl (1998), temperatures on Svalbard increased by 4 °C from 1906 until 1996. Such a temperature increase will have caused changes in vegetation composition that may have been recorded in fossil pollen records.

On Svalbard, the climate is Arctic and the tundra plant community is adapted to the often severe conditions. These adaptations have been used to link changes in vegetation composition to climatic variations, according to climatic and environmental indicator species (cf Isarin 1997; Iversen 1954).

Possible links between changes in the Svalbard pollen records, Ny Ålesund, Blomstrand and Isdammen, and the climatic events just described will now be discussed.

The pollen record from Ny Ålesund, extrapolated to be about 490 (350) years BP old, starts about 1460 (1600) AD (Figure 8) just before or at the start of the Little Ice Age, and it is shown that low-temperature indicating species like *Saxifraga oppositifolia* (van der Knaap 1985, 1988a, b; Birks 1991; Aiken et al. 1999) and *Caryophyllaceae* species (Rønning 1996; Elvebakk 1997), but also *Cassiope tetragona* (indicating dry conditions) dominate the vegetation in that period. Towards the end of the LIA *Saxifraga oppositifolia* and *Caryophyllaceae* decrease and *Salix polaris* has come, reaching a maximum at about 1700 AD, possibly due to summer warming melting of nearby snow fields and glaciers, leading to moist glacier valley soil conditions.

It is unlikely that a glacier was located at the site, since tundra vegetation seems to have been present throughout the entire pollen record and presence and retreat of a glacier would have been indicated by silt in the core.

In the Isdammen pollen record (Figure 9) the vegetation composition seems to have been quite stable during the past centuries. *Salix polaris* dominates throughout the entire record. Also *Cassiope tetragona* occurs throughout the entire period. Brassicaceae species are almost lacking and indicate that the nutrient status of the Isdammen vegetation has been low.

Pollen numbers have been continuously high for the entire period and glaciation of the site during the period covered is unlikely. It is unclear to us why the similarly high time resolution of the Isdammen pollen diagram, likely to cover the LIA, does not clearly show changes of tundra plant species supporting this. Alternatively, no decreased temperatures may have occurred on arctic Svalbard during the LIA.

Obviously, with an improved chronology, based on more C-14 dating of subsamples of the peat profiles studied, conclusions on the relationship between arctic tundra vegetation changes before, during and after the LIA, could be developed more firmly.

Plant climate relationships, as required for a proper implementation of the Climate Indicator Species approach (Zagwijn 1994; Isarin 1997) are insufficiently developed for the Arctic. Arctic plant species distributions in the summer have been studied (Elvebakk 1994; Elvebakk and Prestud 1996; Elvebakk 1997) and mean summer or mean July temperatures may be used to explain changes in proportions of the pollen of *Salix polaris*, *Saxifraga oppositifolia* and *Cassiope tetragona*, as was done in this study. Increased growth of *Cassiope tetragona* obtained with experimental warming (1.5 °C) of tundra plants with Open Top Roofs (Figure 4d) is now being developed as a new biotic proxy of temperatures during the past (Blokker et al. in prep.; Rozema et al. in prep). As a tundra plant species of dry, relatively warm terrestrial environments, *Cassiope* growth responds to summer warming. This evergreen arctic heather shrub is also capable of surviving severe winter frost and dessication. Plant parameters linked with mean winter temperatures may also-help to reconstruct past climate regimes.

Perspective for reconstruction of past UV climate

In the present paper a palynological approach was followed to reconstruct past climate and environment, using pollen species identity and pollen numbers. For at least the last 6000 years *Salix* (cf *polaris*) has been a dominant tundra plant species.We have assumed that pollen identified as *Salix* actually represent *Salix polaris*; at the present day *Salix polaris* is by far the predominant arctic species in the Svalbard tundra, beside *S. reticulata*, *S. arctica* and *S. herbacea*. In warmer periods in the past the latter *Salix* species may have been more abundant.

For the reconstruction of past temperature and solar UV-B irradiance, both the morphological and chemical properties of pollen and macrofossil plant remains of *Salix polaris* will be further studied Rozema et al. 2002.

A proxy to reconstruct historic levels of UV-B irradiance may be the phenolic compounds found in pollen, spores, and preserved leaves and twigs of this tundra shrub (Rozema et al. 2001a, b; Blokker et al. 2005).

Acknowledgments

Field work at Svalbard by J.R. in 2000 was funded by EC contract UVAQTER number ENV-CT97-0580. The field work by P.B and J.R.on Svalbard in 2002 is financially supported by NWO-ALW-NAAP grant number 851.20.010 (UVANTARTIC). We acknowledge the permission for the field work from Sysselmannen, Longyearbyen and the cooperation and support of UNIS. The field work of M.D (2002), J.H. and R.F.(2003) at Svalbard forms part of their MSc research projects, jointly supervised by J.R. P.B. and Dr B. Solheim, which has been funded partially by the Vrije Universiteit, (Dr. K. Kits) which is greatly appreciated. We are grateful to Dr Pim van der Knaap, Bern, Switzerland for discussing peat core collection and analysis at Brøggerhalvøya. We thank Mr. N. van Harlingen, Drs Michel Groen and Mr. Flip de Kriek (Earth Sciences workshop Faculty of Earth and Life Sciences) for developing the motor-driven permafrost soil corer and for supporting the air transport of field equipment. The logistic and scientific support coordinated by base commanders Mr Nick Cox and Mrs Maggie Annat of the National Environment Research Council (NERC) at the Arctic Research Station "Harland Huset", Ny lesund in 2000 and 2002 is greatly acknowledged. The support by and cooperation with Dr B. Solheim and Prof dr Rolf Olsen in preparing the reconnaissance trips at Brøggerhalvøya and Blomstrand appreciated. Dr Dan Yeloff and Dr Hans Cornelissen are greatly acknowledged for constructive comments on the manuscript.

References

Aerts R., Cornelissen, J.H.C. and Dorrepaal E. 2005. Plant performance in a warmer world: general responses of plants from cold, northern biomes and the importance of winter and spring events. Plant Ecol. this volume.

Aiken S.G., Dallwitz M.J., Consaul L.L., Mc Jannet C.L., Ginnespie L.J., Boles R.L., Argus G.W., Gillett J.M., Scott P.J., Elven R., Blanc M.C. le Zamluck A.E. and Brysting A.K. 1999. Flora of the Canadian Arctic Archipelago: Descriptions, Illustrations, Identification and Information retrieval.

Barkman J.J. 1987. Preliminary investigations on the texture of high arctic tundra vegetation. In: Huiskes A.H.L., Blom C.W.P.M. and Rozema J. (eds), Vegetation Between Land and Sea, Dr. W. Junk Publishers, Dordrecht, pp. 120–132.

Birks H.H. 1991. Holocene vegetational history and climatic change in west Spitsbergen – plant macrofossils from Skardtjørna, an Arctic lake. Holocene 13: 209–218.

Birks H.J.B., Jones V.J. and Rose N.L. 2004. Recent environmental change and atmospheric contamination as recorded in lake sediments – an introduction. J. Paleolimnol. 31: 403–410.

Bliss L.C. and Wiegolaski F.E. (eds). 1973. Primary Production and Production Processes, Tundra biome. Tundra Biome Steering Committee, University of Alberta, Edmonton.

Blokker P., Boelen P., Broekman R. and Rozema J. 2005. The potential of p-coumaric and ferulic acid as UV proxies: occurrence in pollen and spores, preservation and pyrolytic analysis. Plant Ecol. (this volume).

Boelen P., de Boer M. K., de Bakker N., Blokker P. and Rozema. J. 2005. Field studies on the effects of solar UV-B on polar bryophytes: overview and methodology. Plant Ecol. (this volume).

Boulton G.S., Dickson J.H., Nichols H., Nichols M. and Short S.K. 1976. Late Holocene glacier fluctuations and vegetation changes at Maktak Fjord, Baffin Island, N.W.T., Canada. Arctic Alpine Res. 8: 343–356.

Clymo R.S. 1998. *Sphagnum*, the peatland carbon economy, and climate change. In: Bates J.W., Ashton N.W. and Duckett J.G. (eds), Bryology for twenty-first Century, Maney Publishing and the British Bryological Society, Leeds, pp. 361–368.

Elvebakk A. 1994. A survey of plant associations and alliances from Svalbard. J. Veg. Sci. 5: 791–802.

Elvebakk A. and Prestrud P. (eds) (1996). A Catalogue of Svalbard Plants, Fungi, Algae and Cyanobacteria. Norsk Polar Institutt Skrifter 198: 1–395.

Elven R. and Elvebakk A. 1996. Vascular plants. In: Elvebakk and Prestrud P. (eds), A Catalogue of Svalbard Plants, Fungi, Algae and Cyanobacteria. Norsk Polar Institutt Skrifter 198: 9–95.

Elvebakk A. 1997. Tundra diversity and ecological characteristics of Svalbard. Ecosystems of the world 3, Polar and Alpine Tundra: 347–359.

Farman J.C., Gardiner B.G. and Shanklin J.D. 1985. Large losses of total ozone in Antarctica reveal seasonal CLO_x/NO_x interaction. Nature 315: 207–210.

Faegri K. and Iversen J. 1989. Textbook of Pollen Analysis. 4th ed. Faegri K., Kaland P.E. and Krzywinski K. (eds), John Wiley & Sons, New York, 328 pp.

Hassol S. J. 2004. Impact of a Warming Arctic. Arctic Climate Impact Assessment. Cambridge Universiry Press. p. 139.

Grimm E.C. 1992. TILIA and TILIA-graph: pollen spreadsheet and graphic programs. Volume of abstracts 8th International Palynological Congress, Aix-en-Provence 1992, p. 56.

Grove J.M. 1988. The Little Ice Age. Methuen, London, 498 pp.

Grove J.M. 2001. The initiation of the "Little Ice Age" in regions round the North Atlantic. Climatic Change 48: 53–82.

Hisdal V. 1985. Geography of Svalbard. Norsk Polar Institutt, Oslo, p. 75.

Hisdal V. 1998. Svalbard Nature and History. Norsk Polar Institutt, Oslo, p. 123.

Houghton J.T., Ding Y., Griggs D.J., Noguer M., van der Linden P.J., Dai X., Maskell K. and Johnson C.A. (eds). IPCC 2001. Climate Change 2001. The Scientific Basis. Cambridge University Press.

Isarin R.F.B. 1997. The climate in north-western Europe during the Younger Dryas. Ph.D. Thesis, Vrije Universiteit, Amsterdam, 159 pp.

Isarin R.F.B. and Bohncke S.J.P. 1999. Mean July temperatures during the Younger Dryas in Northwestern and Central Europe as inferred from Climate Indicator Plant Species. Quaternary Res. 51: 158–171.

Isaaksson E., Hermanson M., Hicks S., Igarashi M., Kamiyama K., Moore J., Motoyama H., Muir D., Pohjola V. and Vaikmae R. 2003. Ice cores from Svalbard-useful archives of past climate and pollution history. Physics Chemistry Earth 28: 1217–1228.

Iversen, J. 1954. The Late-glacial flora of Denmark and its relation to climate and soil. Danmarks Geologiske Undersøgelse Årbog 2: 87–119.

Karl T.R. 1998. Annex A "Regional Trends and Variations of Temperature and Precipitation". The Regional Impacts of Climate Change. IPCC WGII, Cambridge University Press.

van der Knaap W.O. 1985. Human influence on natural Arctic vegetation in the seventeenth century and climatic change since A.D. 1600 in north-west Spitsbergen: a paleobotanical study. Arctic Alpine Res. 17: 371–387.

van der Knaap W.O. 1987. Long-distance transported pollen and spores on Spitsbergen and Jan Mayen. Pollen et Spores XXIX(4): 449–453.

van der Knaap W.O. 1988a. A pollen diagram from Brøggerhalvøya, Spitsbergen: changes in vegetation and environment from ca. 4400 to ca. 800 BP. Arctic Alpine Res. 20: 106–116.

van der Knaap W.O. 1988b. Age and stability of bird-manured vegetation on Spitsbergen. Arctic Botanica Neerlandica 37: 171–179.

van der Knaap W.O. 1989a. Relations between present-day pollen deposition and vegetation in Spitsbergen. Ph.D. Thesis. Laboratory of Paleabotany and Palynology, Utrecht. Arctic centre, Groningen, pp.15–26.

van der Knaap W.O. 1989b. Deposition of long-distance transported pollen and spores since 7900 B.P. studied in peat deposits from Spitsbergen. Ph.D. Thesis. Laboratory of Paleabotany and Palynology, Utrecht. Arctic centre, Groningen. pp. 113–117.

van der Knaap W.O. 1991. Palynology of peat sections from Spitsbergen covering the last few centuries. Nordic J. Botany 11: 213–223.

Mc Peters R., Hollandsworth S., Flynn L., Herman J. and Seftor C. 1996. Long-term ozone trends derived from the 16-year combined nimbus 7/ meteor 3 TOMS version 7 record. J. Geophys. Res. 23: 3699–3702.

Moore P.D., Webb J.A. and Collinson M.E. 1991. Pollen Analysis. Blackwell Scientific, London, p. 216.

Mauquoy D., van Geel B., Blaauw M. and van der Plicht J. 2002. Evidence from northwest European bogs shows Little Ice Age climatic changes driven by variations in solar activity. Holocene 12: 1–6.

Newman P., Gleason J., Mc Peters R. and Stolarski R. 1997. Anomalously low ozone over the arctic. Geophys. Res. Lett. 24: 2689–2692.

Ovenden L. 1988. Holocene proxy-climate data from the Canadian Arctic. Geological Survey of Canada Paper: 88–122.

van der Plicht J. 1993. The Groningen Radiocarbon Calibration Program. Radiocarbon 35: 231–237.

Rozema J., Noordijk A.J., Broekman R.A., van Beem A., Meijkamp B.M., de Bakker N.V., van de Staaij J.W.M., Stroetenga M., Bohncke S.J.P., Konert M., Kars S., Peat H., Smith R.I.L. and Convey P. 2001a. Polyphenolic compounds in pollen and spores of Antarctic plants as indicators of solar UV-B: a new proxy for the reconstruction of past solar UV-B?. Plant Ecol. 154: 11–26.

Rozema J., Broekman R., Blokker P., Meijkamp B.M., de Bakker N.V., van de Staaij J., van Beem A., Ariese F. and Kars S.M. 2001b. UV-B absorbance and UV-B Absorbing Compounds (*para*-coumaric acid) in pollen and sporopollenin: the perspective to track historic UV-B. J. Photochem. Photobiol. 6: 108–117.

Rozema J., van Geel B., Björn L.O., Lean J. and Madronich S. 2002. Towards solving the UV puzzle. Science 296: 1621–1622.

Rozema J., Boelen P., Solheim B., Zielke M., Buskens A., Doorenbosch M., Fijn R., Herder J., Callaghan T.V., Björn L.-O., Gwynn Jones D., Broekman R., Blokker P. and van de Poll W. 2006. Stratospheric ozone depletion: high arctic tundra plant species from Svalbard are not affected by enhanced UV-B after 7 years of UV-B supplementation in the field. Plant Ecol. (this volume).

Rozema J., Boelen P. and Blokker P. 2005. Depletion of stratospheric ozone over the antarctic and arctic: responses of plants of polar terrestrial ecosystems to enhanced UV-B, an overview. Environment. Pollution 137: 428–442.

Rønning O.I. 1965. Studies in Dryadion of Svalbard. Norsk Polarinstitutt Skrifter 134: 1–52.

Rønning O.I. 1996. Svalbards Flora. Norsk Polar Institutt.

Short S.K. and Andrews J.T. 1980. Palynology of six middle and late Holocene peat sections, Baffin Island. Géographie physique et Quaternaire 34: 61–75.

Stolbovoi V. 2002. Carbon in Russian soils. Climatic Change 55: 131–156.

Summerhayes V.S. and Elton C.S. 1923. Contributions to the ecology of Spitsbergen and Bear Island. J. Ecol. 11: 214–286.

Steffensen E.L. 1982. The climate at Norwegian Arctic stations. Klima 5: 1–44.

Svendsen J.I. and Mangerud J. 1992. Paleoclimatic inferences from glacial fluctuations on Svalbard during the last 20,000 years. Climate Dynamics 6: 213–220.

Svendsen J.I. and Mangerud J. 1997. Holocene glacial and climatic variations on Spitsbergen, Svalbard. Holocene 7: 45–57.

Van Geel B., Buurman J. and Waterbolk H.T. 1996. Archaeological and palaeoecological indications of an abrupt climate change in The Netherlands, and evidence for climatological teleconnections around 2650 BP. J. Quaternary Sci. 11: 451–460.

Van Geel B., van der Plicht J., Kilian M.R., Klaver E.R., Kouwenberg J.H.M., Renssen H., Reynaud-Farrera I. and Waterbolk H.T. 1998. The sharp rise of Δ ^{14}C ca. 800 cal BC: possible causes, related climatic teleconnections and the impact on human environments. Radiocarbon 40: 535–550.

Williams M., Dunkerley D., De Deckker P., Kershaw P. and Chapell J. 1998. Quaternary Environments, 2nd ed. Arnold, London.

Zagwijn W.H. 1994. Reconstruction of climate change during the Holocene in western and central Europe based on pollen records of indicator species. Vegetation Hist. Archeobot. 3: 65–88.

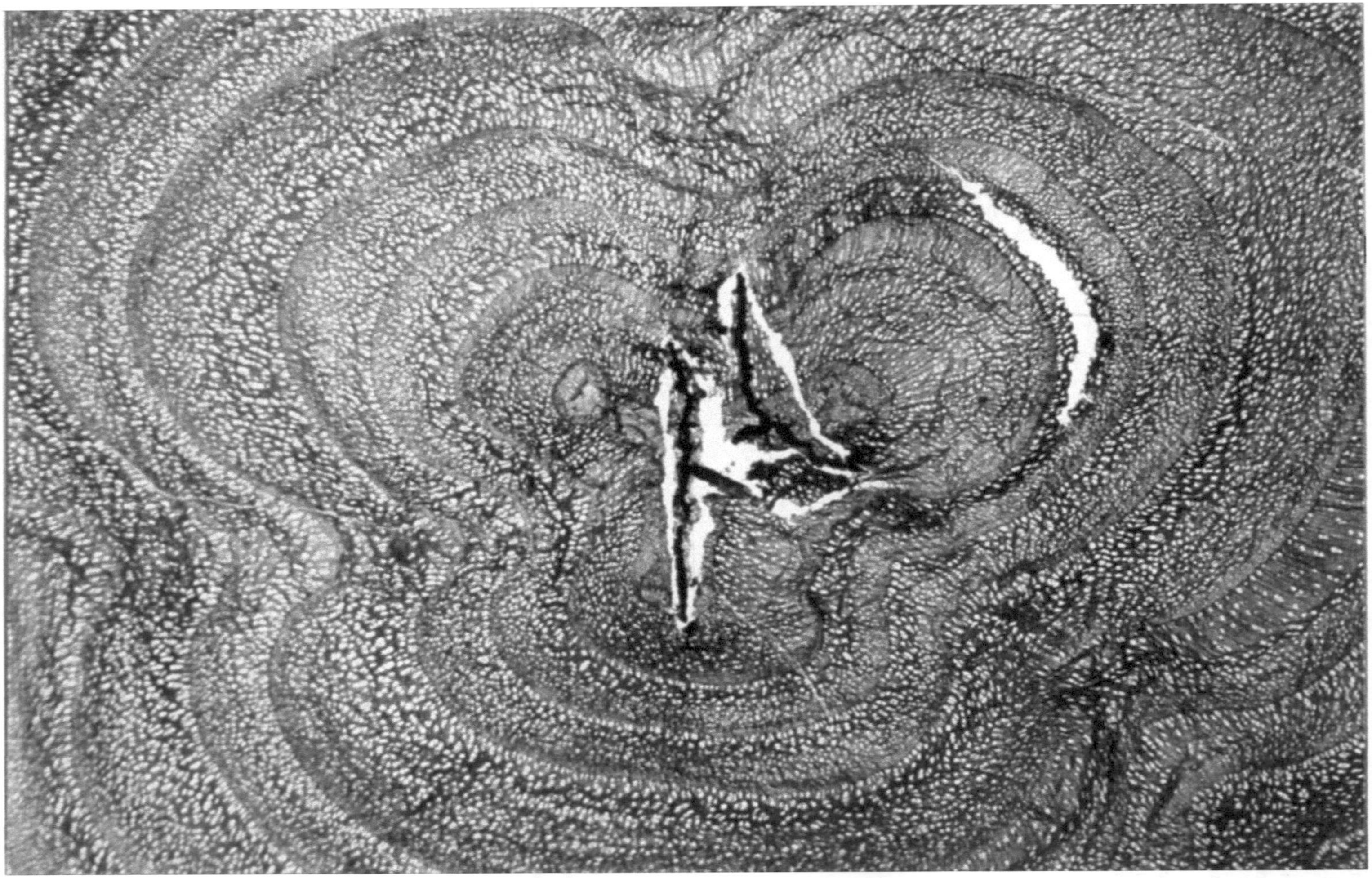

Photo. Transverse section of a piece of permineralised Tertiary wood showing well preserved anatomy classically used in palaeoclimate analysis.

Plant Ecology (2006) 182:175–195
DOI 10.1007/s11258-005-9025-z

Physiognomic and chemical characters in wood as palaeoclimate proxies

Imogen Poole[1,2,3,*] and Pim F. van Bergen[3,4]
*[1]Palaeontological Museum, Oslo University, 1172, Blindern, N0318, Oslo, Norway; [2]National Herbarium of the Netherlands, University of Utrecht Branch, Utrecht University, Utrecht, The Netherlands; [3]Geochemistry, Earth Sciences, Utrecht University, 80021, 3508 TA, Utrecht, The Netherlands; [4]Shell Global Solutions, Flow Assurance, 38000, 1030 BN, Amsterdam, The Netherlands; *Author for correspondence (e-mail: i.poole@geo.uu.nl)*

Received 1 September 2004; accepted in revised form 15 December 2004

Key words: Fossil wood, Lignin, Stable Isotopes, Tree rings

Abstract

Fossil wood is both abundant and ubiquitous through geological time and space. During growth the parent plant was directly influenced by the biotic and abiotic (including climatic-) factors in the surrounding environment. The climate affects wood production in a number of ways and it is the resulting physiognomic and chemical characters that can help retrodict palaeoclimate. Physiognomic characters include those morphological and anatomical characters that in turn have enabled the use of wood characters, tree ring characters and statistical parameters (Mean Sensitivity) to determine seasonality, length and favourability of growing season, growth rates and forest productivity. Potential chemical characters discussed include (i) the preservation of wood-derived compounds (e.g. guaiacyl, syringyl *p*-hydroxyphenyl and resins); (ii) degree of lignin degradation to determine climate induced environmental changes; and (iii) stable isotopes (δD, δ^{13}C and δ^{18}O) to help determine aspects of past climates as derived from environmental changes. The feasibility and methodology of these characters, in both angiosperm and conifer wood, are reviewed in order to establish certain safe guards, or prerequisites, such that interpretations of palaeoclimate can be as unbiased, and thus as reliable, as possible.

Introduction

Due to its omnipresence in sequences through geological time coupled with its robust nature, wood has provided researchers with a potentially rich archive of data impinging on areas as diverse as the evolution of arborescence and biodiversity to chemical taphonomy and palaeoclimate. Such data are used to glimpse past environments often quite unlike those with which we are familiar today. Fossil plants often represent the remains of shrub and canopy-forming organisms thus are particularly important in palaeoecological reconstructions. Palaeoecology, in its loosest sense, circumscribes all biotic and abiotic factors that existed together during geological time. One such abiotic factor, which has caught the attention of the scientific community in recent years, is past (i.e. historical to Quaternary) and palaeo- (i.e. pre-Quaternary) climate (both referred to as 'palaeoclimate' herein for convenience unless otherwise stated). This biological input to our understanding of palaeoclimate presents an important parameter that can be used to help calibrate computer-generated models based on physical data (Taylor et al. 1992; Poole et al. 2003).

Before we undertake a critical review of the potential aspects of wood as a palaeoclimate proxy it is necessary to introduce the concept of variation which underpins data interpretation obtained from living organisms. Unfortunately living organisms do not readily submit to our attempts to neatly categorise them and the natural variation expressed, whether at the chemical or physiognomic level, needs due consideration such that interpretations can be as reliable as currently possible (Poole 2000). It would be misleading to imply that variation has not been considered. In fact variation in observed wood characters has been known since 300 BC through the recordings of Theophrastus (Carlquist 1988). Moreover, disciplines have arisen from acknowledging variation such that variation is used to obtain reliable, rigorous data. For example dendrochronology (see below) utilises anatomical variation whilst forestry science exploits the genetic variation in commercially important technological characters of wood. Other disciplines, including palaeoclimatology, have been slower to recognise and account for variation thus rendering comparisons more tenuous. One of the most important sources of variability in wood is related to its ontogenetic age and growth position within the plant, i.e. whether it is derived from trunk, branch, stump or root material (Jane 1962; Chapman 1994; Gartner 1995; Falcon-Lang 2005a), in addition to chemical variations even at the cell wall layer level (van der Heijden and Boon 1994; van der Heijden et al. 1994). Such intra-tree variability can often be greater than the average values for inter-tree variability from the same or even different sites (Larsen 1967). This highlights the fundamental importance of providing as much information about the fossil material under investigation as possible.

This chapter is subdivided into two parts. Section 'Macroscopic characters of wood' reviews the macroscopic features of wood used in palaeoclimate reconstructions whereas Section 'Chemical characters of wood' focuses on specific chemical characters of wood to this end. When discussing the applicability of the different aspects of wood characters as proxies for palaeoclimate we bear in mind the potential sources of bias and outline means to circumvent them.

Macroscopic characters of wood

Taxonomic based methodology

Latitude is one of many factors that help determine the climate within an area. In turn the climate is one of the major determinants for plant growth. Studies of palaeoclimate have traditionally used a bioclimatic approach (Kershaw and Nix 1988) utilising the nearest living relative (NLR) principal (e.g. Mosbrugger 1999). This relies on the premise that there has been little change in the climatic preference of a particular plant fossil taxon relative to its NLR today. This approach is therefore linked to botanical affinities and nomenclature of identified fossils (Collinson 1986; Mosbrugger 1999). Critics amongst us would be quick to point out that ecological tolerances of plants are unlikely to have remained unchanged during geological time and for isolated organs, such as wood, identification to a living relative is tenuous without attached reproductive structures not to mention problems introduced by sampling strategies and taphonomical biases. Indeed there are very few cases published where the identity of fossil wood can be confirmed by the presence of attached reproductive organs (e.g. Herendeen 1991). To help overcome this Mosbrugger and Utescher (1997) refined this bioclimatic methodology in order to quantify terrestrial climate reconstructions in the Tertiary based on the bioclimatic appraisal of fossil floras. They coined their method the Coexistence Approach (CoA). This approach can be made relatively more rigorous by restricting the extrapolation to the recent past.

Such bioclimatic approaches have usually focused on fruit and seed or leaf floras (e.g. Reid and Chandler 1933; Liang et al. 2003) but there is no reason why it cannot be applied to fossil wood floras (Poole, Cantrill and Utescher 2005). Sceptics may suggest that a wood flora, as opposed to a palynological flora, would not accurately represent the parent vegetation and thus climate. However, a study of trunks greater than 10 cm in diameter lying in palisadas along the river Manu (Peru) showed a statistically highly significant representation of the local vegetation when the material was identified to family or genus level (Poole, Silman and van Bergen unpublished data). This indicates that although not necessarily complete (but certainly complements and supplements data derived from other organ floras) and probably representing a relatively local, rather than regional flora, fossil wood alone goes a long way to providing a good representation of the vegetation from which it was derived and thus, by extrapolation, the palaeoclimate.

As with any organ a bioclimatic approach hinges on a good working knowledge of wood anatomy of both living and fossil material such that the fossil can be identified to a 'nearest living relative'. Applying this method to fossil wood floras is routinely carried out to study former ecosystems and assess changes in vegetational communities over time (Figueiral and Mosbrugger 2000). For example Wheeler and Manchester (2002) published a study of the diverse fossil wood assemblage from the Clarno Nut Beds. They concluded a probable warm temperate to subtropical climate regime existed over Oregon during the Middle Eocene – a conclusion substantiated by the NLRs of the identified woods which are now growing in eastern/southeastern Asia. Poole et al. (2001) had adopted a bioclimatic approach when they determined that the cool temperate Valdivian rainforests of southern Chile are the modern analogue to Tertiary Antarctic vegetation. Independent climate analyses (Hunt and Poole 2003) of the floras corroborated a prevailing cool temperate climate regime. Moreover, an understanding of both the sedimentary environment and the biotic makeup of the vegetation enabled Poole et al. (2001) to determine that ecological disturbance, in this case volcanic activity, was controlling the vegetational shifts and not climate change to which the shifts had been attributed. This illustrates the importance of looking at all available data through interdisciplinary approaches to ensure accurate and precise palaeoecological and thus palaeoclimatic interpretations.

Physiognomic based methodologies

Physiognomic based methodologies to determine palaeoclimate assume that morphological or anatomical characters had the same relationship with climate in the past as they do today without needing reference to systematic status of the plant (Liang et al. 2003) but simply relying on relative differences in the character states. Ideally fossil wood studied would be preserved *in situ* allowing the examination of forest density and productivity as well as analysing morphological and anatomical characters for palaeoclimatic signals. Unfortunately such instances are rare such that isolated wood organs are predominantly the focus of palaeoclimatic studies.

Morphological characters

Most plant fossil assemblages are allochthonous having arrived at their site of incorporation into the geological record after being transported from their parent vegetational community. *In situ* fossil plants, although more rarely preserved, are considerably more valuable in palaeoenvironmental analysis. Morphological characters of *in situ* material can provide indications of the hydrological status of the palaeoenvironment and in particular the seasonality and intensity of precipitation (Fielding and Alexander 2001). Fossil trees in the deposits of variable-discharge rivers are generally low in diversity, comprising pioneer plant assemblages and dominated by individuals of limited diameter reflecting typically high mortality rates of plants in such environments (Fielding et al. 1997; Fielding and Alexander 2001). Moreover, morphological adaptations to periodic submergence may be determined from characters such as trailing roots (conversely thick, spongy roots would be indicative of adaptations to dry seasons), reclined habit, multi-stemmed forms, concentration of individuals in flow-parallel linear grooves, thick, spongy bark (e.g. Hook 1984; Fielding and Alexander 2001; van der Burgh in Poole et al. submitted). In extreme cases following long periods of coverage by standing water or sediment, trunks may have flask-shaped basal portions and/or a second root layer on top of the newly deposited sediment (Stone and Vasey 1968; Jefferson 1982). These morphological specialisations would be in response to the stresses imparted through the need to successfully inhabit a riverbed environment and to ensure access to moisture during the dry season. However, palaeoclimatic interpretation from such assemblages obviously relies on fossil tree morphology in conjunction with analysis of the associated sediments (Demko et al. 1998).

Anatomical characters

Data provided by fossil wood anatomical characters have been used extensively to reconstruct climates and determining climate change during the geological past. Both conifer and angiosperm woods have been employed to this end. The activity of a particular layer of cells, the vascular cambium, located in the bark is responsible for wood formation. Factors influencing cambial activity (see Creber and Chaloner 1990 and references therein) usually mediated by growth hormones (e.g. Aloni

2001) affect the characteristics of the wood cells produced. During unfavourable environments the cessation of cambial cell division leads to the formation of growth rings. Larger diameter cells are produced at the beginning of the growing season (early wood) followed by smaller diameter cells (late wood) as the growing season draws to a close (Figure 1). These anatomical changes help to emphasise the environmentally related growth characteristics within both conifer and angiosperm wood. The wood structure of conifers is relatively simple with a single cell type, tracheid, for both support and water conduction (Figure 1b) whereas dicotyledonous angiosperm (dicot) wood exhibit a division of labour (Wheeler and Baas 1993). Perforate vessel elements undertake water conduction whereas imperforate fibres provide support (Figure 1c; Jane 1962; Wilson and White 1986; Carlquist 2001 and references therein for further details on wood anatomy and function). Dicot wood is further complicated by the diversity in size, arrangement and minute features of vessel elements which have considerable homoplasy (Wheeler and Baas 1993). Based on extensive floristic surveys certain wood anatomical features can be characteristic of certain ecological categories (e.g.

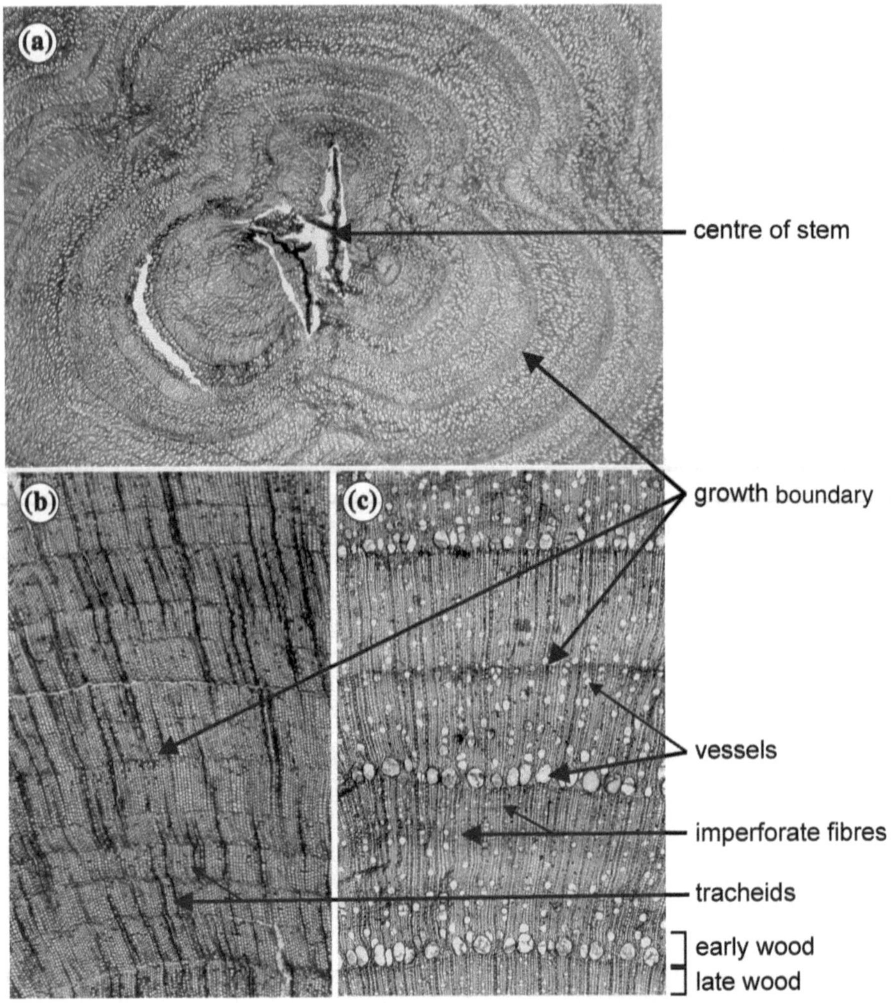

Figure 1. Tertiary fossil wood. (a) Permineralised branch wood from a dicotyledonous angiosperm from Antarctica showing the distinctive growth rings. (b) Mummified conifer wood from Tasmania, Australia showing the relatively simple structure of conifer wood. Growth rings are demarcated by ~2 rows of late wood tracheids. (c) Permineralised angiosperm wood from Antarctica showing growth rings defined by the increase in wall thickening of the imperforate tracheary elements and emphasised by the increased vessel diameter in the early wood relative to the late wood.

Carlquist 1975, 1977, 2001; Baas 1976, 1986; Wiemann et al. 1998, 1999), which in turn are often climate influenced.

Tree rings. Tree rings are probably the most well know anatomical feature used to make generalisations concerning palaeoclimate since tree rings, length and seasonality of the growing season are intimately related. Tree ring analyses for palaeoclimatological inferences rest upon uniformitarian deductions based on observations of (predominantly) conifers from the present boreal temperate realm (Brison et al. 2001) regardless of the incoherent growth rings patterns exhibited by woods growing in warm climates (Jacoby 1989; Worbes 1989, 1995; 1999; Borchert 1999). Understanding sequences of growth ring forms the basis of Dendrochronology which includes the discipline Dendroclimatology. By judicious sampling and the use of rigorous statistical procedures dendroclimatology has provided unique insights into the nature of past climatic variability most significantly from an interannual to centennial time-scale (Briffa et al. 1998) from woods originating from temperate regions. These climate changes reveal themselves as characteristic rings which can be cross matched to overlapping ring series from different trees thereby extending the climate record (ideally) uninterrupted into the past. The success of using growth rings to determine (cool temperate) climate from the past has led palaeoclimatologists to use this methodology on older material even though some researchers consider this approach questionable for pre-Quaternary fossils (e.g. Ammons et al. 1987; Falcon-Lang 2003, 2005b).

Although growth rings in essence appear to be a rich archive of palaeoclimatic information, careful evaluation of the material for study is paramount to ensure that the climate signal is not being confused with expressions of variation which in turn would affect wood growth and growth ring characters (see Creber and Chaloner 1984; Briffa et al. 1998; Tardif et al. 2003). Problems arise from restricted biotic (including genetic) and abiotic effects often surpassing the climate effect. Moreover, not all tree species conveniently lend themselves to such an approach (see Schweingruber 1993 for listings). Within the tropical realm there is no simple correlation between climate and the development of growth rings since the ability to produce rings is primarily determined by the genetic makeup of the individual species (e.g. Chowdhury 1964; Tomlinson and Craighead 1972; Jacoby 1989; Savidge 1996) and exposure to inundation resulting in cessation of growth (Worbes 1985, 1989). Studies of inter-tree variability find that within a given tree many anatomical parameters show significant relationship with (i) taxonomic status (see Schweingruber 1993), (ii) ontogenetic age (fast growing young trees have wider, more variable rings compared with older more slow growing trees which can have missing rings), and (iii) organ of origin: branch wood can be affected by unequal gravitational forces producing 'reaction' and 'compression' wood, and root wood and twig material have wider growth rings than trunk wood (Jane 1962; Zobel and van Buijtenen 1989; Chapman 1994; Falcon-Lang 2003, 2005a). Ring patterns are only meaningful if derived from known organs seen in relation to the whole diameter since ring boundaries can become gradually narrower and fainter when traced around the trunk, some do not even make a full circuit; rings can also 'wedge out' from other rings. This results in confused growth interruptions (see Larson 1956 for physiological explanations) with a growth ring series from one radius not obviously relating to a series complied from another radius from the same cross section (Ash and Creber 1992).

In situ fossil tree stumps provide relatively rigorous, cross-dateable datasets to retrodict palaeoclimate at a particular locality even if such chronologies are "floating" (Jefferson 1982; Ammons et al. 1987; Gregory-Wodzicki 2001). Where such studies are possible, biologically induced growth cycles can be eliminated and sedimentological evidence can be used to help determine (local) environmental (e.g. Falcon-Lang et al. 2001) conditions (Ammons et al. 1987) ensuring more rigorous palaeoclimate interpretation. Since *in situ* material is rare, isolated wood specimens become the focus of attention for palaeoclimate analysis.

When extending the growth ring analysis approach to pre-Quaternary isolated wood material ideally a number of prerequisites need to be met:

(i) assemblages need to be of an adequate sample size (ca. >20 trees) and taxonomically diverse (cf. Wiemann et al. 1998).
(ii) specimens should be identifiable in terms of taxon and ontogenetic age.

(iii) organs should be of complete cross section and have similar origins (with preference for trunk wood) and excluding the inner xylem near the pith to overcome problems of rapid initial growth, confused growth interruptions and unequal gravitational effects (Fritts 1976; Zobel and Buijtenen 1989).
(iv) taxa should be 'reliable' (see Schweingruber 1993; Falcon-Lang 2000a; Falcon-Lang and Cantrill 2000; Brison et al. 2001).
(v) conifer taxa need to have known leaf longevity/retention times (Falcon-Lang 2000a, b).

Unfortunately with many fossil woods all these prerequisites cannot be fulfilled but adequate reference to both the specific taxon and specimen ontogeny, coupled with a significant but not unrealistic sample size from localities with a tight stratigraphic-, and thus temporal, framework are simple additional precautions that can help distinguish climate signals from background noise. This can reduce the chances of introducing fundamental error and ensure the possibility of legitimate comparative data analyses. So with these prerequisites in mind we explore and discuss the various approaches that have been used to interpret palaeoclimatic factors using growth rings.

Mean sensitivity: To overcome the aforementioned biases introduced through intrinsic and extrinsic factors, statistical parameters have been devised for modern and Quaternary material. The common signal of the impact of any external factor, most likely climate, is assumed to influence the growth of all the trees in the same way with the intrinsic variations exerting a different affect on the tree-ring series. One such statistical parameter that has been adopted for palaeoclimatic research is the 'sensitivity' of tree rings (Creber 1977; Creber and Francis 1999) and in particular the Mean Sensitivity (MS) index (Douglass 1928; Fritts 1976, 1991). MS provides a measure of the variability in growth from year to year and thus an indication of the limitations to growth within the environment. In order to use MS, ring width measurements need to be standardized (Fritts 1976, 1991; Monserud 1986) to remove large amounts of the nonclimatic variations (i.e. ring geometry, age, local stand conditions, site factors, individual tree histories, etc; Fritts 1991; Schweingruber 1993) unique to individual trees and sites and preserve a large part of the climatic variation in ring-width measurements common to trees from a particular region (Fritts 1991).

Mean Sensitivity has been used to help interpret palaeoclimates during the Tertiary and Mesozoic (e.g. Jefferson 1982; Francis 1986; Ammons et al. 1987; Morgans 1999; Morgans et al. 1999; Francis and Poole 2002) even though the standardization procedure has only been implemented in one study to our knowledge (Kumagai et al. 1995). Recent work now shows the use of MS to be inappropriate for both modern (Strackee and Jansma 1992) and fossil (Chapman 1994; Brison et al. 2001; Falcon-Lang 2005a,b) material because of the biases outlined above.

Length and favourability of growing season: Ring width patterns in fossil wood have been used to indicate the length and favourability of the growing season. Consistently wide rings in large diameter fossils can be interpreted as having undergone relatively rapid growth under a favourable climate. The early wood-late wood ratio can help interpretations of length of growing season. Longer, more favourable growing seasons lead to the production of a higher percentage of late wood and a wide zone of early wood followed abruptly by a narrow zone of small, dense late wood cells (cf. Creber 1977; Koizumi et al. 2003; Savva et al. 2003). Unfavourable climatic factors such as fire, frost, drought, increased light towards the end of the growth season (but also biotic factors such as defoliation by insects) can result in a characteristic gradual decrease and subsequent increase in cell diameter, which is not always continuous around the circumference of the tree (but see Schulman 1938 for exceptions), manifesting itself as a 'false' or 'traumatic' ring. However, false rings can be highly localised and often restricted to juvenile wood, upper trunk and lateral branches (Chapman 1994) such that a presence/absence record can result in erroneous 'climatic' indications. Additional biases will arise if spurious rings of collapsed, compacted and flattened cells (formed during burial and diagenesis) are not identified (Chaloner and Creber 1973). Cell characters within a tree-ring are also thought to provide information relating to environmental conditions, such as daylight regime within a growing season at high latitudes (Creber and Francis 1999). Analysis of these characters in fossil wood, particularly if they originate from localities where there is no modern analogue for comparative analysis such as the high latitudes, is problematic. For example ring width in fossil woods from the southern high latitudes cannot be related to climatic favourability because a continuous light regime would have ensured elevated growth rates during the growing season (Falcon-Lang et al. 2001).

Seasonality: The markedness of the ring boundary (in particular percentage late wood) has been used as an indicator of climatic seasonality (in precipitation, temperature, etc.) or as a measure of the favourability of growing conditions towards the end of the growing season (Creber and Chaloner 1984; Francis 1986; Ash and Creber 1992; Yao et al. 1994; Keller and Hendrix 1997; Francis and Poole 2002). A high proportion of woods in a fossil assemblage with indistinct growth rings infers a climate that is not highly seasonal (Wheeler and Manchester 2002) and *vice versa*. However, such extrapolations are drawn from similarities between trees growing at (cool temperate) mid-latitudes today (Vetter and Botosso 1989; Worbes 1989; Lindorf 1994). A combination of factors, such as seasonality in precipitation and (freezing) temperature, alongside other biotic factors (including leaf longevity) and abiotic factors mentioned above can all affect ring markedness (Greguss 1972; Woodcock 1994; Woodcock and Ignas 1994; Falcon-Lang 2000a). LaMarche (1982) went as far to say that in modern conifer woods ring markedness appeared to be poorly correlated with climate since they evolved under high carbon dioxide levels and are still genetically adapted to such conditions.

Deciduous angiosperm plants produce woods of a ring- and/or semi ring porous nature today (Figure 1c; Wheeler and Baas 1991). Therefore, deciduousness could be correlated to seasonality in, for example precipitation, temperature or daylength (Wheeler and Manchester 2002). The earliest know ring porous wood is described from the Cretaceous of Antarctica (Poole et al. 2000) suggesting that this may initially have evolved as an adaptation to pronounced seasonality in day length.

Bias can be introduced because there is a continuum in growth ring boundary markedness and porosity (in dicots) in both extant and fossil wood. Consequently there is not always agreement on the assigned category (e.g. 'ring porous' or 'semi ring porous', etc.) especially over the mid range of this continuum (Wheeler and Manchester 2002). Overcoming this problem and ensuring consistency in dicot woods might hinge on using continuous vessel characters rather than assigning discrete categories (Woodcock 1994).

Growth rates and forest productivity: Forest productivity estimates can be obtained from *in situ* stump material using mean ring increment, mean stump diameter and tree density data (see Creber and Francis 1999; Henry and Aarssen 1999) have been successfully applied to *in situ* forests (e.g. Ash and Creber 1992; Falcon-Lang et al. 2001; Gregory-Wodzicki 2001). Conclusions drawn from such studies become more rigorous still if associated palaeosols are present and interpreted. Relative productivity within a forest can be related to the relative solar input provided the palaeolatitude of the fossil forest is known (see Henry and Aarssen 1999; Creber and Francis 1999). Using isolated fossil material and small diameter material to determine growth rates is not realistic since the biases outlined above become overriding.

Additional dicotyledonous wood anatomical characters. Just as sites that have many species in common can be inferred to have similar climates, sites that have plants with similar anatomical characters could also suggest a similar climate. Modern wood anatomical features specific to dicots have been related to a variety of ecological, and thus climatic, factors (e.g. Baas and Schweingruber 1987; Baas 1973, 1986; Carlquist 1975; den Outer and van Veenendaal 1976; Wheeler and Baas 1991, 1993; February 1993; Woodcock 1994; Woodcock and Ignas 1994; Wiemann et al. 1998; Terral and Mengüal 1999; Baas et al. 2004). A survey of the published fossil dicotyledonous wood records reveal considerable potential for using features of dicot woods as an additional tool in the study of palaeoclimates (Wheeler and Baas 1991; Baas et al. 2004). However, this was only realised when a better understanding was gained into the relationship between qualitative and quantitative wood anatomical data from modern wood and prevailing climate. Temperature-related climate variables, particularly mean annual temperature, yield good correlations with anatomy and were best predicted by two or more wood anatomical characters (Wiemann et al. 1998, 1999, 2001). Using the formulas devised by Wiemann et al. (1998) palaeoclimate can been determined (to within accuracy limits) (Wiemann et al. 1999; Wheeler and

Manchester 2002; Francis and Poole 2002; Poole et al. submitted). This approach is applicable to Quaternary and Tertiary material since the evolution of wood anatomical features in angiosperms had already been set (Wheeler and Baas 1991; Baas et al. 2004). Adopting these formulae to Cretaceous material introduces the bias of anatomical adaptations since many dicots had only evolved to the shrub- or small tree habit.

In summary anatomical characters, although seemingly a rich source of climate information, will only provide meaningful climate data when derived from fossil material that have met the sample prerequisites. To date, conifer wood has been extensively used to retrodict palaeoclimate. However, recent evidence (Falcon-Lang 2003, 2005a,b) suggests that the variability in modern conifer wood obscures any palaeoclimatic signal except where sample size is very large and sample taxonomy and ontogenetic age is constrained. Such conditions are unlikely to be met in fossil studies rendering the use of quantitative tree ring parameters as indicators of pre-Quaternary climates highly questionable. The use of angiosperm characters to determine palaeoclimate is in its infancy but there is strong evidence to suggest that the quantitative combination of characters might prove to be successful. However, much more data is still required which quantitatively relates modern dicot anatomy to prevailing climate from different ecological habitats.

Chemical characters of wood

Wood is a complex mixture of distinct chemical compounds that in principal can be used in palaeoclimatic studies. The bulk of wood normally found in the fossil record is secondary xylem and hence this section will focus on the chemical constituents present in this tissue type. Secondary xylem is based primarily on ligno-cellulose complexes. The main constituent is normally the polysaccharide holocellulose (approx. 60–70%) with a lesser amount of lignin (30–40%). In addition, in certain woods solvent extractable compounds, in particular resins, can also contribute up to ca. 5%. It should be noted however, that lignin is not solely derived from secondary xylem but can be found in many other plant organs, including leaves (e.g. van Bergen et al. 1998) and propagules (e.g. Boon et al. 1989; van Bergen et al. 1995; 2004).

Holocellulose can be subdivided further into the cellulose and hemicellulose fractions, with the latter considered to be the bridging units linking cellulose fibres to the lignin macromolecule. Generally, hemicelluloses contribute about 10–20% to the overall ligno-cellulose complex. Cellulose is a large polymeric molecule based solely on glucose monomers, whereas the hemicelluloses are much smaller units (oligomers) based on a variety of monosaccharides including xylose, mannose and glucose. Xylose is normally the most abundant part of hemicellulose in both dicotyledons and conifers. The difference in chemical composition between these two taxa is mainly based on the abundance of mannose which is often abundant in conifers and much less important in dicotyledons.

In contrast to the chemically well-described polysaccharide component, lignin is a relatively less well-defined aromatic macromolecule considered to be present as a more amorphous matrix surrounding the cellulose microfibres. The building blocks of lignin are based on C9 units: a benzene ring (C6) with an attached C3 side chain (Appendix I). Units are linked through an oxygen atom attached to the benzene ring to one of the carbons of the side chain (Appendix I). The most common links are the β-O-4 and the α-O-4. The benzene ring can contain a number of additional methoxyl groups ($-OCH_3$), leading to different basic lignin units. Units with no methoxyl groups are named *p*-hydroxylphenyl (P-units), those with 1 methoxyl adjacent to the O are called guaiacyl (G-units; also 2-methoxyphenol) and those with two methoxyl groups on either side of the O are syringyl (S-units; also 2,6-dimethoxylphenol). During lignin formation the individual monomeric lignin units, [*p*-coumaryl alcohol (P), coniferyl alcohol (G) and sinapyl alcohol (S); Appendix I] present as alcohols with an additional free OH group in the side chain, are linked at random to form three-dimensional networks.

Different taxa form different monomers leading to distinct chemical compositions of the lignin. Lignin in conifer is based solely on the G units, whereas in all angiosperms the lignin contains at least both G and S although the degree of G vs. S differs dependent on the position within either the secondary xylem or even the cell wall. Amongst the angiosperms, variations are also observed:

monocotyledons and legumes contain substantial amounts of P units that are much less important in the non-legume dicotyledons. In addition, monocots and legumes often also contain substantial amounts of cinnamic acids (*p*-coumaric- and ferulic acid) – compounds very closely related to lignin monomers, but which are mainly ester linked.

Although less ubiquitous than ligno-cellulose, resins can also be found as part of fossil wood remains, in particular those of conifers. Resins of most conifers are based on diterpenoids (C_{20} units) whereas the angiosperms contain resins comprising mainly triterpenoids (C_{30} units) (e.g. Otto and Wilde 2001; Langenheim 2003). Resins can lock up, directly or indirectly, information that might be used for palaeoclimatic reconstructions. As with lignin, chemical differences exist mainly at the angiosperm and conifer level, but further variations occur among the conifer taxa (see Otto and Wilde 2001 for a recent review).

Wood-derived compounds as proxies for palaeoclimate

Although differences in the actual molecular composition of ligno-cellulose might be climate dependant, we know of no studies applying changes in chemical composition taken directly from fossil wood material. One major complication for such an approach is that the chemical composition of fossil wood, as with the physiognomic characters, may vary depending on ontogenetic age of the material as well as original positioning within the tree. Nevertheless wood-derived compounds can be used to assess changes in vegetation composition and/or depositional setting which in turn can be linked to climatic changes.

Climate induced vegetation changes

Lignin-derived markers have been used to evaluate changes in vegetation composition which in turn were believed to be climate driven (Hedges et al. 1982; Goñi 1997; Sheng Hu et al. 1999). However, the use of molecular markers only indirectly reflects climate changes. Changes in the ratio of guaiacyl to syringyl (G/S) and *p*-hydroxyphenyl to guaiacyl, (P/G) have been applied to organic matter obtained from various settings including lake sediments and deltaic samples. For example, a trend from a warm to a cold period could lead to a change from an angiosperm dominated vegetation to a conifer dominated vegetation which might subsequently be replaced by a grass dominated ecosystem during the coldest period. In such a case one would expect to find first an increase of G/S (increase of conifers over angiosperms) and subsequently an increase of P/G [increase of monocotyledons – i.e. grasses – relative to conifers].

Climate-induced vegetation changes have also been reported based on wood-derived resins compounds. These were mainly related to changes in the composition of resin-derived markers found in sediment cores. For example, a global cooling trend was inferred from a distinct increase in the relative abundance of retene, a diagenetic compound known to be derived from diterpenoid conifer resins (Simoneit 1998), over cadalene in Jurrasic sediments (van Aarssen et al. 2000; Grice et al. 2001). The increase of retene was suggested to imply larger amounts of conifer derived organic matter entering the marine realm whereby the conifers were considered to be more representative of a cooler climate (van Aarssen et al. 2000). However, an alternative explanation would be that, since this trend co-occurred with a global eustatic sea level rise, larger expanses of conifer dominated hinterland vegetation were being exposed. This illustrates the necessity to work with all available data to ensure accurate palaeoclimatic reconstructions.

Climate induced changes in depositional setting

Another application of wood derived compounds has been to study the degree of lignin degradation which can be related to climate changes. Detailed molecular studies of peat deposits, to determine differences in the level of degradation, have been evaluated in terms of changes in depositional setting (e.g. Kuder and Kruge 1998). These in turn were believed to be related to regional changes from wetter to drier periods (e.g. Kuder and Kruge 1998). The underlying reasoning was that drier periods would result in more enhanced oxic lignin degradation and thus an increase in more oxygenated lignin degradation products. Thus, ratios of various lignin degradation products can indirectly yield palaeoenvironmental information.

Palaeoclimate signals from stable isotopes in wood

In addition to the molecular composition of wood, organic compounds lock up palaeoclimate information through the presence of stable isotopes of Hydrogen, Carbon and Oxygen, i.e. δD, $\delta^{13}C$ and $\delta^{18}O$, respectively. It is beyond the scope of the current paper to review all published data in this research field with the majority focusing on plant material from the recent past. For additional detailed information the reader is referred to the excellent studies for amongst others Epstein, Ehleringer, Leavitt, Loader, Saurer, Schleser (see also http://www.ltrr.arizona.edu/~sleavitt/IsoDendroBib.htm).

Briefly, the stable isotope signals of hydrogen and oxygen measured in plant material are mainly derived from the isotopic composition of the leaf water used by the plant during photosynthesis. This in turn is affected by a number of factors including (1) the isotopic composition of the water that the roots used, which in turn is affected by precipitation and evaporation and the degree to which this water is influenced by surface and ground water sources, (2) evaporative enrichment in the leaf, (3) biological fractionation processes during the incorporation of water into organic molecules, and (4) isotopic exchange of H and O during transfer of the biosynthate from the leaf to the secondary xylem. In contrast, the $\delta^{13}C$ values of plants depend on (1) the $\delta^{13}C$ of the atmospheric CO_2 used, (2) internal and external CO_2 concentration, and (3) biological fractionation processes whereby the enzyme Rubisco fractionates strongly against ^{13}C. Using this information, stable isotopes from plant material in general have provided valuable detailed palaeonvironmental insights.

Wood, and more specifically tree rings, has also been used extensively to this end with the main emphasis being on past climate inferences from modern and Quaternary material (up to several 100,000 years; cf. Loader et al. 2003 and references cited therein). Studies focusing on palaeo-, as distinct from past-, climate are more limited and have concentrated predominantly on Tertiary and Cretaceous material (e.g. Lücke et al. 1999).

Past and palaeoclimate studies using stable isotope analyses of fossil wood have to consider inherent molecular heterogeneity of the wood, i.e. the relative abundance of the lignin and different polysaccharide fractions. It is known that these fractions often have different stable isotope compositions (Benner et al. 1987; Poole and van Bergen 2002; Loader et al. 2003). Therefore, isotope determinations of bulk wood samples alone would be directly affected by the actual molecular composition of the material and the abundance of one fraction relative to another (cf. van Bergen and Poole 2002). To overcome this phenomenon most studies use only a single fraction, in most cases cellulose, although lignin as such has also been evaluated to this end (Benner et al. 1987; Spiker and Hatcher 1987; Loader et al. 2003). However, even within the cellulose differentiations are being made and α-cellulose is suggested to be the most reliable fraction for δD, $\delta^{13}C$ and $\delta^{18}O$ measurements (Loader et al. 1997). Thus, for modern and Quaternary wood material stable isotope measurements of (α-)cellulose are preferred. However, for older specimens cellulose might not be preserved or might be chemically (van Bergen 1994) or isotopically changed (Schleser et al. 1999). In such cases research has focused on bulk wood (cf. Gröcke 1998, 2002; Gröcke et al. 1999; Hesselbo et al. 2000), isolated lignin fractions (Spiker and Hatcher 1987) or alternatively compound specific stable isotope measurements of ligno-cellulose derived markers (see later, Poole et al. 2004). Despite the fact that isolated lignin fraction(s) can provide extremely useful data, it must be noted that preferential lignin degradation within secondary xylem cell wall layers (van der Heijden and Boon 1994; van der Heijden et al. 1994) may affect the isotope composition. Therefore, in most cases it is the relative isotopic change between the individual samples, and not the absolute values, that are of importance.

Despite the numerous processes affecting the final isotope composition of wood, δD and $\delta^{18}O$ data of cellulose have yielded meaningful palaeoclimate information for modern and past specimens (e.g. Lipp et al. 1996; Switsur et al. 1996; Anderson et al. 1998; Waterhouse et al. 2002). The basic premise for this relates to the fact that during evaporation $^{1}H_2^{16}O$ is preferentially lost. Since temperature strongly affects evaporation, the H and O isotope composition can often be linked to temperature. In principal, higher temperatures may cause more evaporation leading to the remaining water, which is subsequently used by the plant, to be D and ^{18}O enriched ($\delta^{18}O$ values less negative). Hence, when less negative values are measured, in cellulose, higher temperatures can be inferred. Alternatively, humidity will affect the $\delta^{18}O$ values. For example, Sauer et al. (1997) used cellulose from a number of different tree species growing at sites with varying soil moisture to show that trees from the drier sites had less ^{18}O depleted values. Once again absolute values are less important than relative changes since the $\delta^{18}O$ values of rainwater varies across the world due to the temperature effect on precipitation.

In addition to the δD and $\delta^{18}O$ results, the $\delta^{13}C$ data of fossil wood can, in principal, provide insights into the $\delta^{13}C$ of the atmospheric CO_2 at the time the tree grew. For example the increased depletion in ^{13}C of atmospheric CO_2 since the industrial revolutions can easily be traced in trees (e.g. Wiesberg 1974). In addition, $\delta^{13}C$ data can provide information on climate-induced physiological changes such as drought-, or inundation-, related stress (Switsur and Waterhouse 1998). This is because stomatal conductance decreases during drought or inundation forcing a reduction in Rubisco fractionation leading subsequently to less ^{13}C depleted primary biosynthates compared with those formed during normal growth.

To date, we know of no studies using δD and $\delta^{18}O$ data of pre-Quaternary wood to provide palaeoclimate information and hence for the older fossil wood specimens we will focus on stable carbon isotopes. The main difference between Quaternary and pre-Quaternary wood samples relates (to a greater extent) to the changes in chemical composition during fossilization which in turn will affect the stable isotopic composition (van Bergen and Poole 2002; Poole et al. 2004). During the fossilization process the holocellulose fraction is preferentially degraded relative to the lignin fraction (Hedges et al. 1985). Even within the holocellulose fraction the hemicelluloses are normally degraded faster than the cellulose. This type of chemical taphonomy (van Bergen and Poole 2002) will dramatically affect bulk wood stable isotope measurements because of their different isotope composition of each of these three moieties (Benner et al. 1987; Loader et al. 2003). With regard to carbon, differences of ca. 5‰ exist between the three fractions, with hemicellulose being least depleted in ^{13}C whereas lignin is most depleted. If only cellulose is used to determine palaeoclimate it could be assumed that chemical taphonomy is of no significance. However, cellulose is normally not preserved in fossils older than a few million years. Moreover, recent data have implied that cellulose under artificial diagenetic conditions does undergo isotopic changes (Schleser et al. 1999). This is in accordance with detailed molecular data of fossil plant material containing cellulose which indicate that the degree of polymerisation of cellulose decreases during fossilisation (van Bergen 1994; Stankiewicz et al. 1997) which in turn affects the stable isotope composition. Furthermore, relatively low yields during cellulose extraction and the distinct yellow, as opposed to white, colour of the cellulose obtained from pre-Quaternary specimens (Schleser and Helle personal communication) imply significant chemical transformations. Hence, cellulose data of pre-Quaternary samples can be used as long as the taphonomic biases which might affect the isotopic signatures are given due consideration.

Based on the aforementioned biases it is not surprising that most studies of older wood specimens have concentrated on stable carbon isotope measurements of bulk wood samples (Gröcke 1998; Gröcke et al. 1999; Hesselbo et al. 2000). From these latter studies it has become clear that the main palaeoclimatic information to be obtained relates to changes in $\delta^{13}C$ of atmospheric CO_2. In particular Hesselbo et al. (2000) showed that large additions of extreme ^{13}C-depleted methane into the atmosphere could be traced in fossilized wood growing at that time. In addition, direct isotopic links between the terrigenous and marine realm have been examined using $\delta^{13}C$ measurements of fossil wood (Gröcke et al. 1999). However, these were less conclusive since positive isotopes excursions in the marine realm were sometimes being reflected by either positive or negative excursions in the wood data.

Another important factor when using pre-Quaternary fossil wood for palaeoclimatic inferences is preservation type (i.e. mummified, permineralised or coalified) since the processes involved during these different fossilization pathways have a major impact on the chemical composition (Stout et al. 1988, 1989; van der Heijden and Boon 1994; van der Heijden et al. 1994) which in turn affects the stable carbon isotope contents (Poole et al. 2004). In general, mummified material mostly contains organic moieties directly related to the ligno-cellulose recognized in modern wood specimens. In contrast, coalified and permineralised specimens are mainly composed of primary and secondary transformation products of lignin, i.e. catechols, phenols and naphthols (Poole et al. 2004). Hence, mummified specimens should in principal yield the most representative isotope data and as such become the preferred fossil type when using wood to this end. However, relative differences in the preserved chemical fractions such as the amount of polysaccharides vs. lignin, or even the lignin composition itself (van der Heijden and Boon 1994; van der Heijden et al. 1994), in fossil mummified wood still need to be determined as they will affect the bulk stable carbon isotope values (van Bergen and Poole 2002; Poole et al. 2004). With respect to permineralised and coalified material comparisons should be made with material of similar preservation to reduce biases introduced through chemical variation.

A further consideration is taxonomic affinity of the specimens. Stable carbon isotope differences are known to exist between modern tree taxa with the main difference usually at the angiosperm vs. conifer level (Stuiver and Braziunas 1987). Angiosperm wood is generally considered to be more ^{13}C depleted relative to conifer wood. However, the stable carbon isotope data of modern bulk wood samples have been found to be more diverse than previously suggested (Table 1) when climate and leaf morphology are not considered. If temperate rain forest conifer species (*Phyllocladus*; *Podocarpus*) are compared with temperate rainforest angiosperm species (*Eucryphia*; *Nothofagus*) then differences in $\delta^{13}C$ are still supported. Moreover, the high $\delta^{13}C$ values of the wet tropical *Iriartea* and *Attalea* may be explained by the low wettability of the leaves of these palm species. Thus, when climate change time series are studied using different taxa, the taxonomic effect may influence interpretations. Therefore, it is preferable to use the same taxon whenever possible. A palaeoclimatic study by Hesselbo et al. (2000) successfully used isotopic date and eliminated taxonomic and preservation biases by focusing on Jurassic jet material. Jet is fossilised conifer wood which acquired the physical appearance of obsidian. Specimens preserved as jet no longer contain ligno-cellulose but are solely composed of phenols (Poole, Kool and van Bergen, unpublished results) which are secondary lignin degradation products. Hence, the absolute isotope data will have been affected by associated chemical transformations. However, since all material preserved as jet will have undergone similar alterations during fossilization, and thus have the same chemical composition, the relative stable carbon isotope changes can provide meaningful palaeoclimatic information showing in this case the large changes in $\delta^{13}C$ of atmospheric CO_2. Where taxonomic status or preservation cannot be kept constant a correction for the chemical composition needs to be made and the $\delta^{13}C$ of the wood and $\delta^{13}C$ of the atmosphere back calculated (cf. Poole et al. 2004).

Table 1. $\delta^{13}C$ bulk wood values from a selection of modern woods to illustrate the range exhibited between the generally $\delta^{13}C$ depleted conifer – vs. the $\delta^{13}C$ enriched angiosperm wood.

Taxon	$\delta^{13}C_{bulk\ wood}$ (‰)
Conifers	
Phyllocladus hypophyllus	−19.3
Pinus cembroides	−21.3
Podocarpus seuoi	−23.3
Pinus cembra	−23.4
Cupressus sempervirens	−24.8
Sequoia sempervirens	−24.8
Araucaria bidwillii	−25.5
Pinus strobus	−27.3
Angiosperms	
Illicium yunnanensis	−23.4
Quercus lobra	−23.7
Eucryphia cordifolia	−25.9
Symplocus sp. (*Symplocus tenuifolia*)	−30.6
Sassafras albidum	−27.1
Nothofagus antarctica	−27.8
Astrocaryum murumuru	−28.6
Attalea excelsa	−29.3
Iriartea deltoidea	−30.2

Palaeoclimatic signals from compound specific stable isotope analyses of wood

An alternative solution to overcome the biases associated with chemical heterogeneity of fossil wood specimens is the use of compound specific stable isotopes. Over the last decade, stable isotope analyses of individual organic compounds for palaeoclimate studies have gained significant momentum. Initially these studies concentrated on $\delta^{13}C$ measurements, but compound-specific δD analyses are now also commonplace (e.g. Xie et al. 2000). The majority of data are derived from solvent soluble lipids, with only few studies evaluating macromolecular material. The main reason being that macromolecules first have to be transformed to yield meaningful smaller building blocks that can be analysed using gas chromatography. For fossil plant material this can be achieved using chemolysis or pyrolysis (van Bergen 1999).

Compound-specific analyses of fossil wood are rare and only focus on stable carbon isotopes (Poole and van Bergen 2002; Poole et al. 2004). This approach subjects the fossil wood specimens to off-line pyrolysis (for analytical details see Poole and van Bergen 2002) which yields characteristic products that can be directly assigned to either polysaccharides (mainly cellulose), lignin or lignin degradation products (Poole and van Bergen 2002; Poole et al. 2004) each with a characteristic isotopic signal. So far fossil wood samples of different taxonomic status and preservation types have been successfully studied using this approach. A typical example of an angiosperm wood specimen is shown in Figure 2a revealing an abundance of different pyrolysis products. However, it should be emphasised that in order for these, relatively polar, products to be analysed,

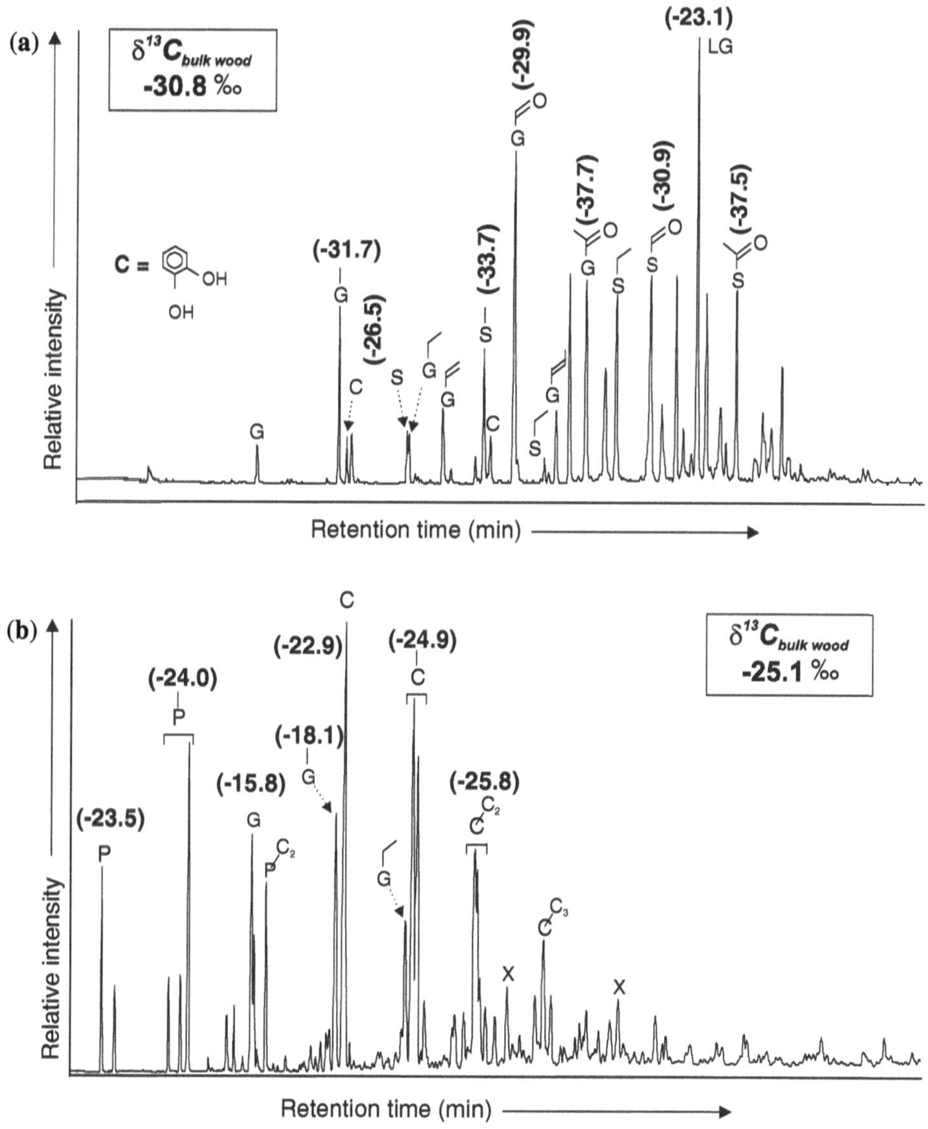

Figure 2. Off-line pyrolysate products, analysed as BSTFA derivatised compounds, of (a) a mummified Peruvian (Pliocene–Miocene) fossil wood specimen showing evidence of polysaccharides (LG) and lignin (G and S) units characteristic of well preserved lignocellulose as well as some degradation products (C); (b) a permineralised fossil (Cretaceous) wood specimen from Antarctica showing evidence of degraded lignin based on the abundance of the degradation products (C, P) and the relatively small amounts of G. Key: LG, levoglucosan; P, phenol; C, catechol; G, guaiacol; S, syringol. Side chains indicated are attached at carbon number 4 *para* to the OH group (cf. Appendix I).

they first need to be derivatised using a carbon-containing reagent, such as BSTFA. Thus the measured data have to be corrected for the carbon added during the derivatisation procedure. An important observation is the distinct isotopic difference between the cellulose-derived product, levoglucosan, and the lignin-derived methoxyphenols, i.e. G and S products. As predicted from the stable carbon isotope data of different ligno-cellulose fractions, the cellulose compounds are ^{13}C enriched relative to the lignin markers. This clearly provides evidence that the off-line pyrolysis method used leaves the isotopic signals intact. This distinct offset has been observed in all the mummified wood specimens studied to date (Poole et al. 2004). Thus for mummified specimens along a time series one could use a single compound, i.e. levoglucosan and/or 2-methoxyphenol (guaiacol) to determine excursions in $\delta^{13}C$ of atmospheric CO_2. This would circumvent variations in bulk wood measurements caused by compositional differences (i.e. relative contribution of polysaccharides over lignin; Poole et al. 2004). With regard to lignin markers the same derivative in each case must be measured because isotopic differences, relating to the number of side-chain carbons, amongst these methoxyphenols can be significant (Figure 2a).

Initially, it had been assumed by the authors that the use of the same organic compound through geological time would provide the most reliable data for palaeoclimate reconstructions. In particular, the distinct lignin marker 2-methoxyphenol (guaiacol) was considered a suitable product since this compound could be found in specimens as old as the middle Cretaceous. The off-line pyrolysis data of these older, permineralised, specimens yielded, apart from the lignin markers, mainly primary and secondary lignin degradation products, i.e. catechols and phenols, respectively, (Figure 2b). The compound specific results of these products yielded intriguing information as significant isotopic differences, of up to 17‰, were found between compound classes (methoxyphenols vs. catechols, phenols). However, in stark contrast to what was expected, i.e. lignin markers being ^{13}C depleted (cf. Figure 2b), the lignin-derived products, (guaiacol) were extremely ^{13}C enriched when compared with the catechols, the primary lignin degradation products. This phenomenon has been observed in all permineralised wood specimens that yielded lignin markers with the majority of the organic material being composed of lignin-degradation products. Lignin degradation to catechols is known to cause an isotopic shift but only of 1‰ causing the catechol to be slightly less ^{13}C depleted (Galimov 1985). Thus this process alone could not explain the guaiacol values measured. Thus, guaiacol cannot be used when the sample has been drastically changed. In contrast, stable isotope results from the primary lignin degradation product, catechol, appears more stable and thus might be a more appropriate target compound. Based on the above the isotopic composition of catechols from a series of permineralised fossil wood specimens from Antarctica was evaluated. The compound specific isotope data revealed large changes in the isotope composition that were similar for catechol as well as its methyl derivatives showing true isotope changes (Figure 3). The main shifts occurred in the middle Eocene and during the K/T boundary implying dramatic alterations of the $\delta^{13}C$ of the atmosphere.

Conclusion

Fossil wood is a unique data store of palaeoclimatic information through geological time (Figure 4). In order to reconstruct palaeoclimate data obtained from physiognomic and chemical characters of fossil material should always be used in conjunction with other evidence, either from sedimentology or in combination with other fossil material such as leaves, wherever possible. Moreover, the safeguards detailed above should be adhered to wherever possible to ensure the data obtained is not only meaningful but also useful for future studies when the understanding of modern responses of trees to climate is more fully understood. Interpretations based on physiognomic (i.e. anatomical and morphological) characters need to acknowledge evolutionary factors coupled with intrinsic natural variation that can bias interpretation. Moreover, care is needed when extrapolating ideas derived from modern systems to palaeo-environments and climates which may have no analogue on Earth today. Compound specific stable carbon isotope analyses of fossil wood specimens can yield new insights with respect to palaeoclimate reconstructions.

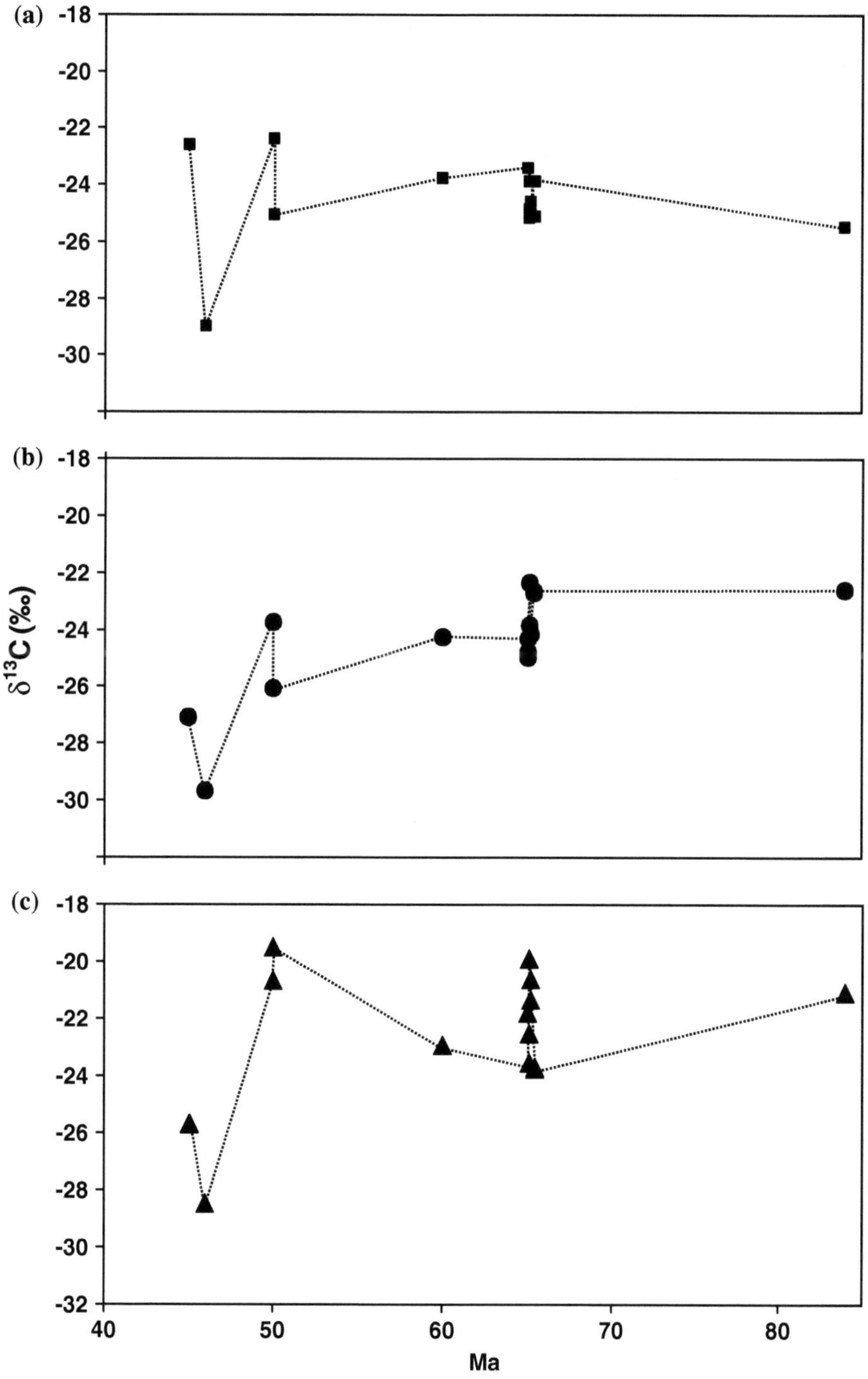

Figure 3. Changes in stable carbon isotope composition of catechol, methylcatechol and bulk organic matter from Antarctic permineralized wood specimens revealing the large shift probably related to shift in $\delta^{13}C$ of atmospheric CO_2, (a) bulk, (b) catechol, (c) methylcatechols.

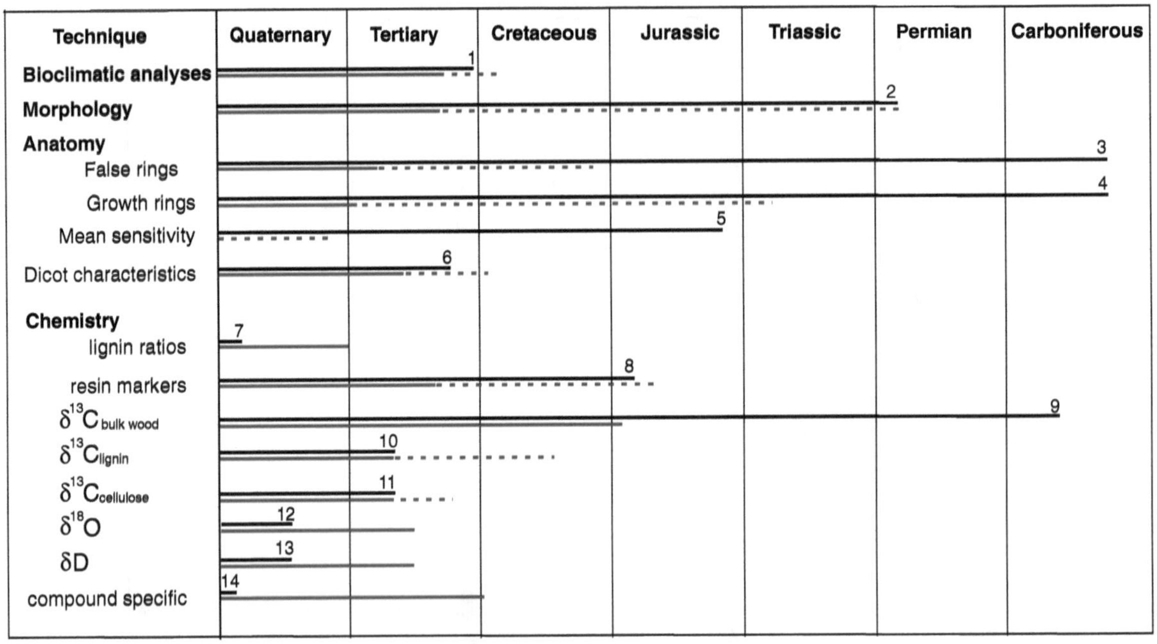

Figure 4. Diagrammatic representation of the applicability of different approaches using fossil wood to determine palaeoclimatic factors through geological time. Solid black line indicates the record from selected published literature (not exhaustive) from: 1 Poole et al. (2001, 2003). 2 Francis (1988), Fielding and Alexander (2001), 3 Falcon-Lang (1999a), Morgans (1999), Morgans et al. (1999), Taylor et al. (1992), 4 Falcon-Lang (1999a, b), Cúeno et al. (2003), Taylor et al. (1992), Francis (1984, 1986), Ash and Creber (1992), Francis et al. (1994), Jefferson and Taylor (1983), Yadav and Bhattacharyya (1996), Poole and Francis (1999), 5 Morgans et al. (1999), Francis (1986), 6 Francis and Poole (2002), Hunt and Poole (2003), Poole et al. (2005), 7 Hedges et al. (1982), Goñi (1997), Kuder and Kruge (1998), 8 van Aarssen et al. (2000), Otto et al. (1994), 9 Jones (1994), Hesselbo et al. (2000), Gröcke et al. (1999), van Bergen and Poole (2002), Poole et al. 2004, 10 Spike and Hatcher (1987), Loader et al. (2003), 11 Spiker and Hatcher (1987), Bates and Spiker (1992), Lücke et al. (1999), 12 Anderson et al. (1998), Sauer et al. (2000), 13 Lipp et al. (1996), Mayr et al. (2003), 14 Poole et al. (2004). Solid grey line indicates *certain* applicability of each methodology, dotted grey line indicates the *possible* applicability if all caveats are acknowledged and discussed (authors' opinion).

However, in addition to the prerequisites outlined for the physiognomic studies, the type of chemical preservation as well at the actual compound measured has to be taken into consideration. Therefore, although wood offers great potential for palaeoclimatic determinations, we recommend a comprehensive, interdisciplinary approach using all biotic and abiotic evidence available when trying to retrodict climates through the geological past.

Acknowledgements

This research was made possible by funding through the Natural Environmental Research Council (UK), NWO (ALW 809.32.004, NL), Percy Sladen Memorial Fund (UK) and a British Council grant JRP472. The authors would like to thank Johan Kool (Utrecht University), D. Makaham (NHN-Utrecht), M. Tabecki (British Antarctic Survey) for help with material preparation over the years. Professor G. Schleser (Research Centre Jülich GmbH), Dr G. Helle (Research Centre Jülich GmbH), Dr U. Sass-Klaassen (Wageningen University), Dr D. Cantrill (Swedish Museum of Natural History) are thanked for valuable discussions. Professor RS Hill, Dr M. Silman and the Bristish Antarctic Survey are thanked for the invitations to collect wood samples from various field locations.

Appendix

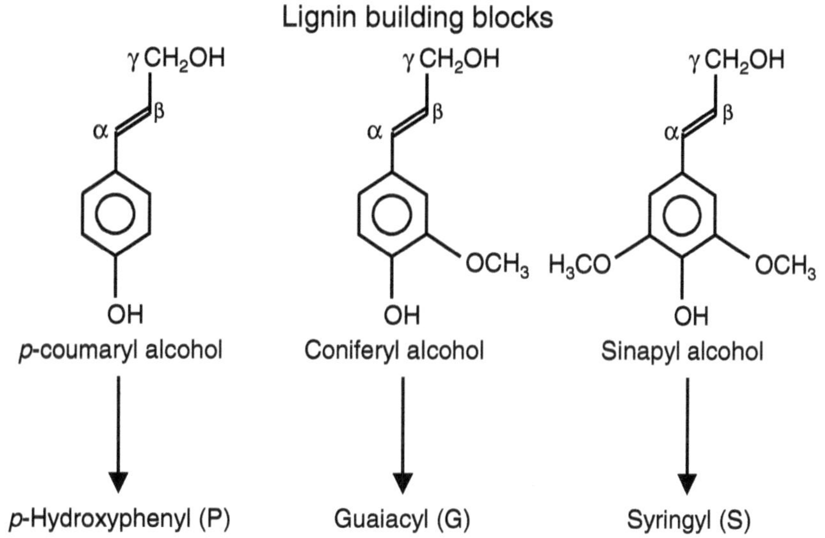

References

Aloni R. 2001. Foliar and axial aspects of vascular differentiation: hypothesis and evidence. J. Plant Growth Regul. 20: 22–34.

Ammons R., Fritz W.J., Ammons R.B. and Ammons A. 1987. Cross-identification of ring signatures in Eocene trees (*Sequoia magnifica*) from the Speciman Ridge locality of the Yellowstone fossil forests. Palaeogeogr. Palaeoclimatol. Palaeoecol. 60: 97–108.

Anderson W.T., Bernasconi S.M., McKenzie J.A. and Saurer M. 1998. Oxygen and carbon isotopic record of climatic variability in tree-ring cellulose (*Picea abies*): an example from central Switzerland (1913–1195). J. Geophys. Res. 103: 635–636.

Ash S.R. and Creber G.T. 1992. Paleaoclimatic interpretation of the wood structures of the trees in the Chinle Formation (Upper Triassic), Petrified Forest National Park, Arizona, USA. Palaeogeogr. Palaeoclimatol. Palaeoecol. 96: 299–317.

Baas P. 1973. The wood anatomical range in *Ilex* (Aquifoliaceae) and its ecological significance. Blumea 21: 193–258.

Baas P. 1976. Some functional and adaptive aspects of vessel member morphology. In: Baas P., Bolton A.J. and Catling D.M. (eds), Leiden Botanical Series 3, Wood Structure in Biological and Technological Research Wood Structure, Leiden University Press, The Hague, pp. 157–181.

Baas P. 1986. Ecological patterns of xylem evolution. In: Givnish J. (ed.), On the Economy of Plant Form and Function, Cambridge University Press, Cambridge, pp. 327–352.

Baas P. and Schweingruber F.H. 1987. Ecological trends in the wood anatomy of trees, shrubs and climbers from Europe. IAWA Bull. n.s. 8: 245–274.

Baas P., Ewers F.W., Savis S.D. and Wheeler E.A. 2004. Evolution of xylem physiology. In: Hemsley A.R. and Poole I. (eds), Linnean Society Symposium Series 21, Evolution of Plant Physiology, Elsevier, Amsterdam, pp. 273–295.

Bates A.L. and Spiker E.C. 1992. Chemical changes and carbon isotope variations in a cross-section of a large Miocene gymnospermous log. Chem. Geol. 101: 247–254.

Benner R., Fogel M.L., Sprague E.K. and Hodson R.E. 1987. Depletion of ^{13}C in lignin and its implications for stable carbon isotope studies. Nature 329: 708–710.

Boon J.J., Stout S.A., Genuit W. and Spackman W. 1989. Molecular paleobotany of *Nyssa* endocarps. Acta Bot. Neerl. 38: 391–404.

Borchert R. 1999. Climatic periodicity, phenology and cambium activity in tropical dry forest trees. Int. Assoc. Wood Anat. J. 20: 239–247.

Briffa K.R., Schweingruber F.H., Jones P.D., Osborn T.J., Harris I.C., Shiyatov S.G., Vaganov E.A. and Grudd H. 1998. Trees tell of past climates: but are they speaking less clearly today?. Phil. Trans. Roy. Soc. London B 353: 65–73.

Brison A-L., Philippe M. and Thevenard F. 2001. Are Mesozoic wood growth rings climate-induced? Paleobiology 27: 531–538.

Carlquist S. 1975. Ecological Strategies in Xylem Evolution. University of California Press, Berkeley.

Carlquist S. 1977. Ecological factors in wood evolution: a floristic approach. Am. J. Bot. 64: 887–896.

Carlquist S. 1988. Comparative Wood Anatomy Systematic Ecological and Evolutionary aspect of Dicotyledon Wood. 1st ed. Springer Verlag, Berlin.

Carlquist S. 2001. Comparative Wood Anatomy Systematic Ecological and Evolutionary aspect of Dicotyledon Wood 2nd ed. Springer Verlag, Berlin.

Chaloner W.G. and Creber G.T. 1973. Growth rings in fossil woods as evidence of past climates. In: Tarling D.H. and Runcorn S.K. (eds), Implications of Continental Drift to Earth Sciences, Academic Press, New York, pp. 425–437.

Chapman J.L. 1994. Distinguishing internal developmental characteristics from external palaeoenvironmental effects in fossil wood. Rev. Palaeobot. Palynol. 81: 19–32.

Collinson M.E. 1986. Use of modern generic names for plant fossils. In: Spicer R.A. and Thomas B.A. (eds), Systematic and Taxonomic Approaches in Palaeobotany, Clarendon Press, Systematics Association Special Volume 31, Oxford pp. 91–104.

Creber G.T. 1977. Tree rings: a natural data storage system. Biol. Rev. 52: 349–383.

Creber G.T. and Chaloner W.G. 1984. Influence of environmental factors on the wood structure of living and fossil trees. Bot. Rev. 50: 357–448.

Creber G.T. and Chaloner W.G. 1990. Environmental influences on cambial activity. In: Iqbal M. (ed.), The Vascular Cambium, John Wiley and Sons Inc., New York, pp. 159–199.

Creber G.T. and Francis J.E. 1999. Fossil tree-ring analysis: palaeodendrology. In: Jones T.P. and Rowe N.P. (eds), Fossil Plants and Spores Modern Techniques, Geological Society London, London, pp. 245–250.

Chowdhury K.A. 1964. Growth rings in tropical trees and taxonomy. J. Indian Bot. Soc. 43: 334–343.

Demko T.M., Dubiel R.F. and Parrish J.T. 1998. Plant taphonomy in incised valleys: implications for interpreting paleoclimate from fossil plants. Geology 26: 1119–1122.

den Outer R.W. and van Veenendaal W.L.H. 1976. Variation in wood anatomy of species with a distribution covering both rainforest and savannah areas of the Ivory Coast, West-Africa. Leiden Bot. Series 3: 182–195.

Douglass A.E. 1928. Climate cycles and tree growth: a study of the annual rings of trees in relation to climate and solar activity. Vol II. Carnegie Institute, Washington Publ. 289.

Falcon-Lang H.J. 1999a. The Early Carboniferous (Courceyan-Arunian) monsoonal climate of the British Isles: evidence from growth rings in fossil woods. Geol. Magazine 136: 177–187.

Falcon-Lang H.J. 1999b. The early Carboniferous (Asbian-Brigantian) seasonal tropical climate of northern Britain. Palaios 14: 116–126.

Falcon-Lang H.J. 2000a. The relationship between leaf longevity and growth ring markedness in modern conifer woods and its implications for palaeoclimatic studies. Palaeogeogr. Palaeoclimatol. Palaeoecol. 160: 317–328.

Falcon-Lang H.J. 2000b. A method to distinguish between woods produced by evergreen and deciduous coniferopsids on the basis of growth ring anatomy: a new palaeoecological tool. Palaeontology 43: 785–793.

Falcon-Lang H.J. 2003. Do tree rings in fossil woods give a palaeoclimatic signal? (Conference abstract). IAWA J. 24: 316.

Falcon-Lang H.J. 2005a. Intra-tree variability in wood anatomy, and its implications for fossil wood systematics and palaeoclimatic studies. Palaeontology 48: 171–183.

Falcon-Lang H.J. 2005b. Global climate analysis of growth rings in woods and its implications for deep time paleoclimate studies. Paleobiology 31: 434–444.

Falcon-Lang H.J. and Cantrill D.J. 2000. Cretaceous (Late Albian) coniferales of Alexander Island, Antarctica. I. Wood taxonomy, a quantitative approach. Rev. Palaeobot. Palynol. 111: 1–17.

Falcon-Lang H.J., Cantrill D.J. and Nichols G.J. 2001. Biodiversity and terrestrial ecology of a mid Creaceous, high latitude floodplain, Alexander Island, Antarctica. J. Geol. Soc. London 158: 709–724.

February E. 1993. Sensitivity of xylem vessel size and frequency to rainfall and temperature implications for palaeontology. Palaeontol. Afr 30: 91–95.

Fielding C.R. and Alexander J. 2001. Fossil tees in ancient fluvial channel deposits: evidence of seasonal and longer-term climatic variability. Palaeogeogr. Palaeoclimatol. Palaeoecol. 170: 59–80.

Fielding C.R., Alexander J. and Newman-Sutherland E. 1997. Preservation of *in situ*, arborescent vegetation and fluvial bar construction in the Burdekin river of north Queensland, Australia. Palaeogeogr. Palaeoclimatol. Palaeoecol. 135: 123–144.

Figueiral I. and Mosbrugger V. 2000. A review of charcoal analysis as a tool for assessing Quaternary and Tertiary environments achievements and limits. Palaeogeogr. Palaeoclimatol. Palaeoecol. 164: 397–407.

Francis J.E. 1984. The seasonal environment of the Purbeck (Upper Jurassic) fossil forests. Palaeogeogr. Palaeoclimatol. Palaeoecol. 48: 285–307.

Francis J.E. 1986. Growth rings in Cretaceous and Tertiary wood from Antarctica and their palaeoclimatic implications. Palaeontology 29: 665–684.

Francis J.E. 1988. A 50-million year old fossil forest from Strathcona Fiord, Ellesmere Island, Arctic Canada: evidence for a warm polar climate. Arctic 41: 314–318.

Francis J.E. and Poole I. 2002. Cretaceous and Tertiary climates of Antarctica: evidence from fossil wood. Palaeogeogr. Palaeoclimatol. Palaeoecol. 182: 47–64.

Francis J.E. Woolfe K.J. Arnott M.J. and Barrett P.J. 1994. Permian climates of the southern margins of Pangea: evidence from fossil wood in Antarctica. Soc. Petrol. Mem. 17: 275–282.

Fritts H.C. 1976. Trees Rings and Climate. Academic Press, New York London.

Fritts H.C. 1991. Reconstructing Large-Scale Climate Patterns from Tree-ring Data. A Diagnostic Analysis. University of Arizona Press, Tuscon.

Gartner B.L. 1995. Patterns of xylem variation within a tree and their hydraulic and mechanical consequences. In: Gartner B.L. (ed.), Plant Stems: Physiology and Functional Morphology, Academic Press, London, pp. 125–149.

Galimov E.M. 1985. The Biological Fractionation of Isotopes. Academic Press, Orlando.

Goñi M.A. 1997. Record of terrestrial organic matter composition in amazon fan sediments. In: Flood R.D., Piper D.J.W., Klaus A. and Peterson L.C. (eds), Proceedings of the Ocean Drilling Program, Scientific Results, Vol. 155, pp. 519–530.

Gregory-Wodzicki K.M. 2001. Paleoclimatic implications of tree-ring growth characteristics of 34.1 Ma *Sequoiaoxylon pearsallii* from Florissant Colorado. Proceedings of the Denver Museum of Nature and Science 4(1), pp. 163–186.

Greguss P. 1972. Xylotomy of Living Conifers. Akademia Kiado, Budapest.

Grice K., van Aarssen B.G.K., Jiang D., Alexander R. and Kagi R.I. 2001. Stable carbon isotpic compositions and distributions of higher plant biomarkers reflecting palaeoclimate changes during the Jurassic. 20th International meeting on organic geochemistry, Abstracts Volume 1. Université Henri Poincaré, Vandoeuvre. p. 175.

Gröcke D.R. 1998. Carbon-isotope analyses of fossil plants as a chemostratigraphic and palaeoenvironmental tool. Lethaia 31: 1–13.

Gröcke D.R. 2002. The carbon isotope composition of ancient CO_2 based on higher-plant organic matter. Phil. Trans. Roy. Soc. London A 360: 633–658.

Gröcke D.R., Hesselbo S.P. and Jenkyns H.C. 1999. Carbon-isotope composition of Lower Cretaceous fossil wood, ocean-atmosphere chemistry and relation to sea-level change. Geology 27: 155–158.

Hedges J.I., Lowie G.L., Ertel J.R., Barbour R.J. and Hatcher P.G. 1985. Degradation of carbohydrates and lignins in buried woods. Geochim. Cosmochim. Acta 49: 701–711.

Hedges J.I., Ertel J.R. and Leopold E.B. 1982. Lignin geochemistry of a Late Quaternary core from Lake Washington. Geochim. Cosmochim. Acta 46: 1869–1877.

Henry H.A.L. and Aarssen L.W. 1999. The interpretations of stem diameter-height allometry in trees: biomechanical constraints, neighbour effects or biased regressions?. Ecol. Lett. 2: 89–97.

Herendeen P.S. 1991. Lauraceous wood from the mid-Cretaceous Potomac group of eastern North America: *Paraphyllanthoxylon marylandense* sp. nov. Rev. Palaeobot. Palynol. 69: 277–290.

Hesselbo S.P., Gröcke D.R., Jenkyns H.C., Bjerrum C.J., Farrimond P., Morgans Bell H.S. and Green O.R. 2000. Massive dissociation of gas hydrate during a Jurassic oceanic anoxic event. Nature 406: 392–395.

Hook D.D. 1984. Adaptations to flooding with fresh water. In: Kozlowski T.T. (ed.), Flooding and Plant Growth, Academic Press, London, pp. 265–293.

Hunt R.J. and Poole I. 2003. Revising Palaeogene West Antarctic climate and vegetation history in light of new data from King George Island. In: Wing S.L., Gingerich P.D., Schmitz B. and Thomas E. (eds), Causes and Consequences of Globally Warm Climates in the Early Paleogene. Geological Society of America Special Paper 369, pp. 395–412.

Jacoby G.C. 1989. Overview of tree ring analysis in tropical regions. IAWA J. 10: 99–108.

Jane F.W. 1962. The Structure of Wood. A&C Black, London.

Jefferson T.J. 1982. The early Cretaceous forests of Alexander Island, Antarctica. Palaeontology 25: 681–708.

Jefferson T.J. and Taylor T.N. 1983. Permian and Triassic woods from the Transantarctic Mountains: paleoenvironmental indicators. Antarct. J. US 1983: 55–57.

Jones T.P. 1994. ^{13}C enriched Lower Carboniferous fossilplants from Donegal, Ireland: carbon isotope constraints on taphonomy, diagenesis and palaeoenvironment. Rev. Palaeobot. Palynol. 81: 53–64.

Keller A.M. and Hendrix M.S. 1997. Palaeoclimatological analysis of a Late Jurassic petrified forest, Southeastern Mongolia. Palaios 12: 282–291.

Kershaw A.P. and Nix H.A. 1988. Quantitative palaeoclimatic estimates from pollen data using bioclimatic profiles of extant taxa. J. Biogeogr. 15: 589–602.

Koizumi A., Takata K., Yamashita K. and Nakada R. 2003. Anatomical characteristics and mechanical properties of *Larix sibirica* grown south-central Siberia. IAWA J. 24: 355–370.

Kuder T. and Kruge M.A. 1998. Preservation of biomolecules in sub-fossil plants from raised peat bogs – a potential paleoenvironmental proxy. Org. Geochem. 29: 1355–1368.

Kumagai H., Sweda T., Hayashi K., Satoru K., Basinger J.F., Shibuya M. and Fukaoa Y. 1995. Growth-ring analysis of Early Tertiary conifer woods from the Canadian High Arctic and its palaeoclimatic implications. Palaeogeogr. Palaeoclimatol. Palaeoecol. 116: 247–262.

La Marche V.C. 1982. Sampling strategies. In: Hughes M.K., Kelly P.M., Pilcher J.R. and La Marche V.C. (ed), Climate from Tree Rings, Cambridge University Press, Cambridge, p. 4.

Langenheim J.H. 2003. Plant Resins: Chemistry, Evolution, Ecology, and Ethnobotany. Timber Press, Cambridge.

Larsen R.R. 1967. Silvicultural control on the characteristics of woods used for furnish. In: ANON (ed), Proceedings of the 4th TAPPI Forest Biology Conference. Pulp and Paper Research Institute of Canada, Quebec, pp. 143–150.

Larson P.R. 1956. Discontinuous growth rings in suppressed slash pine. Trop. Woods 104: 80–99.

Liang M-M., Bruch A., Collinson M.E., Mosbrugger V., Li C.-S., Sun Q.-G. and Hilton J. 2003. Testing the climatic estimates from different palaeobotanical methods: an example from the Middle Miocene Shanwang flora of China. Palaeogeogr. Palaeoclimatol. Palaeoecol. 198: 279–301.

Lindorf H. 1994. Eco-anatomical wood features of species from a very dry tropical forest. IAWA J. 15: 361–384.

Lipp J., Trimborn P., Graf W., Edwards T. and Becker B. 1996. Climate signals in a ^{2}H and ^{13}C chronology (1882–1989) from tree rings of spruce (*Picea abies* L.), Schussbach forest, Germany. In: Dean J.S., Meko D.M. and Swetnam T.W. (eds), Tree Rings, Environment, and Humanity. Radiocarbon. Department of Geosciences, The University of Arizona, Tucson, pp. 603–610.

Loader N.J., Robertson I., Barker A.C., Switsur V.R. and Waterhouse J.S. 1997. An improved technique for the batch processing of small whole wood samples to α-cellulose. Chem. Geol. 136: 313–317.

Loader N.J., Robertson I. and McCarroll D. 2003. Comparison of stable carbon isotope ratios in the whole wood, cellulose and lignin of oak tree-rings. Palaeogeogr. Palaeoclimatol. Palaeoecol. 196: 395–407.

Lücke A., Helle G., Schleser G.H., Figueiral I., Mosbrugger V., Jones T.P. and Rowe N.P. 1999. Environmental history of the German Lower Rhine Embayment during the Middle Miocene as reflected by carbon isotopes of brown coal. Palaeogeogr. Palaeoclimatol. Palaeoecol. 154: 339–352.

Mayr C., Frenzel B., Friedrich M., Spurk M., Stichler W. and Trimborn P. 2003. Stable carbon- and hydrogen-isotope ratios of subfossil oaks of southern Germany: methodology and application to a composite record for the Holocene. Holocene 13: 393–402.

Monserud R.A. 1986. Time-series analysis of tree ring chronologies. Forest Sci. 32: 349–372.

Morgans H.S. 1999. Lower and Middle Jurassic woods of the Cleveland Basin (North Yorkshire), England. Palaeontology 42: 303–328.

Morgans H.S., Hesselbo S.P. and Spicer R.A. 1999. The seasonal climates in the Early–Middle Jurassic, Cleveland Basin, England. Palaios 14: 261–272.

Mosbrugger V. 1999. The nearest living relative method. In: Jones T.P. and Rowe N.P. (eds), Fossil Plants and Spores Modern Techniques, The Geological Society, London, pp. 261–265.

Mosbrugger V. and Utescher T. 1997. The coexistence approach – method for quantitative reconstructions of

Tertiary terrestrial palaeoclimate data using plant fossils. Palaeogeogr. Palaeoclimatol. Palaeoecol. 134: 61–86.

Otto A. and Wilde V. 2001. Sesqui-, di- and triterpenoids as chemosystematic markers in extant conifers – a review. Bot. Rev. 67: 141–238.

Otto A., Walther H. and Püttmann W. 1994. Molecular composition of a leaf- and root-bearing Oligocene oxbow lake clay in the Weisselster Basin, Germany. Org. Geochem. 22: 275–286.

Poole I. 2000. Variation – nature's spanner or an unrecognised tool? Palaios Online 15: 1–2.

Poole I. and van Bergen P.F. 2002. Carbon isotope ratio analysis of organic moieties from fossil mummified wood: Establishing optimum conditions for off-line pyrolysis extraction using GC/MS. Rapid Commun. Mass Spectrom. 16: 1–6.

Poole I. and Francis J.E. 1999. Reconstruction of Antarctic palaeoclimates using angiosperm wood anatomy. Acta Palaeobot. Suppl. 2: 173–179.

Poole I., Hunt R.J. and Cantrill D.J. 2001. A fossil wood flora from King George Island: ecological implications for an Antarctic Eocene vegetation. Ann. Bot. 88: 33–54.

Poole I., Mennega A.M.W. and Cantrill D.J. 2003. Valdivian ecosystems in the late Cretaceous and early Tertiary of Antarctica as evidenced from fossil wood. Rev. Palaeobot. Palynol. 124: 9–27.

Poole I., Richter H. and Francis J.E. 2000. Gondwanan origins for *Sassafras* (Lauraceae)? evidence from Late Cretaceous fossil wood of Antarctica. IAWA J. 21: 463–475.

Poole I., Cantrill D.J. and Utescher T. 2005. Reconstructing Antarctic palaeoclimate from wood floras – a comparison using multivariate anatomical analysis and the Coexistence Approach. Palaeogeogr. Palaeoclimatol. Palaeoecol. 222: 95–121.

Poole I., Dolezych M., Kool J., van der Burgh J. and van Bergen P.F. Do stable carbon isotopes of brown coal woods record changes in Miocene palaeoecology? Palaeogeogr. Palaeoclimatol. Palaeoecol. (submitted).

Poole I., van Bergen P.F., Kool J., Schouten S.S. and Cantrill D. 2004. Molecular isotopic heterogeneity of fossil organic matter: implications for $\delta^{13}C_{biomass}$ and $\delta^{13}C_{palaeoatmosphere}$ proxies. Org. Geochem. 35: 1261–1274.

Reid E.M. and Chandler M.E.J. 1933. The Flora of the London Clay. British Museum (Natural History), London.

Saurer M., Aellen K. and Siegwolf R. 1997. Correlating $\delta^{13}C$ and $\delta^{18}O$ in cellulose of trees. Plant Cell Environ. 20: 1543–1550.

Saurer M., Cherubini P. and Siegwolf R. 2000. Oxygen isotopes in tree rings of *Abies alba*: the climatic significance of interdecadal variations. J. Geophys. Res. Atmos. 105: 12461–12470.

Savidge R.A. 1996. Xylogenesis, genetic and environmental regulation – a review. IAWA Bull. 10: 170–174.

Savva Y.V., Schweingruber F.H., Vaganov E.A. and Milyutin L.I. 2003. Influence of climate change on tree-ring characteristics of Scots pine provenances in southern Siberia (forest-steppe). IAWA J. 24: 371–383.

Schleser G.H., Frielingsdorf J. and Blair A. 1999. Carbon isotope behaviour in wood and cellulose during artificial aging. Chem. Geol. 158: 121–130.

Schulman E. 1938. Classification of false annual rings in Monterey pine. Tree-ring Bull. 4: 4–7.

Schweingruber F.H. 1993. Morphological, Anatomical and Tree-ring Analytical Characteristics of Trees Frequently used in Dendrochronology (Springer Series in Wood science). Springer Verlag, GmbH & Co, Berlin, Heidelberg.

Sheng Hu F., Hedges J.I., Gordon E.S. and Brubaker L.B. 1999. Lignin biomarkers and pollen in postglacial sediments of an Alaskan lake. Geochim. Cosmochim. Acta 63: 1421–1430.

Simoneit B.R.T. 1998. Biomarker PAHs in the environment. In: Hutzinger, O. (ed.), The Handbook of Environmental Chemistry. PAHs and Related Compounds. 3, Part I, A.H. Neilson (ed.), Springer Verlag, Berlin, pp. 175–221.

Spiker E.C. and Hatcher P.G. 1987. The effects of early diagenesis on the chemical and stable carbon isotopic composition of wood. Geochim. Cosmochim. Acta 51: 1385–1391.

Stankiewicz B.A., Mastalerz M., Kruge M.A., van Bergen P.F. and Sadowska A. 1997. A comparative study of modern and fossil cone scales and seeds of conifers: a geochemical approach. New Phytol. 135: 375–393.

Stone E.C. and Vasey R.B. 1968. Preservation of coast redwood on alluvial flats. Science 159: 157–161.

Stout S.A., Boon J.J. and Spackmans W. 1988. Molecular aspects of the peatification and early coalification of angiosperm and gymnosperm woods. Geochem. Cosmochem. Acta 52: 405–414.

Stout S.A., Spackman W., Boon J.J., Kistemaker P.G. and Bensley D.F. 1989. Correlations between the microscopic and chemical changes in wood during peatification and early coalification: a canonical variant study. Int. J. Coal Geol. 13: 41–64.

Strackee J. and Jansma E. 1992. The statistical properties of 'mean sensitivity' a reappraisal. Dendrochronologia 10: 121–135.

Stuiver M. and Braziunas T.F. 1987. Tree cellulose $^{13}C/^{12}C$ isotope ratios and climate change. Nature 328: 58–60.

Switsur R. and Waterhouse J. 1998. Stable isotopes in tree-ring cellulose. In: Griffiths H. (ed.), Stable Isotopes – Integration of Biological, Ecological and Geochemical Processes, Bios Scientific, Oxford, pp. 303–321.

Switsur R., Waterhouse J.S., Field E.M. and Carter A.H.C. 1996. Climatic signals from stable isotopes in oak trees from East Anglia, Great Britain. In: Dean J.S., Meko D.M. and Swetnam T.W. (eds), Tree Rings, Environment, and Humanity Radiocarbon, Department of Geosciences, The University of Arizona, Tucson, pp. 637–645.

Waterhouse J.S., Switsur V.R., Barker A.C., Carter A.H.C. and Robertson I. 2002. Oxygen and hydrogen isotope ratios in tree rings: how well do models predict observed values? Earth Planet. Sci. Lett. 201: 421–430.

Tardif J., Camarero J.J., Ribas M. and Gutierrez E. 2003. Spatiotemporal variability in tree growth in the Central Pyrenees: Climatic and site influences. Ecol. Monogr. 73: 241–257.

Taylor E.L., Taylor T.N. and Cúneo N.R. 1992. The present is not the key to the past: a polar forest from the Permian of Antarctica. Science 257: 1675–1677.

Terral J.-F. and Mengüal X. 1999. Reconstruction of Holocene climate in southern France and eastern Spain using quantitative anatomy of olive wood and archaeological charcoal. Palaeogeogr. Palaeoclimatol. Palaeoecol. 153: 71–92.

Tomlinson P.B. and Craighead F.C. 1972. Growth ring studies on the native trees of sub-tropical Florida. In: Ghouse

A.K.M. and Yunus M. (eds), KA Chowdhury Commemoration Volume, Tata McGraw-Hill, New Delhi, pp. 39–51.

van Aarssen B.G.K., Alexander R. and Kagi R.I. 2000. Higher plant biomarkers reflect palaeovegetation changes during Jurassic times. Geochim. Cosmochim. Acta 64: 1417–1424.

van Bergen P.F. 1994. Paleaobotany of Propagules: An Investigation combining Microscopy and Chemistry. University of London, UK, PhD thesis.

van Bergen P.F. 1999. Pyrolysis and chemolysis of fossil plant remains: applications to palaeobotany. In: Jones T.P. and Rowe N.P. (eds), Fossil Plants and Spores: Modern Techniques, The Geological Society, London, pp. 143–148.

van Bergen P.F. and Poole I. 2002. Stable carbon isotopes in wood: A clue to palaeoclimate?. Palaeogeogr. Palaeoclimatol. Palaeoecol. 182: 31–45.

van Bergen P.F., Collinson M.E., Briggs D.E.G., de Leeuw J.W., Scott A.C., Evershed R.P. and Finch P. 1995. Resistant biomacromolecules in the fossil record. Acta Bot. Neerl. 44: 319–342.

van Bergen P.F., Flannery M.B., Poulton P.R. and Evershed R.P. 1998. Organic Geochemical Studies of Soils From Rothamsted Experimental Station: III Nitrogen-Containing Macromolecular Moieties in Soil Organic Matter from Geescroft Wilderness. In: Stankiewicz B.A. and van Bergen P.F. (eds), ACS Symposium Series, 707, Nitrogen-Containing Macromolecules in the Bio- and Geosphere, Oxford University Press, New York, pp. 321–338.

van Bergen P.F., Blokker P., Collinson M.E., Sinninghe Damsté J.S. and de Leeuw J.W. 2004. Structural biomacromolecules in plants: What can be learnt from the fossil record?. In: Hemsley A.R. and Poole I. (eds), Linnean Society Symposium Series, 21, Evolution of Plant Physiology, Elsevier, Amsterdam, pp. 133–154.

van der Heijden E. and Boon J.J. 1994. A combined pyrolysis mass spectrometric and light microscopic study of peatified *Calluna* wood isolated from raised peat deposits. Org. Geochem. 22: 903–919.

van der Heijden E., Bouman F. and Boon J.J. 1994. Anatomy of recent and peatified *Calluna vulgaris* stems: implications for coal maceral formation. Int. J. Coal Geol. 25: 1–25.

Vetter R.E. and Botosso P.C. 1989. Remarks on age and growth rate determination of Amazonian trees. IAWA Bull. n.s. 10: 135–145.

Wheeler E.A. and Baas P. 1991. A survey of the fossil record for dicotyledonous wood and its significance for evolutionary and ecological wood anatomy. IAWA Bull. 12: 275–332.

Wheeler E.A. and Baas P. 1993. The potential and limitations of dicotyledonous wood anatomy for climatic reconstructions. Paleobiology 19: 487–498.

Wheeler E.A. and Manchester S.R. 2002. Woods of the Eocene nut beds Flora Clarno formation, Oregon USA. IAWA J. Suppl. 3: 1–188.

Wiemann M.C., Dilcher D.L. and Manchester S.R. 2001. Estimation of mean annual temperature from leaf and wood physiognomy. Forest Sci. 47: 141–149.

Wiemann M.C., Wheeler E.A., Manchester S.R. and Portier K.M. 1998. Dicotyledonous wood anatomical characters as predictors of climate. Palaeogeogr. Palaeoclimatol. Palaeoecol. 139: 83–100.

Wiemann M.C., Manchester S.R. and Wheeler E.A. 1999. Paleotemperature estimation from dicotyledonous wood anatomical characters. Palaios 14: 459–474.

Wiesberg L. 1974. Die ^{13}C-Abnahme in Holz von Baumjahresringen, eine Untersuchung zur anthropogenen Beeinflussung des CO_2-Haushaltes der Atmosphäre. Dissertation, RWTH Aachen.

Wilson K. and White D.J.B. 1986. The Anatomy of Wood: Its Diversity and Variability. Stobart & Son Ltd, London.

Woodcock D.W. 1994. Occurrence of woods with a gradation in vessel diameter across a ring. IAWA J. 15: 277–385.

Woodcock D.W. and Ignas C.M. 1994. Prevalence of wood characters in eastern North America: what characters are most promising for interpreting climates from fossil wood? Am. J. Bot. 81: 1243–1251.

Worbes M. 1985. Structural and other adaptations to long-term flooding by trees in Central Amazonia. Amazonia 9: 459–484.

Worbes M. 1989. Growth rings, increment and age of trees in inundation forests, savannahs and a mountain forest in the Neotropics. IAWA Bull. n.s. 10: 109–122.

Worbes M. 1995. How to measure growth dynamics in tropical trees – a review. IAWA J. 16: 337–351.

Worbes M. 1999. Annual growth rings, rainfall dependent growth and long term growth pattern in tropical trees from Caparo Forest Reserve in Venezuela. J. Ecol. 87: 391–403.

Xie S., Nott C.J., Avsejs L.A., Volders F., Maddy D., Chambers F.M., Gledhill A., Carter J.F. and Evershed R.P. 2000. Palaeoclimate records in compound-specific δD values of a lipid biomarker in ombrotrophic peat. Org. Geochem. 31: 1053–1057.

Yadav R.R. and Bhattacharyya 1996. Climatic significance of growth rings in Mesozoic woods from India. Palaeobotanist 45: 57–63.

Yao Z.Q., Lui L.J. and Zhang S.A. 1994. Permian wood from western Henan, China: implications for palaeoclimatological interpretations. Rev. Palaeobot. Palynol. 80: 277–290.

Zobel B.J. and van Buijtenen J.P. 1989. Wood Variation: Its Causes and Controls. Springer-Verlag, Berlin.

Photo. Above: Fossil leaf of *Ginkgo huttoni* mid Jurassic (age 150 million years). Collection: Dr Simon Troelstra, FALW, Vrije Universiteit, Amsterdam; Below: Present-day *Ginkgo biloba* (Photograph: Jelte Rozema).

Plant Ecology (2006) 182:197–207
DOI 10.1007/s11258-005-9026-y

The occurrence of *p*-coumaric acid and ferulic acid in fossil plant materials and their use as UV-proxy

Peter Blokker*, Peter Boelen, Rob Broekman and Jelte Rozema
*Department of Systems Ecology, Faculty of Earth and Life Sciences, Vrije Universiteit Amsterdam, De Boelelaan 1085, 1081, HV, Amsterdam; *Author for correspondence (e-mail: peter.blokker@ecology.falw.vu.nl)*

Received 1 September 2004; accepted in revised form 15 December 2004

Key words: Cuticles, Fossil plant materials, Leaves, Pollen, Scales, Seeds, Spores, Sporopollenin, Stratospheric ozone

Abstract

The applicability of *p*-coumaric acid and ferulic acid concentrations or ratios in (sub)fossil plant remnant as UV-B proxies relies on various aspects, which are discussed in this paper and will be illustrated with some experimental data. A newly developed THM-micropyrolysis–gas chromatography–mass spectrometry method was tested on various spores, pollen and other plant remains, which were analysed for the presence of the UV-absorbing compounds *p*-coumaric acid and ferulic acid. This revealed that these supposed building-blocks of sporopollenin appear to be present in pollen of many plant species but also in moss spores. The development of this micropyrolysis method paved the way for the quantitative analysis of UV-absorbing compounds in case only a small amount of analyte is available, for example for fossil pollen and spores but also other small palynomorphs and plant fossils. The use of this technique will provide a better insight in the plant responses to UV-radiation, the chemistry of pollen and spores, their fossil counterparts and furthermore the means for a further development of a proxy for the reconstruction of past UV-B radiation.

Abbreviations: *p*CA – *p*-coumaric acid; DHP – dehydrogenation polymer; FA – ferulic acid; Fame – fatty acid methyl ester; FTIR – Fourier transform infrared; GC/MS – gas chromatography mass spectrometry; LOD – limit of detection; NMR – nuclear magnetic resonance; py – pyrolysis; SIM – selective ion monitoring; THM – thermally assisted hydrolysis and methylation; TIC – total ion current; TMAH – tetramethyl ammonium hydroxide; UV – ultraviolet

Introduction

In contrast to for example the reconstruction of past atmospheric carbon dioxide concentrations, which has been possible due to the presence of ancient air included in ice cores (Petit et al. 1999), a historic perspective on the natural fluctuations of stratospheric ozone and surface UV-B is virtually absent before 1920 when the first instrumental monitoring started at Arosa (Switzerland) (Staehelin et al. 2002). The formation of a hole in the Antarctic ozone layer in 1974 was detected using a Dobson spectrometer at the Halley Research Station at the Antarctic (Farman et al. 1985).

It is commonly known that a decrease in stratospheric ozone will result in an increase in solar UV-B radiation in the lower atmosphere. However, changes in solar intensity also affect the

UV-B radiation reaching the earth's surface. At solar highs UV-B only slightly increases in contrast to UV-C, which leads to an increase in the amount of ozone in the stratosphere resulting in an anticorrelated solar activity/UV-B signal on the earth's surface (Rozema et al. 2002b).

It is well documented that plants have active defenses against UV radiation (e.g. Bornman 1991; Huang et al. 1997; Krauss et al. 1997; Rozema et al. 1997; Gehrke 1998; Laakso and Huttunen 1998; Olsson et al. 1998; Meijkamp et al. 1999; Meijkamp et al. 2001; Rozema et al. 2001; Sinha et al. 2001; Day and Neale 2002; Rozema et al. 2002a; Rozema et al. 2002b; Sasaki and Takahashi 2002; Feng et al. 2003; Sullivan et al. 2003; Chicaro et al. 2004). Compounds biosynthesised via the phenylpropanoid pathway such as flavonoids, *p*CA and FA are able to shield harmful radiation and are shown to increase in some of the experiments where higher doses of UV-B were applied to the plants (for a recent overview; Rozema et al. 2005). However, it was also shown that not all plants respond similarly to increased UV-B (Rozema et al. 2005). As a consequence, the use of flavonoids or phenols as a UV proxy is not feasible unless it is possible to link concentrations of these compounds to a specific plant. It was recently suggested to use fossil pollen and spores for that purpose (Rozema et al. 2001; Rozema et al. 2002b). Given that these palynomorphs are able to survive millions of years of burial, even dating back to the Ordovician (500 mya) (Edwards 2001) without major morphological or chemical changes, identification on a genus or sometimes even species-level is possible.

It seems plausible that as a consequence of their small size and potential high exposure to solar radiation upon transport from the anthers to the stigma, the parental plant would protect pollen and spores against UV radiation. Since the main protective tissue that composes the exine of pollen and spores, the so-called sporopollenin, remains unaltered upon diagenesis, it was hypothesised that the fossil remains of pollen and spores possibly form a record of past UV-radiation (Rozema et al. 2001). Even so it might be possible to use other fossil plant materials such as cuticles and seed or fruit coats and bud or catkin scales for extracting information about past UV regimes, provided that these plant remnants are diagenetically stable.

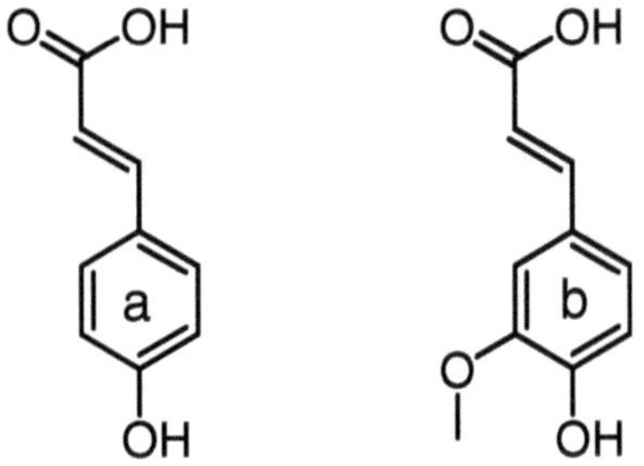

Figure 1. The chemical structure of (a) *p*-coumaric acid; (b) ferulic acid.

The use of chemical information for the reconstruction of past UV-B radiation is hampered by the inconclusive reports on the diagenesis resistant materials composing fossil plant materials, such as sporopollenin. Moreover, the analysis of fossil materials is difficult due to the often low amount of sample material and the lack of analytical instrumentation that can handle these small amounts of material in a reproducible way. This paper describes the potential of using sporopollenin, or more specific, the alleged building-blocks *p*CA and FA as UV-proxies (Figure 1).

Functions of pCA and FA in plants

An important attribute of *p*CA and FA is their function as molecular bridges between lignin and carbohydrates and as cross-linkers between carbohydrates providing rigidity to primary cell walls (Iiyama et al. 1990; Kroon and Williamson 1999; Carnachan and Harris 2000; Lam et al. 2001). Ralph et al. (1992) suggested that these linkages were formed via a catalysed radical reaction by peroxidase comparable to that of lignin formation (Freudenberg and Neish 1968). Although Lam et al. (2001) suggested that such a radical mechanism is possibly unlikely as a consequence of their findings that *p*CA and FA are linked at the benzyl position of a lignin polymer, Boom (2004) shows that a peroxidase/H_2O_2 polymerised DHP of *p*CA can be prepared via a procedure described for artificial lignin (e.g. Hatcher and Minard 1996).

Next to a cross-link function between lignin and carbohydrates, *p*CA and FA are important constituents of unlignified gymnosperm primary cell walls (for an overview of various species Carnachan and Harris (2000)). Comparable to the lignin–carbohydrate bridges, the carboxylic acid group of *p*CA and FA are ester-linked to the hydroxyl groups of carbohydrates. Cross-linking of cell-wall polysaccharides by diferulate was demonstrated to make the primary cell walls in Poaceae more rigid (Tan et al. 1991) suggesting that similar effects may occur in Gymnosperms (Carnachan and Harris 2000). This rigidity may impede their degradation by fungal carbohydrases, thus providing a defence against pathogenic microorganisms.

Pathogenic defence by bound phenolics may not only have its origin in a physical barrier, but may also impair microbial growth by the release of antibiotic phenolics due to the microbial infestation itself (Liakopoulos et al. 2001). The presence of *p*CA esters in the roots of *Tanacetum longifolium* (Asteraceae) might serve a similar function, though (Mahmood et al. 2003) describes these compounds as being plant growth inhibitors. The occurrence of phenolics in the epicuticular layer covering the guard cells of stomata may be related to a defence against biotic and abiotic factors (Kolattukudy 1981; Holloway 1982). Since stomata provide an entry to the leaf interior for pathogens it seems liable that defence metabolites will be deposited in the surface waxes covering the guard cells. Especially *p*CA and FA are effective phenolic components active in the defence against pathogens (Nicholson and Hammerschmidt 1992; Bennett and Wallsgrove 1994).

Despite some controversy, it is widely accepted that both the epicuticular waxes and the cuticular membrane may form a barrier against the penetration of UV into the mesophyll (Barnes and Cardoso-Vilhena 1996). A number of studies revealed that cuticles isolated from various leaves and fruits exhibited a strong absorption in the UV-range (Krauss et al. 1997). UV absorbance of the cuticle seems to be species specific and may differ between the adaxial and abaxial leaf surfaces of the same plant (Bornman and Vogelmann 1988; Robinson et al. 1993; Liakopoulos et al. 2001). *p*CA and FA have been demonstrated to be important constituent of *Prunus persica* and *Olea europaea* leaf waxes. The actual chemical nature of the intermolecular linkages involved in the binding of *p*CA and FA to the leaf waxes are still unknown, though the fact that both compounds are only released by saponification suggests that either the carboxylic acid or the hydroxyl group is involved. Since it was shown that especially the former group is involved in linkages to carbohydrates (Carnachan and Harris 2000) and fatty alcohols (Mahmood et al. 2003) this suggests that most likely *p*CA and FA are linked to the hydroxyl groups of cutin (the solvent insoluble biopolymer in plant cuticles (Tegelaar et al. 1991))-like monomers (see also; Mösle et al. 1997; Mösle et al. 1998; van Bergen et al. 2004). Other examples of esterified *p*CA can be found in flavonol glycosides such as present in the leaves of *Planchonia grandis* (Crublet et al. 2003).

The former examples of the functions of *p*CA and FA in plants illustrate that a polymer composed of these two phenolic compounds, as suggested for sporopollenin, probably provides a manifold protection to pollen and spores: against UV, microbial and physical damage (cf. Rozema et al. 1997).

Sporopollenin

To discern and isolate spores and pollen from other fossil plant remnants and inorganic material acetolysis is often applied in palynology (Erdtman 1960). Due to the fact that the exine of spores and pollen resist this acetolysis treatment it was assumed that it consisted of a chemically resistant organic polymer, conveniently named sporopollenin (Zetzsche and Kälin 1931; Traverse 1988). The actual chemistry of sporopollenin has been a point of debate for years. Varying hypotheses ranging from a polymerised carotenoid network (Brooks and Shaw 1972, 1978) to an aliphatic biopolymer (for reviews: de Leeuw et al. 1991; de Leeuw and Largeau 1993; Largeau and de Leeuw 1995). More recently, it became evident that much of the chemical information obtained on sporopollenins can be ascribed to incomplete removal of other cell constituents or the incorporation of such materials into a more resistant matrix via chemical linkages, greatly biasing the analytical information. Nowadays there seems to be agreement that the resistant material in spores and pollen may have a phenolic and hydrocarbon component (e.g. de Leeuw et al., this issue; Wierman et al. 2001; Ahlers et al. 2003). Although the presence and chemical nature of a hydrocarbon skeleton is still ambiguous, it is evident that the phenolic part of the sporopollenin is primarily composed of *p*CA and FA (Figure 1) (Guilford et al. 1988; Ahlers et al. 1999b; Wierman et al. 2001; van Bergen et al. 2004).

Chemistry of sporopollenin

The amount of techniques that can be used for the chemical analysis of sporopollenin is greatly restricted as a consequence of its resilient nature and insolubility in organic solvents. An important exception is the solubility of the exine material from *Typha angustifolia* L. pollen in 2-aminoethanol commonly up to 5 mg/ml (Ahlers et al. 1999b) allowing the application of nuclear magnetic resonance (NMR) for the analysis of the chemical structure. Derivatisation techniques further increased solubility of these exins allowing more advanced 2D-NMR techniques (Ahlers et al. 2003). Generally it is thought that only insoluble highly cross-linked polymeric or macromolecular structures can resist chemical treatments like acetolysis or survive diagenesis over millions of years (de Leeuw and Largeau 1993). Although the dissolvable sporopollenin mentioned above (Ahlers et al. 1999a; Ahlers et al. 1999b; Ahlers et al. 2003) might be of a unique kind, the conclusions drawn from this work agree with the current ideas about the chemical structure of sporopollenin. Interpretation of NMR results of sporopollenin from *T. angustifolia* (Ahlers et al. 1999b; Ahlers et al. 2000; Ahlers et al. 2003) and *Torreya californica* (Ahlers et al. 1999a) clearly illustrates the presence of aliphatic polyhydroxy compounds next to phenolics suchs as *p*CA. The presence of *p*CA as an intricate part of the sporopollenin structure is supported by Wehling et al. (1989) and Mulder et al. (1992) who described the presence of *p*CA in the sacci of *Pinus* pollen by a detailed study using various pyrolysis techniques combined with mass spectrometry. These authors observe an isoprenoid material next to the phenolic material. Although this could be a species specific characteristic, it is possible that these compounds are remnants of other pollen consituents that were not or incompletely removed by the work-up procedure.

The first unambiguous report of sporopollenin based on *p*CA and FA in spores was published by Boom (2004) who described the chemical structure of the walls of *Isoëtes killipii* megaspores as determined by various pyrolysis techniques combined with GC/MS. Though van Bergen et al. (1993) and Bergen et al. (1995) concluded that the compounds vinylphenol and 2-methoxyvinylphenol amongst the pyrolysis products of extant and fresh spores of *Azolla* and *Salvinia* originated from *p*CA and FA, respectively, other biopolymers like lignin might also generated such pyrolysis fragments. Although it seems unlikely that the benthic spores of *I. killipii* as described by Boom (2004) were exposed to high UV-B radiation or will be during their life-cycle, it illustrates that the use of *p*CA and FA as spore and pollen wall building-blocks also provides other modes of defence.

Other reports on the analysis of sporopollenin confirm the presence of aliphatics and phenolics in sporopollenin for example by degradation with potash fusion (Schulze-Osthoff and Wiermann 1987), by using tracer experiments (Gubatz et al. 1993) and by immunocytochemical experiments (NiesterNyveld et al. 1997).

Next to the fact that certain sporopollenins can be dissolved in organic solvents as 2-amino-ethanol, a synthetic *p*CA/FA DHP produced in a similar manner as that described by Boom (2004) was largely soluble in a 0.1 M phosphate buffer of pH 6. Although this will only give a very crude approximation of the polymer size, dialysis over a 2000 Dalton molecular cut-off membrane demonstrated that the polymer contains more than 10 monomers (Blokker et al. 2005). In general it appears that both *p*CA/FA DHP and sporopollenin are more soluble than lignin, which is composed of their alcoholic counterparts (Freudenberg and Neish 1968). This feature could have serious implications for chemical analysis. Differences in sporopollenin work-up procedures will result in a different dissolution of the phenolic part of the polymer. Since literature data suggests that the aliphatic part, possibly due to a better resistance toward oxidation, is less affected by the chemical treatments used to work-up sporopollenin, different protocols will result in discrepancies between analytical data. For example, Dominguez et al. (1999) describe a work-up method that selectively purifies an aliphatic polymer by removing all ether and ester-bound phenols and fatty acids. A material is obtained that chemically resembles that of an oxidatively polymerised fatty acid network. In contrast, prolonged treatment of *T. angustifolia* sporopollenin with relatively mild acidic methanol

(Bubert et al. 2002) was shown to degrade the sporopollenin in a non-selective way. In other words the ratio between the phenolic and aliphatic part was retained upon further degradation (Bubert et al. 2002).

Acetolysis combines the extracting power of acetic acid anhydride with hydrolysis and derivatisation; most hydroxyl groups will be converted into their corresponding acetals. In theory acetolysis will remove all cell contents under relatively mild conditions, however, chemically altering the sporopollenin by acetylating free hydroxyl groups (Hemsley et al. 1995, 1996). Nevertheless, this chemical alteration does not pose analytical problems when using THM-py-techniques, since the strong hydrolysing environment will replace the acetyl group by a methylene moiety analogous to free hydroxyls. Following the former discussion on the effect of sporopollenin work-up procedures it is likely that acetolysis time, the presence of water in the acetolysis mixture, oxygen or maybe even trace metals will probably affect the chemical composition of the final sporopollenin. Future experiments will reveal if the acetolysis time and conditions will affect the relative quantities of sporopollenin components. This knowledge is of paramount importance when the chemical information retained in fossil pollen and spores are used as a proxy.

The preservation potential of pCA and FA

*p*CA and FA are omnipresent in plants performing various functions. Therefore, it is likely that these phenolic compounds can be found in sedimentary or soil organic matter that underwent limited diagenetic stress. However, to use these UV-screening phenols as a proxy for past UV-B radiation it is necessary to link the chemical information to specific plant parts. The most obvious fossil remnants to investigate the past UV-B are leaves. Both the epicuticular waxes and the cuticular membrane may form a barrier against the penetration on UV into the mesophyll (Bornman and Vogelmann 1988; Robinson et al. 1993; Barnes and Cardoso-Vilhena 1996; Krauss et al. 1997; Liakopoulos et al. 2001). This suggests that maybe a UV-signal might be preserved as chemical information in the fossil leaves. In general plant leaves may preserve in sediments or soils over extremely long periods under the right conditions. It was suggested that this might be due to the presence of resistant aliphatic biopolymeric materials like cutan (e.g. de Leeuw et al., this issue; van Bergen et al. 2004), however, depending on the preservation conditions, it seems that less resilient biomolecular structures survive diagenesis very well. (For a recent review on biomacromolecules and their preservation see: de Leeuw et al., this issue; van Bergen et al. 2004).

The presence of the 4-vinyl phenol in the pyrolysate of fresh and fossil cuticles has been reported in fossil leaf material of *Ginkgo* species (Mösle et al. 1997, 1998). 4-vinyl phenol is thought to be the pyrolytic decarboxylation product of *p*CA (Figure 2). Using the presence of *p*CA in fossil cuticles for the reconstruction of past UV-B might pose difficult if diagenesis affects the chemistry of the cuticle. Huang et al. (1998) shows that upon a 23 years degradation field experiment of *Calluna vulgaris* litter, *p*CA is preferentially removed over the syringyl and guaiacyl moieties as deduced from a decrease in 4-vinyl phenol in the pyrolysate in the time series 0, 0.5, 7 and 23 years. Similar observations were done by Kuder and Kruge (1998), van Bergen et al. (1997) and Karunen and Kalviainen (1988). However, despite a significant loss of *p*CA and FA these authors did not observe a complete removal. This might suggest that a labile fraction of these phenolics, probably from other parts of the leaf than the cuticle, is degraded while a more resilient, probably polymeric, fraction remains. Fossil seeds may also contain *p*CA and FA (e.g. Bergen et al. 1995; van Bergen et al. 1997, 2004, and references therein). These authors state that the polyphenolic constituents in seeds are most likely of a lignin origin and *p*CA and FA are incorporated into such polymeric structure via ester and ether-linkages as indicated by 4-vinylphenol and 4-vinyl-2-meth oxyphenol in the pyrolysis products (for the pyrolysis mechanism see Figure 2).

Pollen and spores are amongst the most resistant organic plant remains (e.g. de Leeuw and Largeau 1993; van Bergen et al. 2004), and will thus be a reliable source for information of past UV-B concentrations. Still, like the chemical information on fresh spores and pollen, not all research seems to agree on the presence of

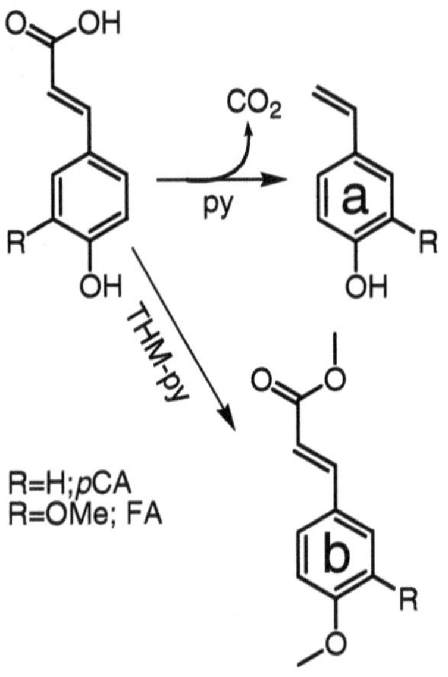

Figure 2. The main pyrolysis pathways of polymeric *p*-coumaric acid (R = H) and ferulic acid (R = OMe) with and without THM. (a) 4-vinyl-phenol R = H and 2-methoxy-4-vinyl-phenol (R = OMe); (b) Methylated *p*-coumaric acid R = H and ferulic acid (R = OMe).

phenolics in fossil spores and pollen. Oxidative conditions will most likely preferentially degrade the polyphenolics in favour of the aliphatic part as demonstrated by Hemsley et al. (1996) using ^{13}C-NMR on *Lycopodium clavatum* spores. Diagenetic modelling by heating *Pinus* sporopollenin furthermore shows that in the early stages the typical polyphenolic signal is replaced by a single aromatic signal, while the aliphatic signal remains. Further heating results in a homogenisation of the signal, resulting in a single aromatic and aliphatic signal without the fine-structure of the original sporopollenin comparable to that of carboniferous megaspores. The most diagenetically altered material consists of only aromatic material as can be seen in fossil samples that are significantly affected by diagenesis (Hemsley et al. 1995, 1996).

The loss of fine-structure and change in the aliphatic vs. aromatic signal was also observed for *Lycopodium clavatum* spores by (Yule et al. 2000) upon simulated diagenesis. They observed a relative intensification of the FTIR aliphatic signal with increasing maturity. The aromatic signal dominated the spores' FTIR spectrum of the most diagenetically altered material.

Development of a micro-scale analytical method for the reconstruction of past UV radiation

The analysis of fossil plant materials from cores is seriously hampered by their low availability. Scarcity of samples requires an analysis method that is non-destructive or uses only limited sample amounts. This is even more crucial for chemical analysis of Herbarium material since destructive sampling of precious collections is unsolicited. Most suitable are these techniques that allow the analysis small samples without laborious pre-treatment, such as pyrolysis-coupled to sensitive GC/MS. It has been demonstrated that such techniques enable the analysis of microscopic entities and would be the analytical method of choice for example fatty acid profiling of flow cytometry isolated microorganisms (Blokker et al. 2002; de Koning et al. 2002). A major drawback of pyrolysis techniques is the often complex fragmentation of macromolecular structures, providing a myriad of compounds that are often unrecognisable by commercial mass spectra-libraries (Haken 2000). In the case of *p*CA and FA conventional pyrolysis will lead to 4-vinyl-phenol and 2-methoxy-4-vinyl-phenol (Figure 2) as primary products (e.g. Boom 2004). Since lignins may also thermally degrade into these compounds amongst others, the presence of 4-vinyl-phenol and 2-methoxy-4-vinyl-phenol does not provide conclusive evidence for the presence of *p*CA and FA in the sample (e.g. Saiz-Jimenez and de Leeuw 1984; van der Hage 1995; Kuroda 2000). However, applying THM-reagents such as TMAH will decrease the degree of fragmentation by simultaneously hydrolysis of macromolecular structures and derivatisation of hydroxyl, carboxylic acid and other functional groups that carry an acidic proton (Challinor 1989). Though the 4-vinyl phenol in the pyrolysate of fresh and fossil *Ginkgo* cuticles (Mösle et al. 1997, 1998) is proposed to originate from *p*CA, only upon using TMAH-pyrolysis this could be unequivocal demonstrated. THM-py on a Paleocene (65–54.8 mya) *Ginkgo adiantoides* leaf (Figure 3) indeed confirmed the presence of pCA next to typical oxidised fatty acid THM-py products. This over 55 million year old fossil cuticle even retained its UV absorbing properties (Figure 4) as compared with absorbance assessed in an extant cuticle of *G. biloba* leaf. These results

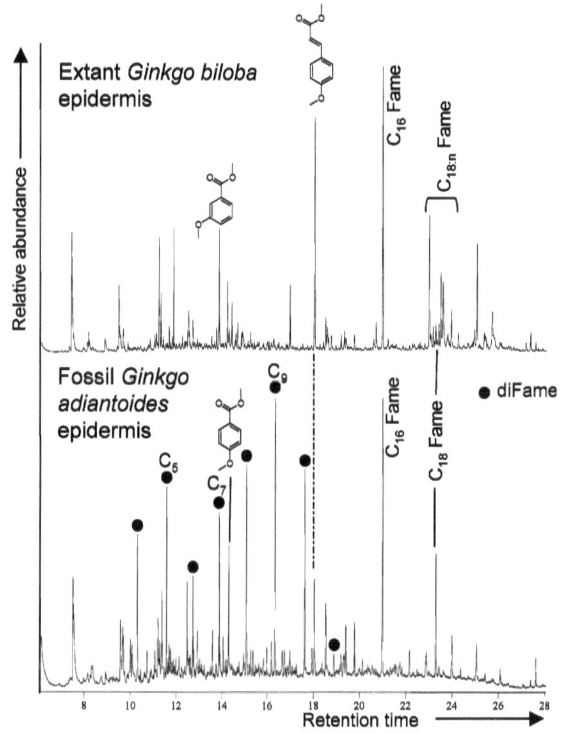

Figure 3. TIC of THM-pyrolysis product mixture of extant and fossil *Ginkgo* cuticles (1 × 3 mm) (Paleocene, ca. 60 mya). diFame – alkyl dicarboxylic acids methyl ester; Fame – fatty acid methyl esters; C_n = carbon number.

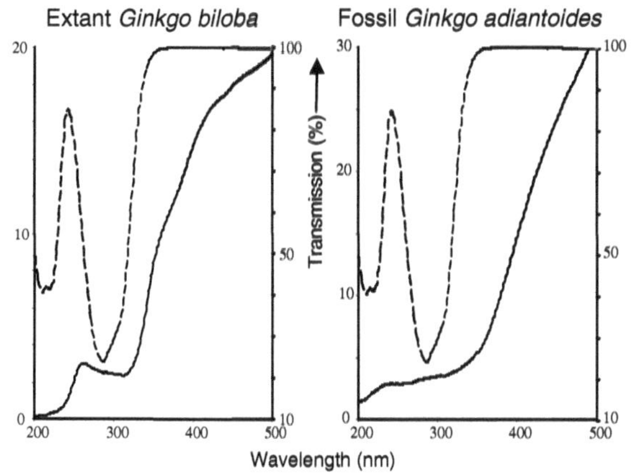

Figure 4. UV-absorption spectra of fresh and fossil *Ginkgo* superimposed on the absorption spectrum of a synthetic *p*-coumaric acid-based sporopollenin (DHP).

indicate that *p*CA is an intricate part of the *Ginkgo* cuticle.

The application of THM reagents and pyrolysis techniques thus supplies the means for the analysis of target compounds in very small sample sizes. To increase the sensitivity of the method, allowing the analysis on a 50–500 pollen grain or moss spore level (depending on the grain size), it was necessary to run the mass spectrometer in selective-ion-monitoring (SIM) mode. This afforded a detection limit of ~60 fresh *Alnus* pollen for pCA and ~10 for FA (S/N = 3) (Blokker et al. 2005).

Though a very low amount of *p*CA and FA can be quantified by using THM-py-GC/MS, absolute quantification of these phenolics as part of a polymeric structure is not possible since a wide variety of different intermolecular linkages will give rise to various pyrolysis products. Therefore, differences observed between samples could be due to differences in linkage types rather than to concentrations. Furthermore, when interpreting results from fossil material it is possible that diagenesis will affect the observed amount of phenolics. Thermal maturation will result in homogenisation of the aromatic signal in NMR and FTIR, possibly due to coalification processes (Hemsley et al. 1995, 1996). Such chemical changes will most likely also be reflected in the pyrograms of such materials and hamper the comparison of fresh and fossil materials.

An alternative to the quantification of pyrolysis products is the use of ratios between different pyrolysis products. Though investigations are ongoing, preliminary results (Figure 5) indicate that next to the concentration of *p*CA and FA, the *p*CA/FA in a THM-pyrolysis mixture of *Vicia faba* pollen is significantly increased by UV-light. THM-py-GC/MS analysis of various fresh spores and pollen (Figure 6) indicate that there is a large variety between the ratio due to interspecies differences of environmental (including radiation regime) conditions. Whether or not UV-B radiation increased the relative amounts of these compounds in the pollen or just affects the degree and/or type of polymerisation remains unknown.

There are still many and large gaps in our knowledge on the effects of UV-B on plants and past UV-radiation. THM-py-GC/MS analysis of *p*CA and FA appears to show perspective for the reconstruction of past UV-B. Especially the analysis of samples from peat deposits and herbarium collections seems to be a fruitful route to gain insight in the impact of fluctuating UV-B on our climate and ecosystems.

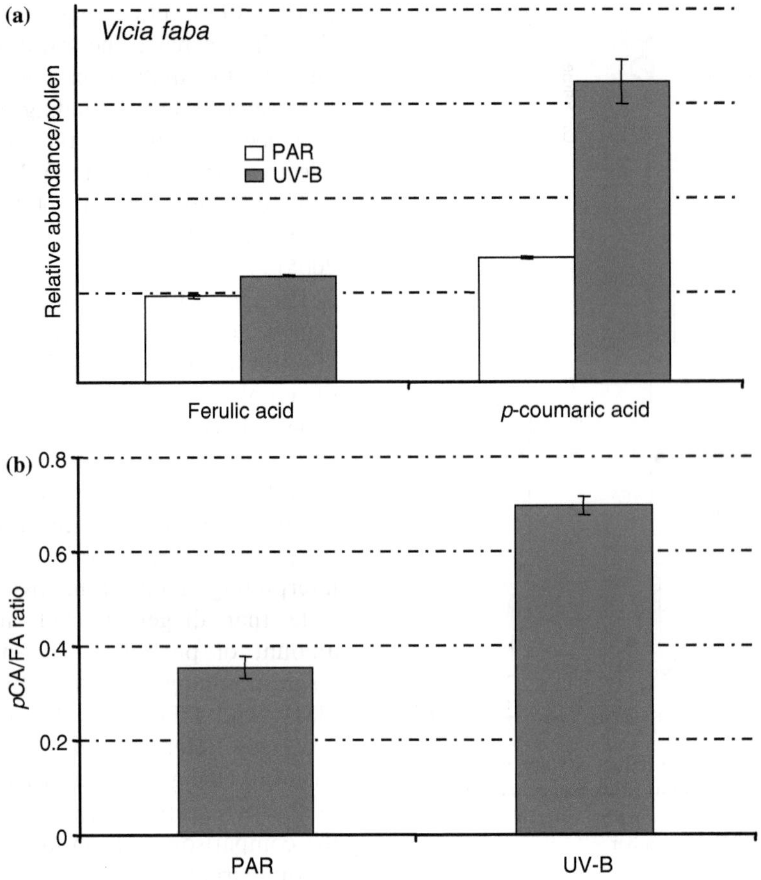

Figure 5. (a) Integrated peak area of *p*-coumaric acid and ferulic acid divided by the number of pollen (between 44 and 68) of unacetolysed *Vicia faba*. Error bars represent the standard deviation. Two-sided *t*-tests (df = 2): ferulic acid, $p = 0.004$; *p*-coumaric acid, $p = 0.007$. (b) Data from (a) plotted as ratio. Error bars represent the standard deviation. Two-sided *t*-tests (df = 2): $p = 0.006$ (Blokker et al. 2005).

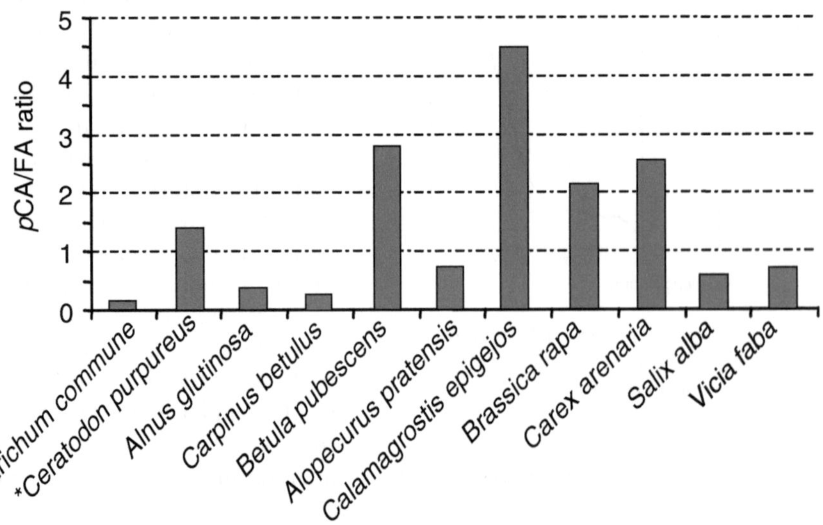

Figure 6. *p*-coumaric acid/ferulic acid ratios of pollen and spores from various higher and lower plants (latter indicated by *).

Experimental

For analysis the samples were placed in a pyrolysis liner (CDS) and 5 μl of a 25% TMAH solution in methanol was added. The sample was allowed to incubate for 2 h at 70 °C and pyrolysed at 550 °C for 5 min. in a CDS AS2500 pyrolysis unit (260 °C interface temperature) coupled to an Agilent 6890 GC equipped with an Agilent 5973 MSD. The GC-oven was programmed from 40 °C (5 min hold time) to 130 °C at 20 °C/min and subsequently to 320 °C at 6 °C/min followed by 10 min isothermal at 320 °C. He was used as a carrier gas at a constant flow of 1.2 ml/min. using a 20:1 splitratio for leaf samples and in splitless mode in case of pollen and spores. The mass spectrometer was operated in full scan mode (m/z 50–800) or in SIM-mode in case of pollen and spore analysis (m/z 192, 222) at 70 eV ionization energy.

The UV spectra were recorded on a Shimatzu UV-1601PC UV–VIS spectrophotometer. The extant leaf was sampled May, 2004 from a 40-year-old *G. biloba* tree on the Vrije Universiteit campus.

Vicia faba was cultured without and with UV-B treatment (biologically active radiation; 10 kJ m^{-2} day^{-1}, representing ~30% ozone depletion) conditions earlier reported by (Meijkamp et al. 2001).

The other pollen and spore materials were collected spring at various places in the north of the Netherlands.

Acknowledgements

Prof Dr Margaret Collinson from the Royal Holloway University in the UK is gratefully thanked for providing the Palaeocene *Ginkgo* leaf. The research of P. Boelen is funded by an ALW-NAAP Grant (851.20.010). The research of P. Blokker was funded by a CLIVAR-Grant (854.00.004).

References

Ahlers F., Lambert J. and Wiermann R. 1999a. Structural elements of sporopollenin from the pollen of *Torreya californica* Torr. (Gymnospermae): Using the ^{1}H-NMR technique. Z. Naturforsch. C: Biosci. 54: 492–495.

Ahlers F., Thom I., Lambert J., Kuckuk R. and Wiermann R. 1999b. 1H NMR analysis of sporopollenin from Typha Angustifolia. Phytochemistry 50: 1095–1098.

Ahlers F., Bubert H., Steuernagel S. and Wiermann R. 2000. The nature of oxygen in sporopollenin from the pollen of *Typha angustifolia* L. Z. Naturforsch. C: Biosci. 55: 129–136.

Ahlers F., Lambert J. and Wiermann R. 2003. Acetylation and silylation of piperidine solubilized sporopollenin from pollen of *Typha angustifolia* L. Z. Naturforsch. C: Biosci. 58: 807–811.

Barnes J.D. and Cardoso-Vilhena J. 1996. Interactions between electromagnetic radiation and the plant cuticle. In: Kerstiens G. (ed.), Plant Cuticles and Intergrated Functional Approach, BIOS Scientific Publishers Ltd., Oxford, pp. 157–174.

Bennett R.N. and Wallsgrove R.M. 1994. Secondary metabolites in plant defense-mechanisms. New Phytol. 127: 617–633.

van Bergen P.F., Bland H.A., Horton M.C. and Evershed R.P. 1997. Chemical and morphological changes in archaeological seeds and fruits during preservation by desiccation. Geochim. Cosmochim. Acta 61: 1919–1930.

van Bergen P., Blokker P., Collinson M.E., Sinninghe Damsté J.S. and de Leeuw J.W. 2004. Structural biomacromolecules in plants: What can be learned from the fossil record?. In: Hemsley A.R. and Poole I. (eds.), Evolution in Plant Physiology, Elsevier, Amsterdam, pp. 133–154.

van Bergen P.F., Collinson M.E. and de Leeuw J.W. 1993. Chemical composition and ultrastucture of fossil and extant salvinialean microspore massulae and megaspores. Grana Suppl. 1: 19–30.

van Bergen P.F., Collinson M.E., Briggs D.E.G., de Leeuw J.W., Scott A.C., Evershed R.P. and Finch P. 1995. Resistant biomacromolecules in the fossil record. Acta Bot. Neerl. 44: 319–342.

Blokker P., Pel R., Akoto L., Brinkman U.A.T. and Vreuls R.J.J. 2002. At-line gas chromatographic–mass spectrometric analysis of fatty acid profiles of green microalgae using a direct thermal desorption interface. J. Chromatogr. A 959: 191–201.

Blokker P., Yeloff D., Boelen P., Broekman R.A. and Rozema J. 2005. Development of a proxy for past surface UV-B irradiation: A THM-py-GC/MS method for the analysis of pollen and spores. Anal. Chem. DOI: 10.1021/ac050696k.

Boom A. 2004. A geochemical study of lacustrine sediments: towards paleo-climatic reconstructions of high Andean biomes in Columbia. Doctorate Thesis, University of Amsterdam, Amsterdam.

Bornman J.F. 1991. UV-Radiation as an environmental-stress in plants. J. Photochem. Photobiol. B 8: 337–340.

Bornman J.F. and Vogelmann T.C. 1988. Penetration of blue and UV-radiation measured by fiber optics in spruce and fir needles. Physiol. Plant 72: 699–705.

Brooks J. and Shaw G. 1972. Geochemistry of sporopollenin. Chem. Geol. 10: 69–87.

Brooks J. and Shaw G. 1978. Sporopollenin: a review of its chemistry, palaeochemistry and geochemistry. Grana 17: 91–97.

Bubert H., Lambert J., Steuernagel S., Ahlers F. and Wiermann R. 2002. Continuous decomposition of sporopollenin from pollen of *Typha angustifolia* L. by acidic methanolysis. Z. Naturforsch. C: Biosci. 57: 1035–1041.

Carnachan S.M. and Harris P.J. 2000. Ferulic acid is bound to the primary cell walls of all gymnosperm families. Biochem. Syst. Ecol. 28: 865–879.

Challinor J.M. 1989. A pyrolysis-derivatisation-gas chromatography technique for the structural elucidation of some synthetic polymers. J. Anal. Appl. Pyrol. 16: 323–333.

Chicaro P., Pinto E., Colepicolo P., Lopes J.L.C. and Lopes N.P. 2004. Flavonoids from *Lychnophora passerina* (Asteraceae): potential antioxidants and UV-protectants. Biochem. Syst. Ecol. 32: 239–243.

Crublet M.-L., Long C., Sevenet T., Hadi H.A. and Lavaud C. 2003. Acylated flavonol glycosides from leaves of *Planchonia grandis*. Phytochemistry 64: 589–594.

Day T.A. and Neale P.J. 2002. Effects of UV-B radiation on terrestrial and aquatic primary producers. Annu. Rev. Ecol. Syst. 33: 371–396.

Dominguez E., Mercado J.A., Quesada M.A. and Heredia A. 1999. Pollen sporopollenin: degradation and structural elucidation. Sex. Plant Reprod. 12: 171–178.

Edwards D. 2001. Early land plants. In: Briggs D.E.G. and Crowther P.R. (eds.), Palaeobiology II, Blackwell Science, Oxford, pp. 63–66.

Erdtman G. 1960. The acetolysis method, a revised description. Sven. Bot. Tidskr. 54: 561–564.

Farman J.C., Gardiner B.G. and Shanklin J.D. 1985. Large losses of total ozone in Antarctica reveal seasonal ClO_x–NO_x interaction. Nature 315: 207–210.

Feng H., An L., Chen T., Qiang W., Xu S., Zhang M., Wang X. and Cheng G. 2003. The effect of enhanced ultraviolet-B radiation on growth, photosynthesis and stable carbon isotope composition ([delta]13C) of two soybean cultivars (Glycine max) under field conditions. Environ. Exp. Bot. 49: 1–8.

Freudenberg K. and Neish A.C. 1968. Lignins-Occurence, Formation, Structure and Reaction. Springer-Verlag, Berlin.

Gehrke C. 1998. Effects of enhanced UV-B radiation on production-related properties of a *Sphagnum fuscum* dominated subarctic bog. Funct. Ecol. 12: 940–947.

Gubatz S., Rittscher M., Meuter A., Nagler A. and Wiermann R. 1993. Tracer Experiments on Sporopollenin Biosynthesis – an overview. Grana 12–17.

Guilford W.J., Schneider D.M., Labovitz J. and Opella S.J. 1988. High-resolution solid-state ^{13}C-NMR-spectroscopy of sporopollenins from different plant taxa. Plant Physiol. 86: 134–136.

van der Hage E.R.E. 1995. Pyrolysis mass spectrometry of lignin polymers. Doctorate Thesis, University of Amsterdam, Amsterdam.

Haken J.K. 2000. Analytical pyrolysis of natural organic polymers. J. Chromatogr. A 877: 241–242.

Hatcher P.G. and Minard R.D. 1996. Comparison of dehydrogenase polymer (DHP) lignin with native lignin from gymnosperm wood by thermochemolysis using tetramethylammonium hydroxide (TMAH). Org. Geochem. 24: 593–600.

Hemsley A.R., Barrie P.J. and Scott A.C. 1995. ^{13}C solid-state N.M.R. spectroscopy of fossil sporopollenins. Variation in composition independent of diagenesis. Fuel 74: 1009–1012.

Hemsley A.R., Scott A.C., Barrie P.J. and Chaloner W.G. 1996. Studies of fossil and modern spore wall biomacromolecules using ^{13}C solid state NMR. Ann. Bot. (Lond.) 78: 83–94.

Holloway P.J. 1982. The chemical constitution of plant cutins. In: Cutler D.F., Alvin K.L. and Price C.E. (eds), The plant Cuticle, Linnean Society Symposium Series 10, Academic Press, London, pp. 45–85.

Huang S., Dai Q., Peng S., Chavez A.Q., Miranda L.L., Visperas R.M. and Vergara B.S. 1997. Influence of supplemental ultraviolet-B on indoleacetic acid and calmodulin in the leaves of rice (*Oryza sativa* L.). Plant Growth Regul. 21: 59–64.

Huang Y., Stankiewicz B.A., Eglinton G., Snape C.E., Evans B., Latter P.M. and Ineson P. 1998. Monitoring biomacromolecular degradation of *Calluna Vulgaris* in a 23 year field experiment using solid state ^{13}C-NMR and pyrolysis-GC/MS. Soil Biol. Biochem. 30: 1517–1528.

Iiyama K., Lam T.B.T. and Stone B.A. 1990. Phenolic acid bridges between polysaccharides and lignin in wheat internodes. Phytochemistry 29: 733–737.

Karunen P. and Kalviainen E. 1988. Lipids and hydroxycinnamic acids in cell-walls of *Eriophorum vaginatum*. Phytochemistry 27: 2045–2048.

Kolattukudy P.E. 1981. Structure, biosynthesis, and biodegradation of cutin and suberin. Annu. Rev. Plant Physiol. Plant Mol. Biol. 32: 539–567.

de Koning J.A., Blokker P., Jungel P., Alkema G. and Brinkman U.A.T. 2002. Automated liner exchange – a novel approach in direct thermal desorption. Chromatographia 56: 185–190.

Krauss P., Markstadter C. and Riederer M. 1997. Attenuation of UV radiation by plant cuticles from woody species. Plant Cell Environ. 20: 1079–1085.

Kroon P.A. and Williamson G. 1999. Hydroxycinnamates in plants and food: current and future perspectives. J. Sci. Food Agric. 79: 355–361.

Kuder T. and Kruge M.A. 1998. Preservation of biomolecules in sub-fossil plants from raised peat bogs – a potential paleoenvironmental proxy. Org. Geochem. 29: 1355–1368.

Kuroda K.-I. 2000. Pyrolysis-trimethylsilylation analysis of lignin: preferential formation of cinnamyl alcohol derivatives. J. Anal. Appl. Pyrol. 56: 79–87.

Laakso K. and Huttunen S. 1998. Effects of the ultraviolet-B radiation (UV-B) on conifers: a review. Environ. Pollut. 99: 319–328.

Lam T.B.T., Kadoya K. and Iiyama K. 2001. Bonding of hydroxycinnamic acids to lignin: ferulic and *p*-coumaric acids are predominantly linked at the benzyl position of lignin, not the β-position, in grass cell walls. Phytochemistry 57: 987–992.

Largeau C. and de Leeuw J.W. 1995. Non-hydrolysable, aliphatic macromolecular constituents of microbial cell walls. In: Jones J.G. (ed.), Adv. Microb. Ecol., Plenum Press, New York, p. 416.

de Leeuw J.W. and Largeau C. 1993. A review of macromolecular organic compounds that comprise living organisms and their role in kerogen, coal and petroleum formation. In: Engel M.H. and Macko S.A. (eds.), Organic Geochemistry, Principles and Applications: Topics in Geobiology, Plenum Press, New York, pp. 23–72.

de Leeuw J.W., Bergen P.F.V., Aarssen B.G.K.V., Gatellier J.-P.L.A., Sinninghe Damsté J.S. and Collinson M.E. 1991. Resistant biomacromolecules as major contributors to kerogen. Philos. Trans.: Biol. Sci. 333: 329–336.

Liakopoulos G., Stavrianakou S. and Karabourniotis G. 2001. Analysis of epicuticular phenolics of *Prunus persica* and *Olea europaea* leaves: evidence for the chemical origin of the

UV-induced blue fluorescence of stomata. Ann. Bot. (Lond.) 87: 641–648.

Mahmood U., Kaul V.K., Acharya R. and Jirovetz L. 2003. *p*-Coumaric acid esters from *Tanacetum longifolium*. Phytochemistry 64: 851–853.

Meijkamp B.B., Aerts R., van de Staaij J., Tosserams M., Ernst W.H.O. and Rozema J. 1999. Effects of UV-B on secondary metabolites in plants. In: Rozema J. (ed.), Stratospheric Ozone Depletion: The Effects of Enhanced UV-B Radiation on Terrestrial Ecosystems, Backhuys publishers, Leiden, pp. 71–99.

Meijkamp B.B., Doodeman G. and Rozema J. 2001. The respons of *Vicia faba* to enhanced UV-B radiation inder low and near ambient PAR levels. Plant Ecol. 154: 135–156.

Mösle B., Finch P., Collinson M.E. and Scott A.C. 1997. Comparison of modern and fossil plant cuticles by selective chemical extraction monitored by flash pyrolysis-gas chromatography–mass spectrometry and electron microscopy. J. Anal. Appl. Pyrol. 40–41: 585–597.

Mösle B., Collinson M.E., Finch P., Stankiewicz B.A., Scott A.C. and Wilson R. 1998. Factors influencing the preservation of plant cuticles: a comparison of morphology and chemical composition of modern and fossil examples. Org. Geochem. 29: 1369–1380.

Mulder M.M., van der Hage E.R.E. and Boon J.J. 1992. Analytical in source pyrolytic methylation electron-impact mass-spectrometry of phenolic-acids in biological matrices. Phytochem. Anal. 3: 165–172.

Nicholson R.L. and Hammerschmidt R. 1992. Phenolic-compounds and their role in disease resistance. Annu. Rev. Phytopathol. 30: 369–389.

NiesterNyveld C., Haubrich A., Kampendonk H., Gubatz S., Tenberge K.B., Rittscher M., Wilmesmeier S. and Wiermann R. 1997. Immunocytochemical localization of phenolic compounds in pollen walls using antibodies against *p*-coumaric acid coupled to bovine serum albumin. Protoplasma 197: 148–159.

Olsson L.C., Veit M., Weissenbock G. and Bornman J.F. 1998. Differential flavonoid response to enhanced uv-b radiation in *Brassica napus*. Phytochemistry 49: 1021–1028.

Petit J.R., Jouzel J., Raynaud D., Barkov N.I., Barnola J.-M., Basile I., Bender M., Chappellaz J., Davis M., Delaygue G., Delmotte M., Kotlyakov V.M., Legrand M., Lipenkov V.Y., Lorius C., PEpin L., Ritz C., Saltzman E. and Stievenard M. 1999. Climate and atmospheric history of the past 420,000 years from the Vostok ice core, Antarctica. Nature 399: 429–436.

Ralph J., Helm R.F., Quideau S. and Hatfield R.D. 1992. Lignin feruloyl ester cross-links in grasses .1. Incorporation of feruloyl esters into coniferyl alcohol dehydrogenation polymers. J. Chem. Soc. Perk. Trans. 1: 2961–2969.

Robinson S.A., Lovelock C.E. and Osmond C.B. 1993. Wax as a Mechanism for protection against photoinhibition – a study of cotyledon-Orbiculata. Bot. Acta 106: 307–312.

Rozema J., van de Staaij J., Bjorn L.O. and Caldwell M. 1997. UV-B as an environmental factor in plant life: stress and regulation. Trends Ecol. Evol. 12: 22–28.

Rozema J., Boelen P. and Blokker P. 2005. Depletion of stratospheric ozone over the Antarctic and Arctic: Responses of plants of polar terrestrial ecosystems to enchanced UV-B, an overview. Environ. Pollut. 137: 428–442.

Rozema J., Broekman R.A., Blokker P., Meijkamp B.B., de Bakker N., van de Staaij J., van Beem A., Ariese F. and Kars S.M. 2001. UV-B absorbance and UV-B absorbing compounds (*para*-coumaric acid) in pollen and sporopollenin: the perspective to track historic UV-B levels. J. Photochem. Photobiol. B 62: 108–117.

Rozema J., Bjorn L.O., Bornman J.F., Gaberscik A., Hader D.-P., Trost T., Germ M., Klisch M., Groniger A. and Sinha R.P. 2002a. The role of UV-B radiation in aquatic and terrestrial ecosystems – an experimental and functional analysis of the evolution of UV-absorbing compounds. J. Photochem. Photobiol. B 66: 2–12.

Rozema J., Van Geel B., Björn L.O., Lean J. and Madronich S. 2002b. Toward solving the UV puzzle. Science 296: 1621–1622.

Saiz-Jimenez C. and de Leeuw J.W. 1984. Pyrolysis-gas chromatography–mass spectrometry of isolated, synthetic and degraded lignins. Org. Geochem. 6: 417–422.

Sasaki K. and Takahashi T. 2002. A flavonoid from *Brassica rapa* flower as the UV-absorbing nectar guide. Phytochemistry 61: 339–343.

Schulze-Osthoff K. and Wiermann R. 1987. Phenols as integrated compounds of sporopollenin from Pinus pollen. J. Plant Physiol. 131: 5–15.

Sinha R.P., Klisch M., Gröniger A. and Häder D.-P. 2001. Reponses of aquatic algae and cyanobacteria to solar UV-B. Plant Ecol. 154: 221–236.

Staehelin J., Mader J., Weiss A.K. and Appenzeller C. 2002. Long-term ozone trends in Northern mid-latitudes with special emphasis on the contribution of changes in dynamics. Phys. Chem. Earth, Parts A/B/C 27: 461–469.

Sullivan J.H., Gitz D.C., Peek M.S. and McElrone A.J. 2003. Response of three eastern tree species to supplemental UV-B radiation: leaf chemistry and gas exchange. Agric. For. Meteorol. 120: 219–228.

Tan K.S., Hoson T., Masuda Y. and Kamisaka S. 1991. Correlation between cell-wall extensibility and the content of diferulic and ferulic acids in cell-walls of *Oryza sativa* coleoptiles grown under water and in air. Physiol. Plant 83: 397–403.

Tegelaar E.W., Kerp H., Visscher H., Schenck P.A. and de Leeuw J.W. 1991. Bias of the paleobotanical record as a consequence of variations in the chemical-composition of higher vascular plant cuticles. Paleobiology 17: 133–144.

Traverse A. 1988. Paleopalynology. Unwin Hyman, Boston.

Wehling K., Niester C., Boon J.J., Willemse M.T.M. and Wierman R. 1989. *p*-coumaric acid – a monomer in the sporopollenin skelleton. Planta 179: 367–380.

Wierman R., Ahlers F. and Schmitz-Thom I. 2001. Sporopollenin. In: Hofrichter M. and Steinbüchel A. (eds.), Biopolymers, Weinheim, Wiley-VCH, pp. 209–227.

Yule B.L., Roberts S. and Marshall J.E.A. 2000. The thermal evolution of sporopollenin. Org. Geochem. 31: 859–870.

Zetzsche F. and Kälin O. 1931. Membranes of spores and pollen. V. Autoxidation of sporopollenin. Helv. Chim. Acta 14: 517–519.

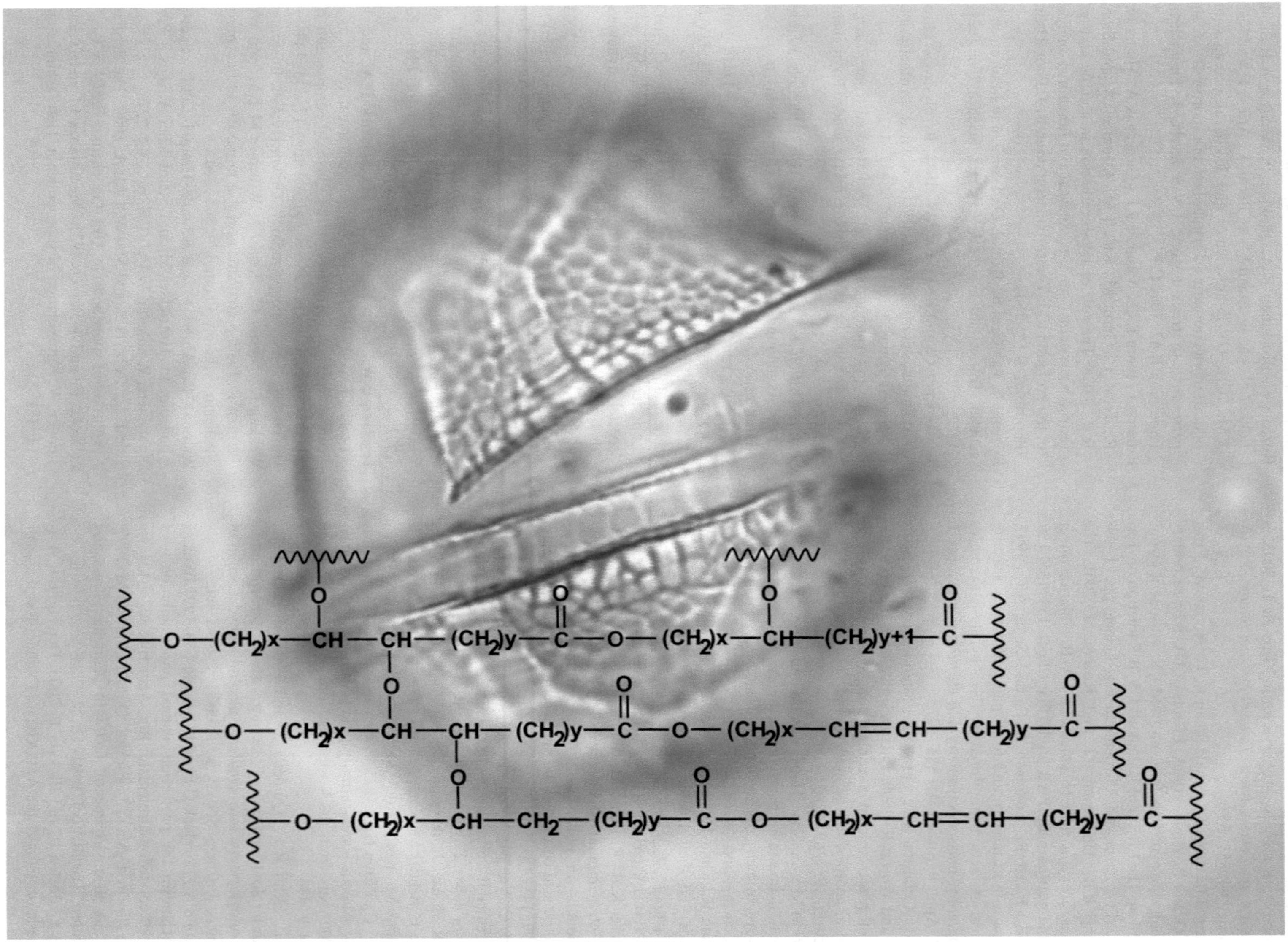

Photo. Theca of a freshwater dinoflagellate with a general chemical structure of chlorophyte algaenan (Photograph: G. Versteegh).

Plant Ecology (2006) 182:209–233
DOI 10.1007/s11258-005-9027-x

Biomacromolecules of algae and plants and their fossil analogues

Jan W. de Leeuw[1,2,3,*], Gerard J. M. Versteegh[1,2,4] and Pim F. van Bergen[5]
*[1]Royal Netherlands Institute for Sea Research, 1797 SZ 't Horntje, Texel, The Netherlands; [2]Organic Geochemistry, Faculty of Geosciences, Utrecht University, Budapestlaan 4, 3584CD, Utrecht, The Netherlands; [3]Palaeoecology, Faculty of Biology, Budapestlaan 4, 3584CD, Utrecht, The Netherlands; [4]Hanse Wissenschaftskolleg, Lehmkuhlenbusch 4, 27753, Delmenhorst, Germany; [5]Shell Global Solutions International, Badhuisweg 3, 1031 CM, Amsterdam, The Netherlands; *Author for correspondence (e-mail: deleeuw@nioz.nl)*

Received 1 September 2004; accepted in revised form 15 December 2004

Key words: Algaenan, Biomacromolecule, Diagenesis, Fossil, Recent, Sporopollenin

Abstract

A review of our current understanding of resistant biomacromolecules derived from present and past algae and higher plants is presented. Insight in the nature of recent and fossil macromolecules is strongly hampered by the difficulties in obtaining the material in pure and unaltered form. For the extant material, avoiding artificial condensation and structural alteration as a result of chemical isolation and purification of biomacromolecules requires constant attention. To date, several types of sporopollenin seem to occur. One type is characterised by oxygenated aromatic building blocks, in particular *p*-coumaric acid and ferrulic acid. The other type is thought to consist predominantly of an aliphatic biopolymer. In this review it is concluded that extant sporopollenin consists of the aromatic type, whereas the aliphatic component of fossil sporopollenin is due to early-diagenetic oxidative polymerization of unsaturated lipids. The cuticles of most higher plants contain the aliphatic biopolyester cutin. Additionally, cuticles of drought-adapted, mostly CAM plants, seem to contain the non-hydrolysable aliphatic biopolymer cutan. Only a very few algae are able to biosynthesize resistant, (fossilisable) cell walls: some Chlorophyta, Eustigmatophyta and Prasinophyta produce the aliphatic biopolymer algaenan. Some Dinophyta are also capable of producing algaenan cell walls. Additionally, some taxa produce highly resistant cyst-walls with a high proportion of aromatic moieties. For the morphologically well-preserved fossil material, contamination by organic particles other than the target taxon is hard to eliminate and can contribute to either the aliphatic or aromatic signal. Furthermore, post-mortem migration of aliphatic moieties into, and their condensation onto the macromolecule might occur, e.g. by oxidative polymerization. These phenomena hamper the evaluation of the aliphatic signature of fossil plant material and may for example explain the preservation of initially cutin-based cuticles or non-algaenan containing algae. The extent to which migration and *in situ* formation of aromatic moieties plays a role in modifying resistant algal macromolecules, notably under elevated temperature and/or pressure conditions, still remains an open question.

Introduction

To extend our understanding of present-day and future natural and anthropogenic climate change, detailed reconstructions of palaeo-environments (with limited or no human factor) and their changes through geological time are crucial. Such reconstructions may be based on instrumental, historical and so-called proxy data. Instrumental data are reliable, but are very limited in time, i.e. at the most 200 years BP. Historical data can be retrieved for millennia but become less reliable with time. Ancient reconstructions rely on proxy data from dated sediment cores, ice cores or fossils. 'Direct' proxies concern environmental entities which are measured as such in the geological record, e.g. measurements of pCO_2 in air bubbles in ice cores. 'Indirect' proxies are entities analysed from sediments, ice cores and fossils formed during the time in question and transfer functions enable these entities to be used to reconstruct past environment and climate. Indirect biomarker proxies should be stable, difficult to degrade or mineralise and have locked-up environmental parameters such as temperature, salinity, humidity, redox potential, etc. Even in the case of partial degradation or mineralisation of biomarkers, ratios of biomarkers may serve as indirect proxies provided that the rate of partial diagenesis of the biomarkers making up the ratio is the same (e.g. Müller et al. 1998; Versteegh et al. 2000; Schouten et al. 2002).

To evaluate existing proxies or to develop new proxies based on organic compounds derived from algae and higher plants, a much better understanding of the origin, the structure, the function and diagenesis of specific low- and high-molecular-weight organic matter is crucial. This chapter therefore focuses on several aspects of molecular structures of resistant biopolymers in algae and higher plants and their fossilised counterparts and can be considered as an extension and up-date of a recent review by van Bergen et al. (2004).

Sporopollenin

Studies concerning the chemical structure of extant and fossilised spore- and pollen-walls have a long history. These studies have been reviewed recently by van Bergen et al. (1995, 2004) and by de Leeuw and Largeau (1993). Based on these reviews and the references cited therein it has been concluded that several types of extant sporopollenins of a variety of ferns, conifers and angiosperms may occur. Although no systematic chemical studies are known regarding relations of these structures with haploid or diploid sporophytes, or about different spore/pollen-wall layers and the different ways different plant groups produce spores and pollen.

One type is characterised by oxygenated aromatic building blocks, in particular *p*-coumaric and ferulic acid. The other type is thought to consist predominantly of an aliphatic biopolymer of unknown structure (e.g. Guilford et al. 1988; Hayatsu et al. 1988; Domínguez et al. 1999). Whether or not two or even more different types of sporopollenins exist – and the extent to which both types may occur in the same pollen-grain or spore, in separate layers or mixed – is not clear and difficult to determine because a multitude of different, incomparable methods have been used to separate the spore- or pollen-walls from the contents of the spores and pollen consisting of lipids, proteins, polysaccharides, etc. In virtually all cases this separation has been performed chemically, sometimes under very harsh conditions. The chemical methods applied may alter the structure of sporopollenin, may remove specific parts of the sporopollenin or may not be complete in removing non-sporopollenin constituents. This problem is approached by assuming that during burial and diagenesis sporopollenin behaves conservatively and that other non-sporopollenin constituents are selectively degraded and mineralised. This assumption is justified on the basis of the presence of perfectly morphologically preserved spore- and pollen-walls in sediments without microscopically recognisable remains of the contents of spores and pollen. In other words it was assumed that diagenetic processes represent the best (geo)chemical pathway to purify sporopollenin.

However it must be noted that the diagenetic 'purification' of sporopollenin probably depends on the preservation conditions. Yule et al. (2000) demonstrated through micro-FTIR that the spores of *Lycopodium clavatum* go through various phases of chemical degradation characterised by an initial relative increase in aliphatics followed by a FTIR signal more dominated by an aromatic signal. Especially oxidative conditions might affect the final chemistry of the remaining material as illustrated by Gabarayeva et al. (2003) who demonstrated that by successive chemical treatments starting with a relatively mild glacial acetic acid followed by oxidative potassium permanganate degradation isolates the exine of pollen but progressively erodes it specifically with the duration of the treatment. Though the permanganate conditions may not have an equivalent under 'natural' conditions it illustrates that the sporopollenins investigated by Gabarayeva et al. (2003) can be regarded as inhomogeneous on a molecular level probably with respect to polymerisation grade of the monomeric components. So the degree of diagenetic purification and degradation, together with the wide variety of analytical techniques and samples analysed over time results in a extremely blurred insight into the actual chemistry of sporopollenin. Furthermore, there is no elaborate information on the effect of climatic conditions on the chemistry of sporopollenin, making it difficult to interpret the chemical data obtained from fossil materials solely.

To date, however, virtually all fossilised spore- and pollen-walls analysed consist of both aromatic and highly aliphatic moieties as revealed by pyrolytic and spectroscopic data (e.g. Schenck et al. 1981; van Bergen et al. 1993, 1995, 2004, subm.), even if their recent counterparts are almost exclusively of the aromatic type. To illustrate this phenomenon the pyrolysates of extant and fossil *Salvinia* megaspores are compared (van Bergen et al. 1993; Figure 1). In this particular example the extant megaspore material, exine and perispore, was dissected from mature plants and extracted extensively with organic solvents to remove low-molecular-weight organic matter, thus representing one of the very few extant spore wall isolations without extensive chemical treatments.

The gas chromatogram of the pyrolysate of the extant spore wall consists almost exclusively of peaks representing aromatic compounds that can be produced from a polymeric form of *p*-coumaric acid (see hereafter). Aliphatic compounds, i.e. C_{16} and C_{18} fatty acids (FAs), are present as minor compounds in the pyrolysate. It is not clear whether these aliphatic compounds originate from sporopollenin or represent membrane or other lipids associated with the material. In sharp contrast the pyrolysate of the fossil megaspore

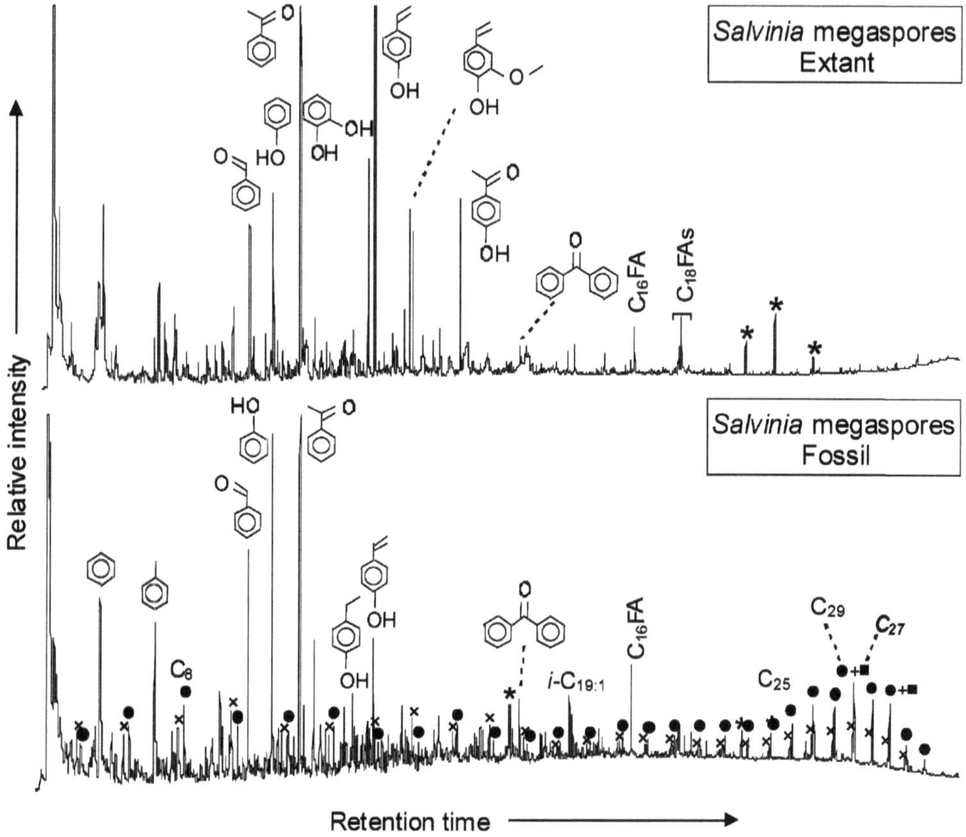

Figure 1. Comparison of pyrolysates of extant and fossil *Salvinia* megaspores. x = *n*-alk-1-enes, ● = *n*-alkanes, ■ = *n*-methylketones, * = contaminants (van Bergen et al. 1993).

wall material yields a strong aliphatic signal as indicated by the homologous series of alkane/alkene doublets. A number of aromatic compounds similar to those recognised in the modern material are still present, in particular, acetophenone and 4-vinylphenol; the latter most probably is a pyrolysis product of *p*-coumaric acid.

To explain the clear and sometimes dominant occurrence of aliphatic constituents in these and other fossilised spore- and pollen-walls it has been assumed that the corresponding or related extant sporopollenins do contain a (very) small portion of aliphatic building blocks that are selectively preserved at the cost of the aromatic constituents, i.e. cinnamic acids such as *p*-coumaric and ferulic acid, during burial and diagenesis. However, it cannot be excluded that the aliphatic constituent represents an aliphatic geopolymer produced during burial and diagenesis from low-molecular-weight lipids such as plant waxes (cf. Collinson et al. 1998) consisting of saturated and unsaturated hydrocarbons, alcohols FAs or from membrane lipids, consisting of saturated and unsaturated FAs, or, most likely, from unsaturated lipids present in the original spores and pollen that have become attached to the original sporopollenin serving as a matrix through oxidative cross linking.

Strong circumstantial evidence of such an oxidative cross linking of low-molecular-weight lipids is presented by a recent study of Versteegh et al. (2004) through a very detailed analysis of a dominantly present aliphatic constituent in morphologically very well preserved fossil algae (dinoflagellates). These algae did not contain a microscopically recognisable cell wall, thus indicating that the aliphatic constituent could not represent algaenan, a biopolymer present in several algae (van Bergen et al. 2004 for a review) and this paper. Based on extensive and detailed microscopical, chemical and spectroscopical analyses of these preserved dinoflagellates, Versteegh et al. (2004) had to conclude that a 'post-mortem polymerisation' of FA moieties from phospholipids, glycolipids or glycerol esters originally present in the cytoplasm by means of cross linking through ether bonds had taken place. Analogous to the migration and condensation of aliphatic lipids described for cavities in coals (Zhang et al. 1993), migration of additional aliphatic moieties from the surrounding environment taking part in the polymerisation process forming the preserved dinoflagellates can not be excluded either. Further detailed evidence for this oxidative cross-linking during diagenesis is described by Kuypers et al. (2002), in a study of Cretaceous black shales. In that study it has been evidenced that a substantial part of the kerogen of these shales consists of ether-linked isoprenoids derived from archaeal membranes as indicated by advanced pyrolysis GC-MS studies in combination with chemical degradation using RuO_4 oxidation. Furthermore, post-mortem formation of an ether cross-linked aliphatic biopolymer been observed in fossil arthropod cuticles (Briggs et al. 1995; Stankiewicz et al. 2000). In this case, the original chitin-based cuticles had been preserved very well morphologically, whereas the chitin was completely replaced by a full aliphatic non-hyrolysable geopolymer. Moreover, there are other indications that this oxidative lipid polymerisation during burial and diagenesis has been underestimated. For example, the oxidative polymerisation of triacylglycerols is a long known process, well known from the drying of linseed oil-based paints triggered by the autooxidation of unsaturated FAs (e.g. Blom, 1936; Fjällström et al. 2002), causes problems in vegetable oil storage and their use for diesel fuel engines (e.g. Srivastava and Prasad 2000) and is easily induced in the laboratory by keeping vegetable oil in sunlight (Versteegh et al. 2004). Finally, the absence of cutan in living *Ginkgo* suggests that fossil *Ginkgo* cuticles originate from oxidative cross linking of the less resistant, saponifiable organic entities originally suggested to be cutin (Mösle et al. 1997), later believed to be derived from cuticular lipids (Collinson et al. 1998).

The above clearly illustrates that the assumption that well-preserved fossil spore- and pollen-walls are highly representative for the chemistry of extant sporopollenin may be invalid, although further studies including FTIR and ^{13}C NMR spectroscopy are required to substantiate this. During burial and diagenesis the original biopolymer can change and/or newly generated geopolymers may be generated and may become closely associated with the original biopolymer or may even completely replace the original biopolymer without a change in morphology. Thus, these types of processes also indicate that chemical treatments of extant spore- and pollen-walls to separate sporopollenin from the cytoplasm constituents are rather problematic and have to be kept in mind when interpreting results of sporopollenin studies based on the chemical separation of sporopollenin.

In a very recent study (Boom 2004) isolated megaspore walls of *Isoetes killipi* C. Morton were treated relatively mildly by sulphuric acid to remove proteins and polysaccharides and after extractions to remove low-molecular-weight lipids. The resistant material obtained was analysed using a series of complementary analysis methods such as Direct Temperature-resolved Mass Spectrometry (DT-MS), Curie-point pyrolysis-GC/MS (Py-GC/MS) and Fourier Transform-IR (FTIR). The data showed that the sporopollenin of these spores consists of polymerised *p*-coumaric acid because the analysis data were very similar to those of a synthetic *p*-coumaric acid-based dehydrogenation polymer (DHP) (Figure 2). The only difference between the synthetic polymer and the sporopollenin was explained by a small additional contribution of ferulic acid moieties in the *Isoetes* spore walls. The chemical treatment used in this particular case had not produced an aliphatic polymer.

Based on this and earlier results (e.g. Wehling et al. 1989; Mulder et al. 1992) and on re-inspection of other literature data concerning the oxidised aromatic type sporopollenin it may be concluded that sporopollenin consists of DHP type polymers based on *p*-coumaric acid and some ferulic acid, predominantly through ether linkages at the alpha, beta and C_4 carbon atoms and that the aliphatic constituent in fossilised spore- and pollen-walls is not part of the original

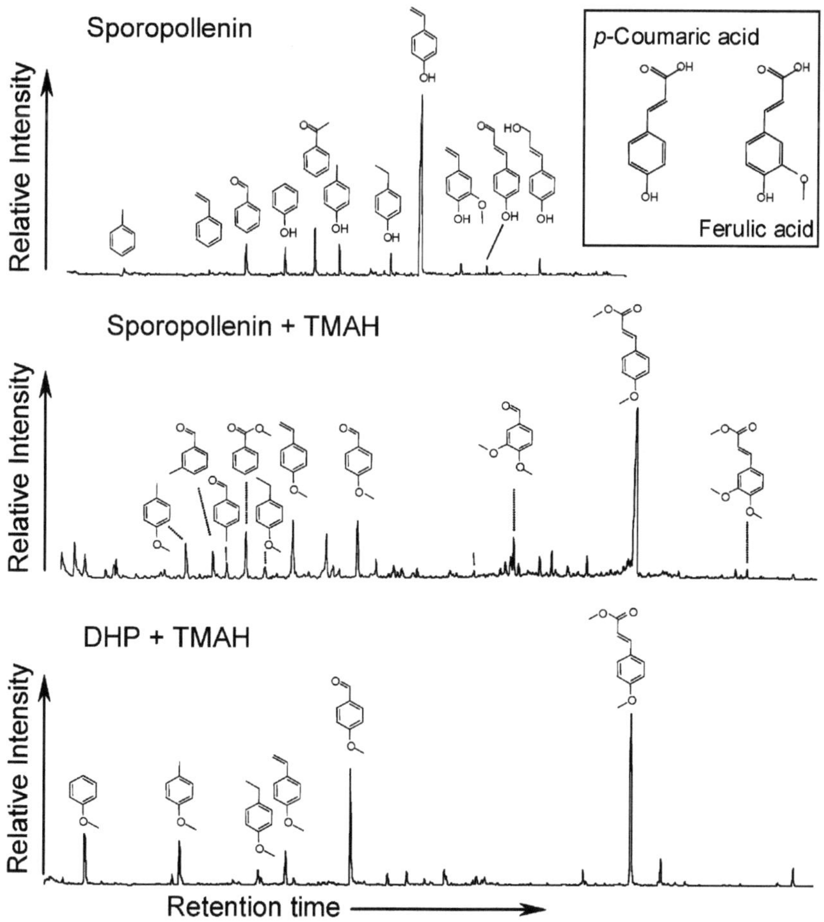

Figure 2. a. Pyrolysate of *Isoetes killipii* megaspore walls. b. Pyrolysate of *Isoetes killipii* megaspore walls with tetramethylammonium hydroxide (TMAH) coating. Upon thermolysis the TMAH methylates the oxygen radicals preventing secondary reactions of the pyrolysis products. c. Pyrolysate of a synthetic, *p*-coumaric acid-based, dehydrogenation polymer (DHP). The similarity of this pyrolysate with that of the *Isoetes killipii* megaspore walls confirms the high proportion of *p*-coumaric acid in the latter. (Adapted from (Boom 2004).

sporopollenin structure. Its highly variable presence and signature in many fossil spore- and pollen-walls may be due to the co-occurrence of the aromatic constituent representing the original sporopollenin, i.e. polymeric *p*-coumaric acid present as such or transformed during diagenesis, and an allochtonously-derived aliphatic component, i.e. a diagenetically produced geopolymer as indicated above. Depending on its availability, this aliphatic component may be derived from the entity itself and/or, from the outside, i.e. directly from the sediment.

This hypothesis implies that the relative proportion of *p*-coumaric acid or, more generally, cinnamic acids or their diagenetic counterparts present in fossil spore- and pollen-walls can not be used straightforwardly as a proxy for UV-B exposure as has been suggested earlier (Rozema et al. 2001, 2002). It is, however, interesting to investigate whether ratios of cinnamic acids or even the degree of esterified *p*-coumaric acid over ether-bound *p*-coumaric acid, can be used as UV-B proxies.

Cutin and cutan

The dominant, sometimes exclusive, presence of a non-hydrolysable aliphatic polymer in well-preserved and morphologically recognisable fossil cuticles led to the assumption that such a polymer occurs as a biopolymer labelled cutan, in a variety of extant plant cuticles (Figure 3; de Leeuw and Largeau 1993; van Bergen et al. 2004). This cutan should not be confused with cutin, a long and well-known biopolyester present in almost every plant cuticle (see hereafter). In several cases, however, the non-hydrolysable aliphatic polymer thought to represent the preserved and selectively preserved biopolymer cutan was only present in the fossil cuticles and not in the extant counterparts consisting of cutin only (Mösle et al. 1997; Collinson et al. 1998). As is the problem with sporopollenin (see above) the genesis of an aliphatic geopolymer replacing the cutin in the original cuticles has to be assumed (cf. Tegelaar et al. 1991; Collinson et al. 1998). In this particular case the most likely candidates to be transferred diagenetically into an aliphatic polymer are cutin (cf. Tegelaar et al. 1991) and/or leaf waxes (cf. Collinson et al. 1998), since these aliphatic constituents are part of the cuticle or very closely associated with it, respectively. Pyrolysates of fossil cuticles consist mostly of homologous series of alkanes and alkenes with carbon chain lengths up to C_{30} to C_{35} (Figure 3) indicating that leaf waxes are the preferred

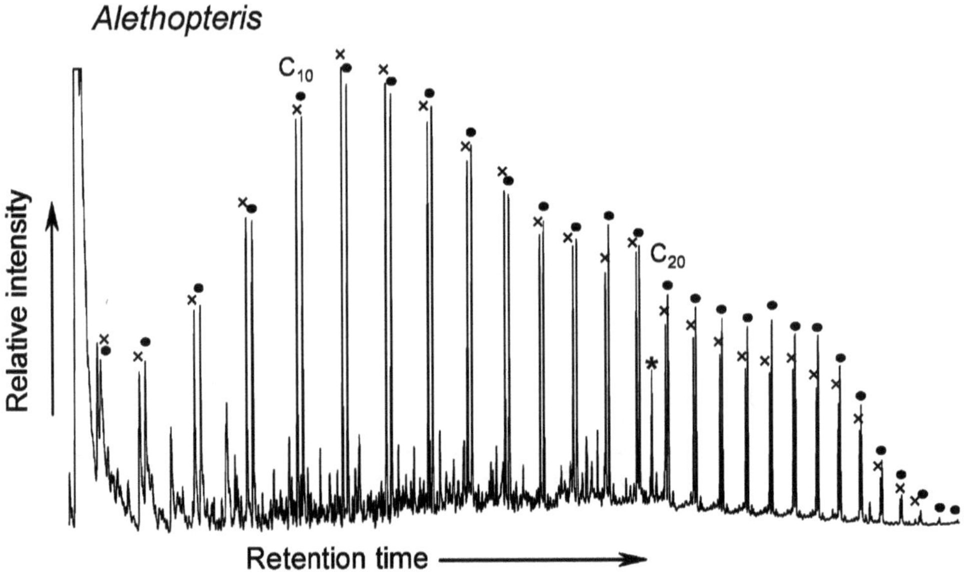

Figure 3. Pyrolysate of *Alethopteris lesquereuxi* pinnule cuticle (Carboniferous). x = *n*-alk-1-enes, ● = *n*-alkanes, * = contaminants (after Collinson et al. 1994).

candidates since the FA building blocks of cutin are C_{16} and C_{18} saturated and unsaturated FAs (e.g. Holloway 1982).

Once again, the problem is whether or not there exists a non-hydrolysable biopolymer cutan next to the biopolyester cutin in extant cuticles, also taking into consideration that chemical treatments performed to purify cutan from the cuticle matrix may lead to artefacts similarly to those mentioned above for sporopollenin. Until recently, rigourous studies of extant cuticles showed that the biopolymer cutan is indeed present in the cuticle of *Agave americana* and *Clivia miniata*, but that the presence of cutan previously reported in a number of other plant cuticles could not be confirmed (Collinson et al. 1998). Very recently however, cutan has been reported as a significant component of the cuticles in drought-adapted, mostly CAM plants (Boom 2004). Although these cutans were isolated chemically by extractions and acid- and base treatments no artificial aliphatic polymer was produced since other, non-drought-adapted plant cuticles treated identically, thus serving as blanks, yielded no cutan. Boom et al. (2005) speculate that the presence of cutan in the thick cuticles of these CAM plants serves as a physiological and chemical adaptation to survive drought conditions. Further studies have to show whether the presence of cutan in fossil plants cuticles can be used as indicators, i.e. proxies, for drought. These data imply that the selectively preserved biopolymer cutan must be discriminated from the diagenetically produced aliphatic geopolymer in fossil cuticles.

Several studies have indicated the presence of cinnamic acids, in particular *p*-coumaric and ferulic acid, as constituents of isolated extant and fossil cuticles from leaves and seeds from several plants (e.g. Deshmukh et al. 1964; Kolattukudy 1981; Holloway 1982; van Bergen 1994; McKinney et al. 1996; Collinson and van Bergen 2004; van Bergen et al. 2004). These aromatic components may be derived from the cell wall of the epidermis directly underlying the cuticle. Further studies have to reveal if these aromatic components play a role in the shielding of UV-B and if they can serve as proxies of UV-B irradiation in the palaeo-environment using well-dated fossilised cuticles (Rozema et al. 2001, 2002).

Resistant algal biomacromolecules

Micro-algae are diverse and abundant in aquatic environments. Most of them have no, or only very limited preservation potential. Nevertheless, a diverse and rich micro-algal fossil record of aquatic palynomorphs exists. This record is an important source of information for solving stratigraphic, palaeoenvironmental (e.g., palaeoclimatological) and evolutionary questions. For this purpose, almost solely the morphological characteristics of the fossils have been used, whereas their potential at the molecular level hardly has been exploited. As a result our knowledge of the structure, chemical diversity, function and diagenesis of the cell walls of algae is still very fragmentary. Improving this knowledge addresses directly the formation, transport and degradation of organic matter, which is central to our understanding of biogeochemical cycles, the formation of petroleum, gas and coal and the history of life and its environment. Below, an overview will be given on our current knowledge of cell wall biomacromolecules from micro-algae and their fossil geomacromolecular analogues.

Cell walls from extant micro-algae

Only the biomacromolecules of cell walls that are considered to fossilise will be discussed here. For cell walls of extant algae for which fossils may be unknown this implies that they resist the methods that palynologists have used to isolate fossil palynomorphs from sediments. In organic geochemical practice this means that the walls of extant algae must be able to resist base and acid hydrolysis. However, this criterion does not seem to hold since several fossil palynomorphs do not resist such harsh hydrolytic treatments.

There has been considerable confusion on the relation between wall ultrastructure and the occurrence of such ‘fossilisable’ algal walls. The walls of extant and fossil algae often consist of several layers. For the Chlorophyta, the presence of a trilayered (trilaminar) outer wall is often associated with the presence of hydrolysis resistant walls. However, there is no strict relationship between both features. Species that lack a hydrolysis resistant wall may have a trilaminar wall and *vice versa* (for a discussion see, de Leeuw and Largeau 1993).

Furthermore, the different methodologies of isolating the walls of extant algae have led to considerable confusion on the chemical wall structure. This is largely related to artificial polymerisation of the cell contents induced by some isolation methods (Brunner and Honegger 1985; Gelin et al. 1997; Allard et al. 1998). Isoprenoids (e.g. carotenoids), which are highly abundant in many algae, easily become incorporated in these artefacts with the result that isoprenoids have been claimed to be important building blocks of resistant algal cell walls (e.g. Burczyk 1987a, b; Derenne et al. 1996; Kokinos et al. 1998). Despite these problems, there are also early studies e.g. on *Botryococcus braunii* (Berkaloff et al. 1983) and *Tetraedron minimum* (Goth et al. 1988), suggesting that such walls are highly aliphatic, i.e. consist of non-cyclic carbon chains which have been cross-linked. Careful re-examination, with more recent technology avoiding the co-analysis of condensed cytoplasm (e.g. by breaking the cell walls prior to chemical treatment), (Brunner and Honegger 1985; Allard et al. 1998; Blokker et al. 1998a) did not confirm this but instead demonstrated either the absence of a resistant wall or the presence of a highly aliphatic cell wall composed of unbranched, but cross-linked carbon atoms, termed algaenan. Consequently, claims in the older literature that a given taxon produces resistant macromolecules, should be evaluated with due consideration. Although not all algae have been subject to such re-examination yet, all evidence suggests that resistant isoprenoid-based biomacromolecules are not produced by modern plants (algae and higher plants).

To date, only two biochemical pathways seem to lead to resistant algal walls (and resistant plant macromolecules in general): (I) the acetate–malate pathway (leading via lipid-synthesis to algaenans, cutin and cutan), and (II) the phenylpropanoid pathway (leading to e.g. sporopollenin, and probably dinosporins). However, we have to acknowledge that only a very limited portion of the living and fossil algae has been studied for the presence and composition of acid and base resistant cell walls. Most species belong to the Chlorophyta and most are from fresh water environments. The marine realm, with the richest and longest fossil record, has hardly been exploited. New pathways leading to fossilisable biomacromolecules may therefore still await discovery. Despite this bias it is clear that only a few living taxa are capable of producing such walls (Table 1).

Table 1. Microalgae and their fossil palynomorphs.

Bacillariophyta	– (but + + record of opal skeletons)
Chlorarachniophyta	–
Chloromonadophyta	–
Chlorophyta	+ (mainly freshwater Chlorococcales and mesospores of Zygnematales)
Cryptophyta	–
Dinophyta	+ + (mainly marine, very abundant long and diverse record)
Euglenophyta	+ (but very rare)
Eustigmatophyta	–
Haptophyta	– (but + + for record of $CaCO_3$ skeletal elements)
Prasinophyta	+
Xanthophyta	–
Acritarcha	+ + (mainly marine, many Palaeozoic and earlier taxa. Polyphyletic)

Fossilised walls of micro-algae

From the above it is clear that the presence of cell contents hampers the isolation and analysis of cell walls of living algae. This problem is absent for analysis of fossil palynomorphs. However, three other difficulties complicate the evaluation of fossil algal walls. First, isolating pure, monotypic assemblages is difficult. The unusually high abundance of *Pediastrum* fossils (60% of the palynomorphs) in a late Miocene sediment led to the conclusion that the *Pediastrum* walls contain aromatic or partly aromatic compounds (Sinninghe Damsté et al. 1993) contradicting later analyses indicating an aliphatic wall for this taxon (Blokker et al. 2000). Second, the lack of recent counterparts, as is the case for the Acritarcha. Third, the transformation of the original biomacromolecule into a geomacromolecule which would provide an alternative explanation to contamination for explaining the results on the Miocene *Pediastrum* walls. At least, if we assume that aromatic moieties 'invaded' the originally aliphatic biomacromolecule *post-mortem*. Despite these problems, very close morphological *and* chemical correspondence does occur between fossil algae and their living counterparts, e.g. for *T. minimum* (Goth et al.

1988). This demonstrates that algaenans can survive relatively unchanged in sediments for millions of years.

Compared with algal diversity in present time, the fossil record is severely incomplete. Some algal groups seem to leave no, or almost no, fossils, e.g. like the Euglenophyta (Gray and Boucot 1989). Others leave lipids (e.g. Eustigmatophyta) and/or biominerals (e.g., Bacillariophyta and Haptophyta) but are unknown as fossil palynomorphs. However, some extant Eustigmatophyta and Chlorophyte taxa without known micro-fossil record have been shown to produce hydrolysis resistant walls. It seems likely that their fossil cell walls are present in the sediments but have not been recognised for a lack of morphological characteristics. The ubiquitous presence of 'ultralaminae' with an aliphatic nature in sediments corroborates with this idea (Derenne et al. 1991). Of the algal groups that have been identified from the fossil record, the palynomorphs of Prasinophyta and Chlorophyta are known from the Proterozoic (Knoll 1992, 1996) and continue to the present day. Their fossil records are, however, not very diverse (Batten and Grenfell 1996; Batten 1996; Guy-Ohlson 1996; van Geel and Grenfell 1996; Wicander et al. 1996). Only one, still existing, group of algae, the dinoflagellates, is known to have given rise to a diverse record of largely marine microfossils since the early Mesozoic (Fensome et al. 1999). A second rich and diverse group of fossil palynomorphs is the Acritarcha. This group consists entirely of taxa with an unknown biological affinity but is considered to include a large proportion of palynomorphs from micro-algae. Acritarchs have been found in Precambrian and younger strata. They reached particularly high diversity during the Palaeozoic (Strother 1996). It must be noted that the fossil record from lacustrine environments is much more limited that that from marine environments so that the preservation potential of fresh water algae (notably Chlorophyta) must be underestimated.

Algaenans

Algaenans of extant micro-algae

Algaenans (Tegelaar et al. 1989) represent a series of acid and base-resistant aliphatic biomacromolecules. It is important to note that other compounds may be associated with the algaenan e.g. isoprenoids in the case of *B. braunii* race L (Bertheas et al. 1999) or sugars in the case of *Coelastrum sphaericum* (Rodríguez et al. 1999) but they are removed upon hydrolysis. The aliphatic nature suggests that algaenans are biosynthesised via the acetate/malate pathway which leads to FAs. Algaenans appear widespread in Chlorophyta and have been detected in some Eustigmatophyta and a member of the Dinophyta but have not been detected in Bacillariophyta or Haptophyta (Table 2). The walls of some Prasinophyta and the pellicles of several Dinophyta have also been reported to be resistant (e.g. Morrill and Loeblich III 1981; Aken and Pienaar 1985) but information on the wall chemistry is too sparse to infer that they consist of algaenan. For the few more closely analysed algae, three general algaenan structures have been proposed. For most Chlorophyta for which structural information is available, the building blocks consist of linear C_{22} to C_{34} even-numbered carbon chains with functional groups at the α, ω, ω^9 and sometimes ω^{18} positions. In the algaenan the functional groups cross-link the monomers with ether and ester bonds (Blokker et al. 1998a, 1999). However, the algaenan of the Chlorophyte *B. braunii*, (at least race A) seems to be based on unsaturated aliphatic aldehydes and unsaturated hydrocarbons with on average 40 carbon atoms. Here, the monomers are cross-linked by acetal and ester bonds (Simpson et al. 2003). The third algaenan type proposed is produced by the Eustigmatophyta. Here, the building blocks are probably mid-chain (ω^{15} to ω^{18}) C_{28} to C_{36} diols and C_{30} to C_{32} alkenols, as well as C_{25}, C_{27}, and C_{29} (poly)unsaturated free hydrocarbons which in the algaenan are cross-linked with mid-chain ether bonds (Gelin et al. 1997). Algaenans are among the most frequently studied resistant algal macromolecules. They are mostly isolated from actively growing cultures. The active metabolism implies that the algaenan walls must contain pores to exchange compounds with the outer environment. Despite such pores, algaenan walls have been shown to form an effective barrier for extracellular enzymes (e.g. Atkinson Jr et al. 1972; Syrett and Thomas 1973) and detergents (Biedlingmaier et al. 1987; Corre et al. 1996).

During a phase of quiescence, the reduced metabolic activity allows for much less exchange with the outer environment. These cyst walls could

Table 2. Occurrence and composition of resistant biomacromolecules in recent algae.

Taxon	Biomacromolecule	Latest Reference(s)
Bacillariophyta (Diatoms)		
Chaetoceros calcitrans (CCAP1010/5)	–[a]	(Gelin et al. 1999)
Chaetoceros muelleri (CS-176)	–	(Gelin et al. 1999)
Nitzschia palea (UTEX1813)	–	(Gelin et al. 1999)
Skeletonema costatum (CCAP 1077/1b)	–	(Gelin et al. 1999)
Chlorophyta (green algae)		
Bulbochæte sp.	? non-sugar-based resistant 'chitinous' outer wall layer (filamentous taxon)	(Tiffany 1924)
Botryococcus sudeticus (braunii A race, UTEX572)	algaenan: non-isoprenoid polyaldehyde network (poly-botryal) C_{40} average chain length	(Metzger and Largeau 2002; Simpson et al. 2003)
Botryococcus braunii (SAG 30.81)	algaenan: C_{32} di-unsaturated α, ω-dialdehyde polymer	(Blokker et al. 2000)
Botryococcus braunii (B race)	algaenan: aliphatic polymer $< C_{31}$ chains	(Kadouri et al. 1988)
Botryococcus braunii (L race Yamoussoukro strain)	algaenan similar to A race, but isoprenoids attached to algaenan backbone	(Bertheas et al. 1999)
Brachiomonas submarina (CCAP7/2b)	–	(Gelin et al. 1997)
Chlorella[b] *ellipsoidea* (CCAP211/1a)	–	(Burczyk et al. 1999)
Chlorella emersonii (CCAP211/8p)	algaenan: C_{26} & C_{28} FA & OH; C_{30} & C_{32}-OH-FA & α, ω diols	(Allard and Templier 2001)
Chlorella fusca (several strains)	+[c]	(Burczyk et al. 1999)
Chlorella nana (type strain)	+ acetolysis resistant, multiple sporopollenin layers	(Rascio et al. 1979)
Chlorella marina (CCAP211/27)	–	(Allard and Templier 2000)
Chlorella minutissima (Utex2219)	algaenan	(Allard and Templier 2000)
Chlorella minutissima marina (Utex2341)	–	(Allard and Templier 2000)
Chlorella pyrenoidosa (A-24)	–	(Burczyk et al. 1999)
Chlorella saccharophila (1 str. + CCAP211/1a)	–	(Burczyk et al. 1999)
Chlorella sorokiniana (2 str. + CCAP211/8k)	–	(Burczyk et al. 1999)
Chlorella sorokiniana (Utex1230)	algaenan	(Zelibor Jr et al. 1988)
Chlorella spaerckii (CCAP211/29A)	–	(Gelin et al. 1997)
Chlorella vacuolatus (CCAP211/8b)	algaenan: up to C_{27}polymethylenic chains	(Derenne et al. 1992a; Burczyk et al. 1999)
Chlorella vulgaris (4 str. + CCAP211/1e)	–	(Burczyk et al. 1999)
Chlorococcum sp. (CCAP11/62)	–	(Gelin et al. 1997)
Cladophora glomerata	? non-sugar-based resistant 'chitinous' outer wall layer (filamentous taxon)	(Wurdack 1923)
Coccomyxa dispar	+ acetolysis, K_2CrO_4 (30%) resistant (free living taxon)	(Honegger and Brunner 1981)
Coccomyxa glaronensis	+ (lichen phycobiont)	(Brunner and Honegger 1985)
Coccomyxa tirolensis	+ (lichen phycobiont)	(Brunner and Honegger 1985)
Coelastrum proboscideum var.*dilatatum* (UTEX282)	+ acetolysis resistant	(Marchant 1977)
Coelastrum reticulatum (SAG8.81)	algaenan: mainly polymer of $\Delta\omega^9$ $C_{30}\omega$-OH FA	(Blokker et al. 2000)
Coelastrum sphaericum var. *dilatatum*	algaenan: polymethylenic chains (with amido groups ?)	(Rodríguez and Cerezo 1996)
Cylindrocapsa geminella	– filamentous taxon	(Tiffany 1924)

Draparnaldia plumosa	– filamentous taxon	(Wurdack 1923)
Dunaliella tertiolecta	algaenan	(Zelibor Jr et al. 1988)
Dysmorphococcus globosus (UTEX65)	– lorica vegetative cells	(Porcella and Walne 1980)
Elliptochloris bilobata	+ (lichen phycobiont)	(Brunner and Honegger 1985)
Elliptochloris sp.	+ (lichen phycobiont)	(Brunner and Honegger 1985)
Microsopra willeana	– filamentous organism	(Tiffany 1924)
Myrmecia biatorellae	– (lichen phycobiont)	(Brunner and Honegger 1985)
Myrmecia reticulata	– (lichen phycobiont)	(Brunner and Honegger 1985)
Myrmecia sp.	– (lichen phycobiont) the resistant walls reported by König and Preveling (1980) may be considered artefacts	(Brunner and Honegger 1985)
Nannochloris sp. (CCAP 251/2)	–	(Gelin et al. 1997)
Nanochlorum eucaryotum	algaenan: up to C_{30} polymethylenic chains (salt water taxon)	(Derenne et al. 1992a)
Oedogonium crassum amplum	? unidentified non-sugar-based resistant outer wall layer (filamentous taxon)	(Wurdack 1923)
Oedogonium irregulare	? non-sugar-based resistant 'chitinous' outer wall layer (filamentous taxon)	(Wurdack 1923)
Oocystus solitaria (SAG 83.80)	–	(Blokker 2000)
Pediastrum braunii (SAG 43.85)	algaenan: polymer of mainly C_{30} & $C_{32}\omega$-OH FA	(Blokker 2000)
Pediastrum boryanum (SAG N.N.)	algaenan: polymer of mainly $\Delta\omega^9$ C_{30} & C_{32}, $\Delta\omega^{9,19}$ C_{30}, $\Delta\omega^{9,18}$ C_{32} ω-OH-FA	(Blokker et al. 1998b)
Pediastrum duplex ()	+ 'sporopollenin-like'	(Gunnison and Alexander 1975)
Pediastrum kawraiskyi (SAG35.81)	algaenan: C_{30} & C_{32} ω-OH-FA	(Blokker 2000)
Phycopeltis epiphyton	+ acetolysis and K_2CrO_4 (30%) resistant (subaerial taxon)	(Good and Chapman 1978)
Prototheca wickerhami	+ (but terpene units need verification)	(Puel et al. 1987)
Prototheca moriformis (CCAP263/2)	–	(Atkinson Jr et al. 1972)
Prototheca portoricensis (CCAP263/3b)	+	(Atkinson Jr et al. 1972)
Prototheca zopfii (CCAP263/5)	+	(Atkinson Jr et al. 1972)
Pseudochlorella pyrenosidosa	–	(Brunner and Honegger 1985)
Pseudochlorella sp.	–	(Brunner and Honegger 1985)
Pseudodidymocystis planctonica (SAG40.98)	+ acetolysis resistant	(Hegewald and Deason 1989)
Pseudodidymocystis fina	+ acetolysis resistant	(Hegewald and Deason 1989)
Pseudoschroederia punctata (UTEX LB2490)	– not acetolysis resistant	(Hegewald and Deason 1988)
Pseudotrebouxia corticola (UTEX909)	+ reported but probably –, see *Trebouxia* (taxa are lichen phycobionts)	(König and Peveling 1984)
Pseudotrebouxia impressa (UTEX893)	+ reported but probably –, see *Trebouxia* (taxa are lichen phycobionts)	(König and Peveling 1984)
Scenedesmus armatus (Paris)	algaenan	(Allard and Templier 2000)
Scenedesmus longus (Indiana Cult. Coll. 614)	+ acetolysis and boiling 6N NaOH resistant	(Staehelin and Picket-Heaps 1975)
Scenedesmus communis (CCAP 276/4b)	algaenan: C_{26} & C_{28} FA & OH; C_{30} & C_{32}-OH-FA & α, ω diols	(Allard and Templier 2001)
	polymer of mainly $\Delta\omega^9$ C_{30} ω-OH-FA	(Blokker et al. 1998b)
Scenedesmus obliquus (3+ strains)	algaenan	(Zelibor Jr et al. 1988; Burczyk et al. 1999)
Scenedesmus pannonicus (Utex77)	algaenan (=CCAP276/4a)	(Allard and Templier 2000)
Scenedesmus subspicatus (Göttingen)	algaenan	(Allard and Templier 2000)
Staurastrum sp.	+ but lignin-like	(Gunnison and Alexander 1975)
Sorastrum spinulosum (SAG B40.81)	algaenan: $\Delta\omega^9$ C_{30} & $C_{32}\omega$-OHFA	(Blokker 2000; Blokker et al. 2000)
Stichococcus bacillaris (CCAP379/32)	–	(Gelin et al. 1997)
Trebouxia spp.	– the resistant walls reported by König and Preveling (1980, 1984) may be considered artefacts (taxa are lichen phycobionts)	(Honegger and Brunner 1981) (Brunner and Honegger 1985)

Table 2. (Continued).

Taxon	Biomacromolecule	Latest Reference(s)
Trentepholia sp. (UTEX1227)	– (both free living and as lichen phycobiont)	(Brunner and Honegger 1985)
Tribonema bombycina	– filamentous organism	(Tiffany 1924)
Tribonema urticulosa	– filamentous organism	(Tiffany 1924)
Tribonema minus	– filamentous organism	(Tiffany 1924)
Tetraedron minimum (Göttingen)	algaenan: C_{26}, C_{28} FA; some C_{30}-C_{34} ω-OH FA, minor aromatics	(Goth et al. 1988; Allard and Templier 2001)
	polymer of mainly $\Delta\omega^9$ C_{32} & C_{34} ω-OH-FA	(Blokker et al. 1998b)
Vaucheria geminata	– filamentous taxon	(Wurdack 1923)
Zygnemataceae	– filaments of *Debarya decussata, Mougeotia* (5 spp.), *Spirogyra* (23 spp.) species names are listed below in section 'spores and resting stages'	(Tiffany 1924)
Zygnema cruciatum	– wall of filaments made of pectose and cellulose	(Wurdack 1923)
Zygnema sp.	? NMR spectrum does not support algaenan (filamentous organism)	(Zelibor Jr et al. 1988)
Zygnema spp.	– species are *collinsianum, decussatum, ericetorum, insignis, pectinatum* and *stellinum*	(Tiffany 1924)
Dinophyta (Dinoflagellates)		
Alexandrium acatenella (ML524)	–	(Morrill and Loeblich III 1981)
Alexandrium catenella (ML497; ML525)	–	(Morrill and Loeblich III 1981)
Alexandrium monilatum (ML487)	–	(Morrill and Loeblich III 1981)
Alexandrium tamarense (6 strains)	–	(Morrill and Loeblich III 1981)
Alexandrium tamarense (ML440)	+ acetolysis resisant	(Morrill and Loeblich III 1981)
Amphidinium carterae (UTEX1561)	–	(Morrill and Loeblich III 1981)
Amphidinium corpulentum (UTEX1652)	–	(Morrill and Loeblich III 1981)
Amphidinium operculatum (ML137; ML143)	–	(Morrill and Loeblich III 1981)
Ceratocorys horrida (ML496)	–	(Morrill and Loeblich III 1981)
Crypthecodinium cohnii (UTEX1649)	+ acetolysis resistant	(Morrill and Loeblich III 1981)
Ensiculifera loeblichii (UTEX1595)	+ acetolysis resistant	(Morrill and Loeblich III 1981)
Fragilidinium heterolobum (ML491)	–	(Morrill and Loeblich III 1981)
Heterocapsa illdefina (ML495)	–	(Morrill and Loeblich III 1981)
Heterocapsa niei (UTEX1654)	+ acetolysis resistant	(Morrill and Loeblich III 1981)
Heterocapsa pygmaea (4 strains)	+ acetolysis resistant	(Morrill and Loeblich III 1981)
Heterocapsa triquetra (ML532)	–	(Morrill and Loeblich III 1981)
Heterocapsa triquetra (NEPCC89)	+ acetolysis resistant	(Morrill and Loeblich III 1981)
Gloeodinium montanum (UTEX1651)	+ acetolysis resistant	(Morrill and Loeblich III 1981)
Gonyaulax diegensis (UTEX1992)	–	(Morrill and Loeblich III 1981)
Gonyaulax grindleyi (UTEX1956)	–	(Morrill and Loeblich III 1981)
Gonyaulax grindleyi (ML511)	+ acetolysis resistant	(Morrill and Loeblich III 1981)
Gonyaulax sphaeroidea (UTEX1947)	–	(Morrill and Loeblich III 1981)
Gonyaulax spinifera (ML502)	–	(Morrill and Loeblich III 1981)
Gymnodinium catenatum (CS301-theca)	algaenan: up to C_{35} *n*-alkyl units, maximum at C_{14}	(Gelin et al. 1999)
Gymnodinium sp. (UTEX1654)	–	(Morrill and Loeblich III 1981)

Gymnodinium sp. (FCRG47)	+ acetolysis resistant	(Morrill and Loeblich III 1981)
Gymnodinium sp. (LG161)	–	(Morrill and Loeblich III 1981)
Gyrodinium resplendens (UTEX1655)	–	(Morrill and Loeblich III 1981)
Gyrodinium dorsum (LG21)	–	(Morrill and Loeblich III 1981)
Lingulodinium polyedrum (red tide; ML492)	+ acetolysis resistant but may be contaminated; strain not tested	(Morrill and Loeblich III 1981)
Ostreopsis siamensis (ML538)	–	(Morrill and Loeblich III 1981)
Oxyrrhis marina (LG91)	–	(Morrill and Loeblich III 1981)
Peridinium balticum (UTEX1563)	+ acetolysis resistant	(Morrill and Loeblich III 1981)
Peridinium cinctum (CCAP1134/2)	+ acetolysis resistant	(Morrill and Loeblich III 1981)
Peridinium cinctum f.*ovoplanum* (UTEX1336)	+acetolysis resistant	(Morrill and Loeblich III 1981)
Peridinium foliaceum (UTEX1688 & 2006)	+ acetolysis resistant	(Morrill and Loeblich III 1981)
Peridinium gatunense (UTEX2051)	+ acetolysis resistant	(Morrill and Loeblich III 1981)
Peridinium volzii (UTEX2175)	+ acetolysis resistant	(Morrill and Loeblich III 1981)
Peridinium willei (ML488)	+ acetolysis resistant	(Morrill and Loeblich III 1981)
Peridinium willei (UTEX2028)	–	(Morrill and Loeblich III 1981)
Prorocentrum cassubicum (UTEX1596)	–	(Morrill and Loeblich III 1981)
Prorocentrum mexicanum (CS28-theca)	–	(Gelin et al. 1999)
Prorocentrum micans (CS28-theca)	–	(Gelin et al. 1999)
Prorocentrum minimum (ML5 & 403)	–	(Morrill and Loeblich III 1981)
Pyrocystis lunula Schütt	+ acid, base and acetolysis resistant	(Swift and Remsen 1970)
Pyrocystis pseudonocticula Murray	– (in fact not explicitly mentioned)	(Swift and Remsen 1970)
Pyrocystis fusiformis Murray (ML493)	–	(Swift and Remsen 1970; Morrill and Loeblich III 1981)
Scrippsiella gregaria (UTEX1948)	+ acetolysis resistant	(Morrill and Loeblich III 1981)
Scrippsiella trochoidea (FCRG72; ML540)	+ acetolysis resistant	(Morrill and Loeblich III 1981)
Woloszynskia coronata (CCAP1117/2)	+ acetolysis resistant	(Morrill and Loeblich III, 1981)
Zooxanthella micro-adriatica (ML395 & 406)	–	(Morrill and Loeblich III, 1981)
Eustigmatophyta		
Nannochloropsis sp. (CCAP849/7)	similar to *N. salina* (also functionality at C_9 ?)	(Gelin et al. 1997)
Nannochloropsis granulata (CCMP529)	algaenan	(Gelin et al. 1999)
Nannochloropsis oculata (CS179)	algaenan	(Gelin et al. 1999)
Nannochloropsis salina (CCAP849/4)	C_{28}-C_{36} alkyl diols & alkenols, $C_{25,\ 27,29}$ alkenes; functionalities around C_{15}	(Gelin et al. 1997)
Haptophyta		
Emiliania huxley (NIOZ)	–	(Gelin et al. 1999)
Phaeocystis sp.(North Sea)	–	(Gelin et al. 1999)
Prasinophyta		
Tetraselmis chui (CS28)	–	(Gelin et al. 1999)
Spores and resting stages		
Chlorophyta		
Chlamydomonas monoica (UTEX220)	algaenan: C_{22}-C_{30} *n*-OH & *n*-FA & α, ω OH-FA & di-FA polymer	(Blokker et al. 1999)
Chlamydomonas geitleri (Ettl 1966/3,	+ acetolysis resistant	(Zárský et al. 1985)
Debarya decussata	? see *M. calcarea*	(Tiffany 1924)

Table 2. (Continued).

Taxon	Biomacromolecule	Latest Reference(s)
Dunaniella sp. (aplanospores)	algaenan	(Blokker 2000)
Dysmorphococcus globosus (culture?)	+ acetolysis resistant	(Porcella and Walne 1980)
Haematococcus pluvialis (akinetes)	algaenan: similar to *C. monoica*	(Blokker 2000; Montsant et al. 2001)
Mougeotia calcarea	? but, wall contains discontinuously non-sugar-based 'chitinous'-deposits	(Tiffany 1924)
Mougeotia laevis	+ acetolysis, HF (40%) and K_2CrO_4 (30%) resistant	(de Vries et al. 1983)
Mougeotia genuflexa	? see *M. calcarea*	(Tiffany 1924)
Mougeotia quadrangulata	? see *M. calcarea*	(Tiffany 1924)
Mougeotia robusta	? see *M. calcarea*	(Tiffany 1924)
Mychonastes desiccatus	+ acetolysis resistant	(Hull et al. 1985)
Sirogonium melanosporum	+ acetolysis, HCl (3%)–KOH (2%)–H_2SO_4 (65%) resistant	(Hull et al. 1985)
Spirogyra acanthomorpha	+ acetolysis, HF (40%) and K_2CrO_4 (30%) resistant	(de Vries et al. 1983)
Spirogyra calospora	+ acetolysis resistant	(Ashraf and Godward 1980)
Spirogyra crassa	+ acetolysis resistant	(Ashraf and Godward 1980)
Spirogyra gracilis	+ acetolysis, HF (40%) and K_2CrO_4 (30%) resistant	(de Vries et al. 1983)
Spirogyra hassallii	+ acetolysis, HF (40%) and K_2CrO_4 (30%) resistant	(de Vries et al. 1983)
Spirogyra jatobae	+ acetolysis, HCl (3%)–KOH (2%)–H_2SO_4 (65%) resistant	(Hull et al. 1985)
Spirogyra longata	+ acetolysis, HF (40%) and K_2CrO_4 (30%) resistant	(de Vries et al. 1983)
Spirogyra sp. (zygospores)	algaenan: (presence of suberin/cutin like monomers needs confirmation)	(Blokker 2000)
Spirogyra spp. (7 unnamed species)	+ acetolysis resistant	(Ashraf and Godward 1980)
Spirogyra submarginata	+ acetolysis resistant	(Ashraf and Godward 1980)
Spirogyra submaxima	+ acetolysis, HCl (3%)-KOH (2%)-H_2SO_4 (65%) resistant	(Hull et al. 1985)
Spirogyra weberi	+ acetolysis, HF (40%) and K_2CrO_4 (30%) resistant	(de Vries et al. 1983)
Spirogyra spp.	? see *M. calcarea*, species are *S. catenaeformis, circumlineata, crassa, daedala, ellipsopora, flavescens, fluviatilis, fovealata, hydrodictya, illinoiensis, inflata, insignis, irregularis , laxa, majuscula, novae-angliae, protecta, rectangularis, stictica, tenuitissima, varians, velata* and *weberi*	(Tiffany 1924)
Zygnema (spp.)	? see *M. calcarea*. Species are *Z. collinsianum, decussatum, ericetorum, insignis, pectinatum* and *stellinum*	(Tiffany 1924)
Dinophyta (dinoflagellates)		
Lingulodinium polyedrum (GpES19-cyst)	aromatic	(Kokinos et al. 1998), this study
Scrippsiella ramonii (Naples)	aromatic + aliphatic	This study
Scrippsiella sp.	complex including aromatic and aliphatic moieties	(Hemsley et al. 1994)
Prasinophyta		
Prasinocladus marinus	–	(Shaw 1971)
Pyramimonas pseudoparkeae	acetolysis resistant walls supposed to be 'sporopollenin': needs confirmation	(Aken and Pienaar 1985)

[a]For all taxa with a '–' indication a resistant biomacromolecule could not be evidenced.[b]*Chlorella* taxonomy based on Huss et al. (1999). The status of *C. fusca* and *C. pyrenoidosa* in Burczyk et al. (1999) however remains unclear.[c]For all taxa with a '+' indication, assessment of the presence of a resistant wall is based on old methodology and the presence of such a wall needs to be verified.

help to survive dryness during aerial dispersion or dry periods (e.g., *Haematococcus pluvialis*), and/or they could be more resistant to attack if the cysts form a 'seed bank'. Clearly, the constraints set to cell walls of metabolically active and quiescent algae may be very different. Nevertheless, for the Chlorophyta, this does not seem to have had an effect on the wall chemistry. All resting stages analysed also have algaenan walls. The possible presence of aromatic moieties in the algaenan of *Spirogyra* sp. mesospores, however, needs further investigation (Blokker 2000). Interestingly, in extant algae, algaenans have almost exclusively been detected for fresh water species. The highly aliphatic (plastic like) algaenan may function as a relatively water-proof layer. Such a layer might be an important adaptation for fresh water species or species that live in smaller enclosed habitats; it enables them to spread from one place to another (by wind, birds etc.) and to resist periods of dryness. Most marine species would not necessarily need this.

Fossil algaenans and algaenan-likes

Only a few fossil algal-walls have been shown to closely resemble their Modern algaenan counterparts. These are all derived from the Chlorophyta (Table 3). To what extent did, the hypothesised migration of aliphatic moieties into the algaenan and the process of oxidative polymerisation attaching the aliphatic moieties to the algaenan (Versteegh et al. 2004) (analogous to the process proposed for changing sporopollenin and cutan above) turn the original algal wall into its present state? Due to the aliphatic nature of the algaenan itself, these processes usually remain unnoticed with current methodology. Seen in this light, the mainly aliphatic nature reported for some Proterozoic acritarchs (Arouri et al. 1999) should be interpreted with care. The additional presence of aromatic and (unusual) amide groups in the walls of these acritarchs, generate questions on the purity of the material, the extent to which the acritarch geomacromolecule still represents the original biomacromolecule and the degree to which this original biomacromolecule was aromatic. For this very old material, a second process, like the removal of aliphatic moieties from the acritarch wall at elevated temperature and pressure conditions may additionally have influenced the wall composition. For obvious reasons, this second process has been subject to intense study in relation to understanding oil and gas formation (e.g. Combaz 1971; Rullkötter, 1993) and will therefore not be discussed further.

The acritarch *Gloeocapsomorpha prisca* (see, Wicander et al. 1996 for an overview) is the principal component of middle Ordovician marine oil shales (kukersites). The taxonomic position of *G. prisca* has been a matter of considerable debate. On the basis of morphological comparison with modern organisms, it has been assigned mostly to the Cyanobacteria and Chlorophyta, notably the fresh water species *B. braunii* which when cultured under salt stress produces structures that are morphologically similar to those of *G. prisca* (Derenne et al. 1992b). Recent chemical analyses (Blokker et al. 2001; Lille 2003) indicate an aliphatic wall composed of 1,3-benzediol (resorcinol) building blocks with mainly C_{15}, C_{17}, and C_{19} alkyl side chains. Resorcinols are known from a variety of higher plants, mosses, fungi, bacteria. They have also been identified as free lipids from *Botryococcus* (with C_{25}–C_{29} alkyl chains) (Metzger and Largeau 1994). Upon salt stress the contribution of phenols to *Botryococcus* pyrolysates increases (Derenne et al. 1992b) but if these phenols were derived from resorsinols or not, remains to be investigated. Fossil *Botryococcus* differ from *G. prisca* in lacking evidence for an aliphatic macromolecular wall structure and the shorter chain lengths of the *n*-alkyl resorcinol building blocks of the fossil macromolecule (Blokker et al. 2001).

Moreover, in the light of the widespread occurrence of resorcinols in organisms their occurrence in *Botryococcus* forms no argument to consider *G. prisca* a member of the Chlorophyta. Resorcinols are produced via the polyketide or acetogenic pathway (Kozubek and Tyman 1999). Depending on the phylogenetic position of *G. prisca* this could represent a third pathway in algae for the biosynthesis of resistant macromolecules. However, Blokker et al. (2001) indicate that *G. prisca* may have produced resorcinols for UV and/or microbial protection. They further suggest that the high reactivity of the resorcinols favours their polymerisation, e.g. catalysed by oxygen or trace metals, and that the fossilised macromolecule may have been formed *post-mortem*. Instead of representing a third biosynthetic pathway, this would provide a

Table 3. Composition of resistant biomacromolecules in fossil algae.

Taxon	Biomacromolecule	Age	Latest Reference(s)
Acritarchs			
Reduviasporonites	mainly aliphatic and minor aromatic components filamentous taxon	P/T boundary	(Foster et al. 2002)
Gloeocapsomorpha prisca	algaenan: mainly C_{21} & C_{23} *n*-alkyl resorcinol polymer	Ordovician	(Blokker et al. 2001; Lille, 2003)
Leiosphaeridia	aliphatic	Silurian and Ordovician	(Kjellström 1968)
Multifronsphaeridium peliorum	manly aliphatic, also aromatic and amide groups	Neoproterozoic	(Arouri et al. 1999)
Species A	manly aliphatic, also aromatic and amide groups	Neoproterozoic	(Arouri et al. 1999)
1.*Tanarium* sp. A	1–3 acanthomorph, multilayered fibrillar wall		
2. *Hocosphaeridium scaberfacium*	4–6 no wall layering		
3. *Alicesphaeridium medusoidum*	1–6, no pyrolysates available, but IR and Raman	Neoproterozoic	(Arouri et al. 2000)
4. Species C2	Spectroscopy indicates a highly ordered aromatic		
5. *Chuaria circularis*	biopolymer which can also be ascribed to over-		
6. *Leiosphaeridia* sp.	maturity of the organic matter.		
Chlorophyta			
Botryococcus braunii	aliphatic, like recent representatives (isoprenes questionable)	Permian to Recent	(Derenne et al. 1994; 1997)
Tetraedron minimum	aliphatic, like recent representative	Eocene (Messel 45 Ma)	(Blokker et al. 2000)
Pediastrum	aliphatic, like recent *P. boryanum*	Early Holocene (10 ka)	(Blokker et al. 2000)
Coelastrum reticulatum (inferred)	aliphatic, like recent representative	Early Holocene (10 ka)	(Blokker et al. 2000)
Prasinophyta			
Tasmanites	polymer of aliphatics, (+ isoprenoids and aromatics?) Permian		(Collinson et al. 1994; Greenwood et al. 2000)
Dinophyta	Biomacromolecule purity[a]	Age	Reference
Brigantedinium spp.	aliphatic/aromatic 83	± E/Oboundary	(Dammers 2003)
Chiropteridium spp.	aliphatic/aromatic 95(97)	Oligocene	This study
Deflandrea sp.	aliphatic/aromatic 74(75)	Middle Eocene	(Warnaar 2001)
Enneadocysta sp.	aliphatic/aromatic 89(92)	Middle Eocene	(Warnaar 2001)
Nematosphaeropsis labyrinthus	aromatic 63	± 77500 y	(Dammers 2003)
Manumiella druggii	aliphatic/aromatic > 75	± K/T boundary	This study
Palaeoperidinium spp.	aliphatic/aromatic 88	± K/T boundary	(Dammers 2003)
Polysphaeridium zoharii	aromatic 76	± 50000 y	(Dammers 2003)
Spiniferites sp.(D3)	aliphatic/aromatic (> 50%)	± K/T boundary	(van Mourik 2000)
Thalassiphora sp.	aliphatic/aromatic 42(60)	Middle Eocene	(Warnaar 2001)
Trithyrodinium evittii (D2; D4)	aliphatic/aromatic (95; 90%)	± K/T boundary	(van Mourik 2000)
Mixed gonyaulacoids[b] (D1; GMB-A)	aliphatic/aromatic? 73	± K/T boundary	(van Mourik 2000; Warnaar 2001)
'Dinocasts'	aliphatic C_{16} and C_{18} 100 fatty-acid-based geopolymer	Eocene	(Versteegh et al. 2004)

[a]Purity indication denotes the percentage of cysts of the indicated taxon on organic particles (between brackets the percentage of dinoflagellate cysts on organic particles).
[b]The mixture includes *Areoligera senonensis, Cribroperidinium* sp., *Glaphyrocysta perforata, Hystrichosphaeridium* sp., *Spiniferites ramosus.*

second example of oxidative polymerisation of algae, analogous to the recent findings of Versteegh et al. (2004). The possibility that the organism excreted the resorcinols to form a protective layer, by oxidative polymerisation, can not be ruled out though.

Tasmanites, represents the only fossil member of the Prasinophyta studied for its wall composition. Pyrolysis of an extracted Tasmanite coal, almost entirely consisting of *Tasmanites*, by Collinson et al. (1994) and nuclear magnetic resonance (NMR) analysis of HF treated and picked *Tasmanites* specimens (Hemsley et al. 1993) indicate that these fossils predominantly feature alkane/alkene doublets in the GC-FID trace and thus, that the *Tasmanites* walls could be made of algaenan, confirming the early suggestion of Kjellström (1968) that the walls consist of long chains of aliphatic saturated hydrocarbons. However, Greenwood et al. (2000) report that the solvent extracted walls are 'comprised of ubiquitous *n*-alkane/alkene, parent and alkyl aromatics, and tricyclic terpenoids' on the basis of GC-MS analysis. They also noted that the untreated parent Tasmanite oil shale has the same composition as *Tasmanites*. We attribute this discrepancy to the higher ionisation efficiency of aromatics and isoprenoids compared to aliphatics, resulting in underrepresentation of the latter upon mass spectrometry. Therefore, we infer that *Tasmanites* indeed largely consists of aliphatic moieties. Moreover, the formation of aromatic moieties from isoprenoids since the Permian may explain the aromatic signature of the material. The absence of acid or base hydrolysis treatments of the *Tasmanites* furthermore leaves the possibility that the isoprenoids became attached to the wall macromolecule *post mortem* via ester bonds. Analogous to all other fossil and recent walls of algae analysed we suggest that the isoprenoids did not form cell wall constituents. Greenwood et al. (2000) suggest on the basis of the same composition of parent Tasmanite and *Tasmanites* that the isoprenoids are wall derived, we rather would like to propose the opposite, i.e. that the isoprenoid moieties have become attached to the *Tasmanites* wall. Based on the results of Kuypers et al. (2002), it can be speculated that apart from *Tasmanites*-derived also the degradation resistant archaeal isoprenoid membrane lipids have become oxygen cross-linked and contributed to the macromolecular isoprenoid constituent in these samples.

Dinosporins

Dinosporin of extant dinoflagellates

In contrast to the algaenan wall of the motile stage of the dinoflagellate *Gymnodinium catenatum*, (Gelin et al. 1999) dinoflagellates seem to be able to produce a completely different kind of wall for their resting cysts, called dinosporin (Fensome et al. 1993). There is only very limited information on recent dinoflagellate cyst walls. The walls of *L. polyedrum* are reported to be relatively condensed and predominantly aromatic, compositionally distinct from 'sporopollenin', unrelated to the walls of green algae (algaenan) and that the isoprenoid tocopherol as an important monomer (Kokinos et al. 1998). Upon re-analysis using the methodology of Blokker et al. (1998a; Figure 4) these conclusions are confirmed with the exception that no evidence of isoprenoids is observed in the pyrolysates. The discrepancy may be attributed to the phosphoric acid treatment (3 weeks) during the cyst wall isolation procedure by Kokinos et al. (1998) which has been shown to produce artefacts (see for a review, Allard et al. 1998; Allard and Templier 2000). Our pyrolysates show few, short chain (C_6-C_{10}), carbon chains and a dominance of mono- to poly-methylated aromatic fragments. Longer chain alkyl benzenes (e.g., butyl-benzene) are virtually absent as is any evidence of alkane/alkene doublets which is so characteristic for algaenans. Clearly, the cyst walls consist of a dense aromatic network in which few short alkyl chains are intercalated (dinosporin) which is, in agreement with Kokinos et al. (1998), totally different from the aliphatic algaenans. Considering the highly aromatic nature of the cyst biomacromolecule, we propose the phenylpropanoid pathway as the most logical one for the synthesis of its precursors. NMR analysis of *Scrippsiella* sp. cysts suggests for this taxon a very complex cyst wall macromolecule with also a substantial aliphatic component (Hemsley et al. 1994). Our preliminary analysis of the transparent cyst-walls from a culture of the peridinioid *Scrippsiella ramonii* confirms this. Both aromatic and aliphatic moieties are observed. Isoprenoid moieties are, however, absent (Figure 5). A series of alkane/alkyl doublets (up to C_{18}) and their corresponding ketones (not shown) are present. Interestingly, the aliphatic and aromatic moieties overlap in the sense

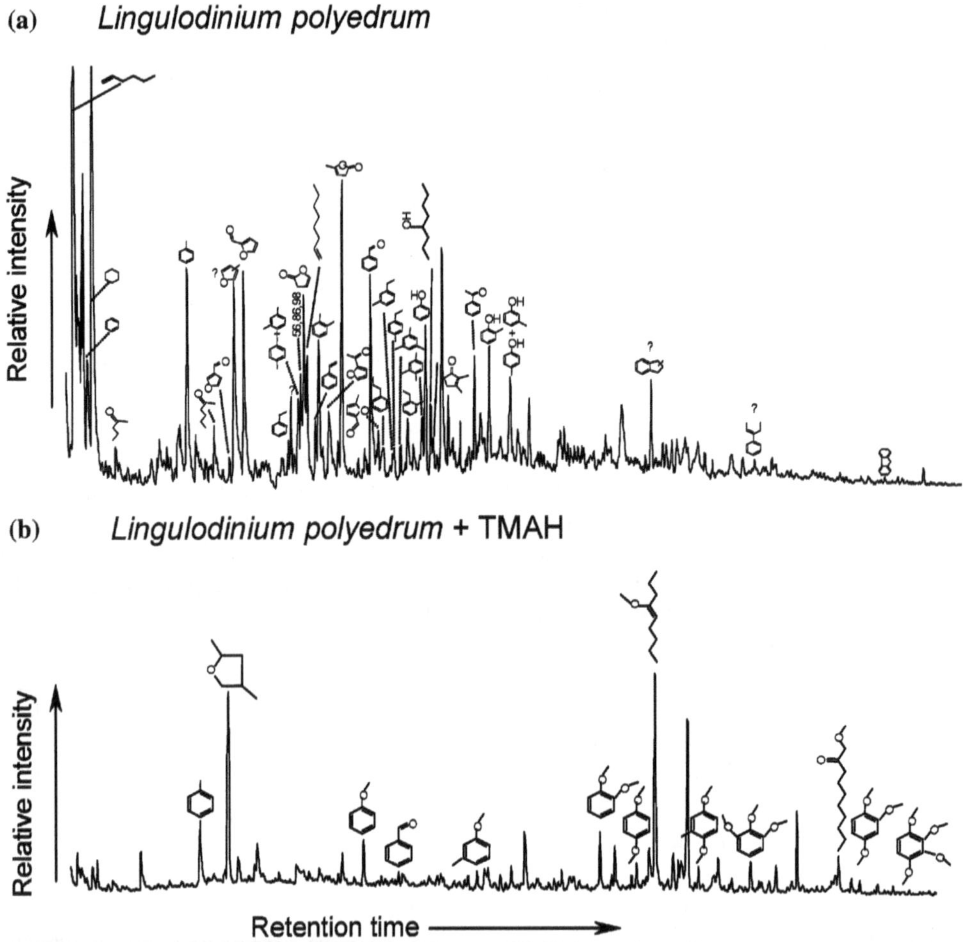

Figure 4. a. Pyrolysate of *Lingulodinium polyedrum* cyst walls from cultured material. b. Pyrolysate of *Lingulodinium polyedrum* cyst walls using tetramethylammonium hydroxide (TMAH). Upon thermolysis the TMAH methylates the oxygen radicals preventing secondary reactions of the pyrolysis products. Both panels demonstrate the aromatic nature of the walls. Note also the absence of isoprenoid moieties.

that also a series of alkyl-benzenes (up to C_7) is present but there is no trace of resorcinols. These latter observations suggest that the aliphatic and aromatic moieties are not organised in separate layers of the cyst wall but are mixed. However, the extent to which the alkyl-benzenes can be formed from aromatic and aliphatic pyrolysis products still remains to be investigated and thus can not be excluded either. Finally, C_{14} and C_{16} FAs are prominent in the pyrolysate whereas the C_{18} FA is absent. This absence of the C_{18} FA makes contamination unlikely as well as a contribution of free, dinoflagellate derived, lipids or their salts (Hartgers et al. 1995) since the C_{18} FA is also prominent in the free lipid extract. Upon pyrolysis with TMAH the C_{14} and C_{16} FAs reappear prominently. However, a series of distally unsaturated and α, ω-dicarboxylic acids are also present suggesting that the aliphatic moieties are part of the biomacromolecular network and represent FA monomers bound to the macromolecule.

The chemical and geological stability of the cysts informs us on a different aspect of dinoflagellate cyst walls. In marine palynology, strong acids, acetolysis and base treatment are avoided as much as possible upon processing of dinoflagellate assemblages from sediments. The reason is that although most gonyaulacoid cysts (the group to which *Lingulodinium* belongs) resist such treatments, they destroy many protoperidinioid cysts (the group to which *Scrippsiella* belongs), notably the brown-walled taxa (e.g. Dale

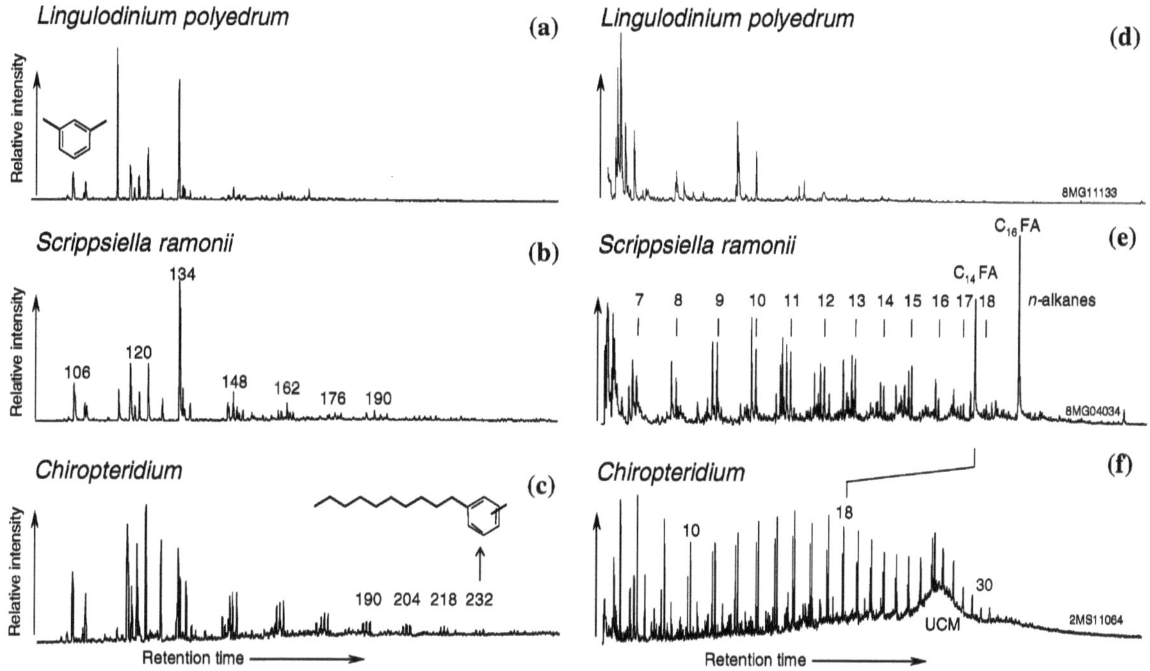

Figure 5. Pyrolysate mass chromatograms of a, b. *Lingulodinium polyedrum* cysts from cultures, c, d. *Scrippsiella ramonii* cysts from cultures, e, f. *Chiropteridium* cysts from Oligocene sediments. *m/z* 105 (a,c,e) showing the methyl-, alkyl-benzenes and *m/z* 55 + 57 (b,d,f) showing the aliphatic moieties, notably alkene/alkane doublets. Numbers in panels c, e refer to the M^+ of the corresponding components. FA = fatty acid, UCM = unresolved complex mixture.

1976; Turon 1984; Schrank 1988; Marret 1993; Hopkins and McCarthy 2002; and pers. obs.). This suggests that the gonyaulacoid macromolecule consists of a high proportion of carbon and ether-linked building blocks whereas the building blocks of protoperidinioid macromolecules are much more ester linked. Interestingly, in the sediments, the resistance of cysts to oxidation parallels their resistance to chemical treatment in the laboratory (Zonneveld et al. 1997, 2001; Versteegh and Zonneveld 2002).

On the basis of the above we conclude that there are probably two variables influencing dinosporin composition, the proportion of aliphatic *vs.* aromatic moieties and the proportion of ether- and carbon-bonds vs. ester-bonds. We further hypothesise that these two variables covary and relate to cyst preservation such that the chemically and/or geologically more resistant cysts (c.f. *L. polyedra*) have more aromatic moieties and more ether-bonds connecting these moieties than the less resistant ones (c.f. *S. ramonii*). However, this hypothesis urgently needs further testing. A major problem in achieving this is that many organic-cyst forming dinoflagellates are notoriously difficult to culture and that inducing cyst formation in culture in sufficient is technically challenging.

We speculate that the presence of aromatic moieties in the cyst wall relates to the fact that they are resting cysts. Some taxa are known to survive in the sediments for more than a decade (Lewis et al. 1999). The cysts are metabolically almost completely inactive (Binder and Anderson 1990) but simultaneously have to protect themselves against bacterial and fungal attack. The aromatic moieties may function as toxins, similar to flavins and tannins in higher plants. Interestingly, this implies that a positive correlation may exist between the survival period of the encysted organism in the sediment, the chemical composition of the cyst wall and its fossilisation potential.

Fossil dinosporin

In recent years several attempts have been made to obtain pure dinoflagellate cyst fractions from sediments, starting in the early 70s (Combaz 1971). In a few cases, high purity was obtained (Table 3).

Pyrolysis of the purest, base- and acid-hydrolysed, fraction with 96% of the Gonyaulacoid *Chiropteridium* displays a mixture of aliphatic and aromatic moieties and no isoprenoids. The pyrolysate resembles that of *S. ramonii* but the aliphatic fragments continue to longer chain lengths; alkane/alkene doublets and their corresponding ketones to C_{30} and the alkylbenzenes up to C_{11}. Furthermore, the FAs are absent. The purity of the sample strongly suggests that the cyst walls are composed of both aliphatic and aromatic moieties. Upon pyrolysis with TMAH, a series of saturated and unsaturated FA moieties is formed, sharply dropping of at C_{18}. The importance of a high purity is illustrated by the observation that analysis of less pure samples yielded highly aliphatic products with the aromatic contribution being strongly dependant on the amount of little black blocky entities in the sample (van Mourik 2000; Warnaar 2001). These black bits yielded predominantly phenols and not only simple aromatics. Clearly, more analyses are needed on pure fossil cyst fractions before and after artificial maturation to elucidate the nature of the fossil cyst walls.

Since the only extant micro-algae known to produce aromatic walls are dinoflagellates, the basically aromatic wall composition of several acantomorph Neoproterozoic Acritarcha has been used to argue that these Acritarcha are related to the dinoflagellates (Arouri et al. 2000). This corroborates with earlier suggestions, based on acritarch morphology (e.g. Sarjeant 1978; Butterfield and Rainbird 1998; Leppig and Montenari 2000), the fossil record of dinosteroids (e.g. Moldowan et al. 1996; Talyzina et al. 2000) and molecular phylogeny (Javaux et al. 2003) that the dinoflagellates originate in the Neoproterozoic.

However, it must be stressed that alternative interpretations are still feasible since the highly aromatic nature of the Neoproterozoic acritarchs may also be related to maturity levels (Arouri et al. 2000), or even convergence.

Concluding remarks

It is clear that the analysis of recent and fossil resistant biomacromolecules requires extreme care especially with respect to the purification procedures and maturation. For the extant material, avoiding artificial condensation and oxidative polymerisation of cytoplasm and ester-bound moieties requires constant attention. Notably addition of aliphatic moieties may occur. For the fossil material, contamination by organic particles other than the target taxon is hard to eliminate and can contribute to either the aliphatic or aromatic signal. Furthermore, *post-mortem* migration of aliphatic moieties into, and their condensation onto the macromolecule might occur as well. These things hamper the evaluation of the aliphatic signature of fossil plant material. The extent to which migration and *in situ* formation of aromatic moieties plays a role in modifying resistant algal macromolecules, notably under elevated temperature and/or pressure conditions, still remains an open question.

Acknowledgements

We thank Peter Blokker (Free University, Amsterdam) and two anonymous referees for useful suggestions for improvement of the manuscript. Peter Blokker is also thanked for processing and analysing the *Lingulodinium* and *Scrippsiella* cysts; Arnoud Boom for providing Figure 2; Arne van Mourik, Jeroen Wanaar and Niels Dammers, Utrecht University for purification and analysis of several fossil dinoflagellate cyst fractions. Jane Lewis and Richard Hallett (University of Westminster, London), are thanked for providing the *Lingulodinium* cysts, Stefano Torricelli (AGIP ENI, Milan) for providing the purified *Chiropteridium* sample, and Marina Montresor (Stazione Zoologica, Naples) for providing the *Scrippsiella* cysts. This is NSG paper 2004.05.08

References

Aken M.E. and Pienaar R.N. 1985. Preliminary investigations on the chemical composition of the scale-boundary and cyst wall of *Pyramimonas pseudoparkeae* (Prasinophyceae). S. Afr. J. Bot. 51: 408–416.

Allard B. and Templier J. 2000. Comparison of neutral lipid profile of various trilaminar outer cell wall (TLS)-containing microalgae with emphasis on algaenan occurrence. Phytochemistry 54: 369–380.

Allard B. and Templier J. 2001. High molecular weight lipids from the trilaminar outer wall (TLS)-containing microalgae *Chlorella emersonii, Scenedesmus communis* and *Tetraedron minimum*. Phytochemistry 57: 459–467.

Allard B., Templier J. and Largeau C. 1998. An improved method for the isolation of artifact-free algaenans from microalgae. Org. Geochem. 28: 543–548.

Arouri K., Greenwood P.F. and Walter M.R. 1999. A possible chlorophycean affinity of some Neoproterozoic acritarchs. Org. Geochem. 30: 1323–1337.

Arouri K.R., Greenwood P.F. and Walter M.R. 2000. Biological affinities of Neoproterozoic acritarchs from Australia: microscopic and chemical characterisation. Org. Geochem. 31: 75–89.

Ashraf M. and Godward M.B.E. 1980. Ultrastructure and chemistry of the zygospore wall of *Spirogyra*. Ann. Bot. 46: 485–487.

Atkinson A.W. Jr, Gunning B.E.S. and John P.C.L. 1972. Sporopollenin in the cell wall of Chlorella and other algae: ultrastructure, chemistry and incorporation of ^{14}C-acetate, studied in synchronous cultures. Planta 107: 1–32.

Batten D.J. 1996. Green and blue-green algae. Colonial Chlorococcales. In: Jansonius J. and McGregor D.C. (eds.), Palynolog: Principles and Applications, AASP Foundation, Salt Lake City, pp. 191–203.

Batten D.J. and Grenfell H.R. 1996. Green and blue-green algae. Botryococcus. In: Jansonius J. and McGregor D.C. (eds.), Palynology: Principles and Applications, AASP Foundation, Salt Lake City, pp. 205–214.

Berkaloff C., Casadevall E., Largeau C., Metzger P., Peracca S. and Virlet J. 1983. The resistant polymer of the walls of the hydrocarbon-rich alga *Botryococcus braunii*. Phytochemistry 22: 389–397.

van Bergen P.F. 1994. Palaeobotany of Propagules: An Investigation combining Microscopy and Chemistry. Ph.D. Thesis. University of London, London.

van Bergen P.F., Blokker P., Collinson M.E., Sinninghe Damsté J.S. and de Leeuw J.W. 2004. Structural biomacromolecules in plants. What can be learnt from the fossil record?. In: Hemsley A.R. and Poole I. (eds.), Evolution of Plant Physiology, Elsevier, Amsterdam, pp. 133–154.

van Bergen P.F., Collinson M.E., Blokker P., Van Moerkerken P., Barrie P.J. and de Leeuw J.W. Chemical characterization of (fossil) sporopollenin: a molecular comparison of Azolla microspore massulae and Kurtzipites pollen with a review of previous literature, in press.

van Bergen P.F., Collinson M.E., Briggs D.E.G., de Leeuw J.W., Scott A.C., Evershed R.P. and Finch P. 1995. Resistant biomacromolecules in the fossil record. Acta Bot. Neerl. 44: 319–345.

van Bergen P.F., Collinson M.E. and de Leeuw J.W. 1993. Chemical composition and ultrastructure of fossil and extant salvinialean microspore massulae and megaspores. Grana Suppl. 1: 18–30.

Bertheas O., Metzger P. and Largeau C. 1999. A high molecular weight complex lipid, aliphatic polyaldehyde tetraterpenediol polyacetal from *Botryococcus braunii* (L. race). Phytochem. 50: 85–96.

Biedlingmaier S., Wanner G. and Schmidt A. 1987. A correlation between detergent tolerance and cell wall structure in green algae. Z. Naturforsch. 42: 245–250.

Binder B.J. and Anderson D.M. 1990. Biochemical composition and metabolic activity of *Scripsiella trochoidea* (Dinophyceae) resting cysts. J. Phycol. 26: 289–298.

Blokker P. 2000. Structural analysis of resistant polymers in extant algae and ancient sediments. Geol. Ultratrajectina 193: 1–145.

Blokker P., Schouten S., de Leeuw J.W., Sinninghe Damsté J.S. and van den Ende H. 1999. Molecular structure of the resistant biopolymer in the zygospore cell walls of *Chlamydomonas monoica*. Planta 207: 539–543.

Blokker P., Schouten S., de Leeuw J.W., Sinninghe Damsté J.S. and van den Ende H. 2000. A comparitive study of fossil and extant algaenans using ruthenium tetroxide degradation. Geochim. Cosmochim. Acta 64: 2055–2065.

Blokker P., Schouten S., van den Ende H., de Leeuw J.W., Hatcher P.G. and Sinninghe Damsté J.S. 1998a. Chemical structure of algaenans from the fresh water algae *Tetraedron minimum*, *Scenedesmus communis* and *Pediastrum boryanum*. Org. Geochem. 29: 1453–1468.

Blokker P., Schouten S., van den Ende H., de Leeuw J.W. and Sinninghe Damsté J.S. 1998b. Cell wall specific ω-hydroxy fatty acids in freshwater green microalgae. Phytochem. 49: 691–695.

Blokker P., van Bergen P., Pancost R., Collinson M.E., de Leeuw J.W. and Sinninghe Damsté J.S. 2001. The chemical structure of *Gloeocapsamorpha prisca* microfossils: Implications for their origin. Geochim. Cosmochim. Acta 65: 885–900.

Blom A.V. 1936. Quelques remarques sur le mécanisme de séchage des peintures à l'huile. Peintures, Pigments, Vernis 13: 156–162.

Boom A. 2004. A Geochemical Study of Lacustrine Sediments: Towards Palao-climatic Econstructions of High Andean Biomes in Colombia. University of Amsterdam, Amsterdam, p. 125.

Boom A., Sinninghe Damsté J.S. and de Leeuw J.W. 2005. Cutan, a common aliphatic biopolymer in cuticles of drought-adapted plants. Org. Geochem. 36: 596–601.

Briggs D.E.G., Kear A.J., Baas M., de Leeuw J.W. and Rigby S. 1995. Decay and composition of the hemichordate *Rhabdopleura*: implications for the taphonomy of graptolites. Lethaia 28: 15–23.

Brunner U. and Honegger R. 1985. Chemical and ultrastructural studies on the distribution of sporopollenin like biopolymers in six genera of lichen phycobionts. Can. J. Bot. 63: 2221–2230.

Burczyk J. 1987a. Biogenic relationships between ketocarotenoids and sporopollenins in green algae. Phytochemistry 26: 113–119.

Burczyk J. 1987b. Cell wall carotenoids in green algae which form sporopollenins. Phytochemistry 26: 121–128.

Burczyk J., Śmietana B., Terminska-Pabis K., Zych M. and Kowalowski P. 1999. Comparison of nitrogen content amino acid composition and glucosamine content of cell walls of various chlorococcalean algae. Phytochemistry 51: 491–497.

Butterfield N.J. and Rainbird R.H. 1998. Diverse organic-walled fossils, including 'possible dinoflagellates' from the early Neoproterozoic of arctic Canada. Geology 26: 963–966.

Collinson M.E., Mösle B., Finch P., Scott A.C. and Wilson R. 1998. Structure, biosynthesis and biodegradation of cutin and suberin. Ancient Biomol. 2: 251–265.

Collinson M.E. and van Bergen P.F. 2004. Evolution of angiosperm fruit and seed physiology: anatomical and chemical evidence from fossils. In: Hemsley A.R. and Poole I. (eds.), Evolution of Plant Physiology, Elsevier, Amsterdam, pp. 343–377.

Collinson M.E., van Bergen P.F., Scott A.C. and de Leeuw J.W. 1994. The oil-generating potential of plants from coal and coal-bearing strata through time: a review with new evidence from Carboniferous plants. Geol. Soc. Spec. Publ. 77: 31–70.

Combaz A. 1971. Themal degradation of sporopollenin and genesis of hydrocarbons. In: Brooks J. (ed.), Sporopollenin, Academic Press, London, pp. 621–653.

Corre G., Templier J., Largeau C., Rousseau B. and Berkaloff C. 1996. Influence of cell wall composition on the resistance of two *Chlorella* species (Chlorophyta) to detergents. J. Phycol. 32: 584–590.

Dale B. 1976. Cyst formation, sedimentation, and preservation: factors affecting dinoflagellate assemblages in recent sediments from Trondheimsfjord. Norway. Rev. Palaeobot. Palynol. 22: 39–60.

Dammers N. 2003. Chemical characterization of the organic-walled dinoflagellate cyst taxa *Nematosphaeropsis labyrinthus*, *Polysphaeridium zoharii*, *Brigantedinium* spp. and *Palaeoperidinium* spp. and a review on algaenan and dinosterol. Masters Thesis, Department of Geochemistry, Utrecht University, 43 pp.

de Leeuw J.W. and Largeau C. 1993. A review of macromolecular compounds that comprise living organisms and their role in kerogen, coal and petroleum formation. In: Engel M.H. and Macko S.A. (eds.), Organic Geochemistry. Principles and Applications, Plenum Press, New York, pp. 23–72.

de Vries P.J.R., Simons J. and van Beem A.P. 1983. Sporopollenin in the spore wall of Spirogyra (Zygnemataceae, Chlorophyceae). Acta Bot. Neerl. 32: 252–258.

Derenne S., Largeau C. and Behar F. 1994. Low polarity pyrolysis products of Permian to Recent Botryoccus-rich sediments: first evidence for the contribution of an isoprenoid algaenan to kerogen formation. Geochim. Cosmochim. Acta 58: 3703–3711.

Derenne S., Largeau C. and Berkaloff C. 1996. First example of an algaenan yielding an aromatic-rich pyrolysate. Possible geochemical implications on marine kerogen formation. Org. Geochem. 24: 617–627.

Derenne S., Largeau C., Berkaloff C., Rousseau B., Wilhelm C. and Hatcher P.G. 1992a. Non-hydrolysable macromolecular constituents from outer walls of *Chlorella fusca* and *Nanochlorum eucaryotum*. Phytochemistry 31: 1923–1929.

Derenne S., Largeau C., Casadevall E., Berkaloff C. and Rousseau B. 1991. Chemical evidence of kerogen formation in source rocks and oil shales via selective preservation of thin resistant outer walls of microalgae: origin of ultralaminae. Geochim. Cosmochim. Acta 55: 1041–1050.

Derenne S., Largeau C., Hetényi M., Brukner-Wein A., Conan J. and Lugardon B. 1997. Chemical structure of the organic matter in a Pliocene maar-type shale: implicated *Botryococcus* race strains and formation pathways. Geochim. Cosmochim. Acta 61: 1879–1889.

Derenne S., Metzger P., Largeau C., van Bergen P.F., Gatellier J.P., Sinninghe Damsté J.S., de Leeuw J.W. and Berkaloff C. 1992. Similar morphological and chemical variations of *Gloeocapsomorpha prisca* in Ordovician sediments and cultured *Botryococcus braunii* as a response to changes in salinity. Org. Geochem. 19: 299–313.

Deshmukh A. P., Simpson A.J. and Hatcher P.G. 1964. Evidence for cross-linking in tomato cutin using HR-MAS NMR spectroscopy. Phytochemistry 64: 1163–1170.

Domínguez E., Mercado J.A., Quesada M.A. and Heredia A. 1999. Pollen sporopollenin: degradation and structural elucidation. Sex. Plant Reprod. 12: 171–178.

Fensome R.A., et al., 1993. A Classification of Modern and Fossil Dinoflagellates. Sheridan Press, Hanover, p. 351.

Fensome R.A., Saldarriaga J.F. and Taylor F.J.R. 1999. Dinoflagellate phylogeny revisited: reconciling morphological and molecular based phylogenies. Grana 38: 66–80.

Fjällström P., Andersson B., Nilsson C. and Andersson K. 2002. Drying of linseed oil paints: a laboratory study of aldehyde emissions. Indust. Crop. Prod. 16: 173–184.

Foster C.B., Stephenson M.H., Marshall C., Logan G.A. and Greenwood P.F. 2002. A revision of Reduviasporonites Wilson 1962: description, illustration, comparison and biological affinities. Palynology 26: 35–58.

Gabarayeva N.I., Blackmore S. and Rowley J.R. 2003. Observations on the experimental destruction and substructural organisation of the pollen wall of some selected Gymnosperms and Angiosperms. Rev. Palaeobot. Palynol. 124: 203–226.

van Geel B. and Grenfell H.R. 1996. Green and blue-green algae. Spores of Zygnemataceae. In: Jansonius J. and McGregor D.C. (eds.), Palynology: Principles and Applications, AASP Foundation, Salt Lake City, pp. 173–179.

Gelin F., Boogers I., Noordeloos A.A.M., Sinninghe Damsté J.S., Riegman R. and de Leeuw J.W. 1997. Resistant biomacromolecules in marine microalgae of the classes Eustigmatophyceae and Chlorophyceae: Geochemical applications. Org. Geochem. 26: 659–675.

Gelin F., Volkman J.K., Largeau C., Derenne S., Sinninghe Damsté J.S. and de Leeuw J.W. 1999. Distribution of aliphatic nonhydrolyzable biopolymers in marine microalgae. Org. Geochem. 30: 147–159.

Good B.H. and Chapman R.L. 1978. The ultrastructure of Phycopeltis (Chroolepidaceae: Chlorophyta). I. Sporopollenin in the cell walls. Am. J. Bot. 65: 27–33.

Goth K., de Leeuw J.W., Püttmann W. and Tegelaar E.W. 1988. Origin of messel oil shale kerogen. Nature 336: 759–761.

Gray J. and Boucot A.J. 1989. Is Moyeria a euglenoid?. Lethaia 22: 447–456.

Greenwood P.F., Arouri K.R. and George S.C. 2000. Tricyclic terpenoid composition of *Tasmanites* kerogen as determined by pyrolysis GC-MS. Geochim. Cosmochim. Acta 64: 1249–1263.

Guilford W.J., Schneider D.M., Labovitz J. and Opella S.J. 1988. High resolution solid state ^{13}C NMR spectroscopy of sporopollenin from different plant taxa. Plant Physiol. 86: 134–136.

Gunnison D. and Alexander M. 1975. Basis for the resistance of several algae to microbial decomposition. Appl. Microbiol. 29: 729–738.

Guy-Ohlson D. 1996. Green and blue-green algae. Prasinophyceanalgae. In: Jansonius J. and McGregor D.C. (eds.), Palynology: Principles and Applications, AASP Foundation, Salt Lake City, pp. 181–189.

Hartgers W.A., Sinninghe Damsté J.S. and de Leeuw J.W. 1995. Curie-point pyrolysis of sodium salts of functionalized fatty acids. J. Anal. Appl. Pyrolysis 34: 191–217.

Hayatsu R., Botto R.E., Mcbeth R.L., Scott R.G. and Winans R.E. 1988. Chemical alteration of a biological polymer 'sporopollenin' during coalification: origin, formation, and transformation of the coal maceral sporinite. Energy Fuel 2: 843–847.

Hegewald E. and Deason T.R. 1988. *Pseudoschroederia punctata* (Chlorophyta, Chlorococcales), a new species with an unusual cell wall. Arch. Hydrobiol. Suppl. 78: 465–473.

Hegewald E. and Deason T.R. 1989. *Pseudodidymocystis*, a new genus of Scenedesmaceae (Chlorophyceae). Arch. Hydrobiol. Suppl. 82: 119–127.

Hemsley A.R., Barrie P.J., Chaloner W.G. and Scott A.C. 1993. The composition of sporopollenin and its use in living and fossil plant systematics. Grana Suppl. 1: 2–11.

Hemsley A.R., Barrie P.J. and Chaloner W.G. 1994. Studies of fossil and modern spore and pollen wall biomacromolecules using 13C solid state NMR. NERC Spec. Publ. 94: 15–19.

Holloway P.J. 1982. The chemical constitution of plant cutins. In: Cutler D.F. (ed.), The Plant Cuticle, Elsevier, Amsterdam, pp. 45–85.

Honegger R. and Brunner U. 1981. Sporopollenin in the cell walls of Coccomyxa and *Myrmecia phycobionts* of various lichens: an ultrastructural and chemical investigation. Can. J. Bot. 59: 2713–2734.

Hopkins J.A. and McCarthy F.M.G. 2002. Post-depositional palynomorph degradation in Quaternary shelf sediments: a laboratory experiment studying the effects of progressive oxidation. Palynology 26: 167–184.

Hull H.M., Hoshaw R.W. and Wang J.-C. 1985. Interpretation of zygospore wall structure and taxonomy of *Spirogyra* and *Sirogonium* (Zygnemataceae, Chlorophyta). Phycologia 24: 231–239.

Huss V.A.R., Frank C., Hartmann E.C., Hirmer M., Kloboucek A., Seidel B.M., Wenzeler P. and Kessler E. 1999. Biochemical taxonomy and molecular phylogeny of the genus *Chlorella sensu* lato (Chlorophyta). J. Phycol. 35: 587–598.

Javaux E., Knoll A.H. and Walter M. 2003. Recognizing and interpreting the fossils of early eukaryotes. Origin. Life Evol. Biosph. 33: 75–94.

Kadouri A., Derenne S., Largeau C., Casadevall E. and Berkaloff C. 1988. Resistant biopolymer in the outer walls of *Botryococcus braunii*, B race. Phytochemistry 27: 551–557.

Kjellström G. 1968. Remarks on the chemistry and ultrastructure of the cell wall of some Palaeozoic leiospheres. Geol. Föreningens Stockholm Förhandl. 90: 221–228.

Knoll A.H. 1992. The early evolution of eukaryotes: a geological perspective. Science 256: 622–627.

Knoll A.H. 1996. Archean and Proterozoic paleontology. In: Jansonius J. and McGregor D.C. (eds.), Palynology: Principles and Applications, AASP Foundation, Salt Lake City, pp. 51–80.

Kokinos J.P., Eglinton T.I., Goñi M.A., Boon J.J., Martoglio P.A. and Anderson D.M. 1998. Characterisation of a highly resistant biomacromolecular material in the cell wall of a marine dinoflagellate resting cyst. Org. Geochem. 28: 265–288.

Kolattukudy P.E. 1981. Structure, biosynthesis and biodegradation of cutin and suberin. Ann. Rev. Plant Physiol 32: 539–576.

König J. and Peveling E. 1980. Vorkommen von sporopollenin in der zellwand des phycobionten Trebouxia. Z. Pflanzenphysiol. 98: 459–464.

König J. and Peveling E. 1984. Cell walls of the phycobionts *Trebouxia* and *Pseudotrebouxia*: constituents and their localisation. Lichenology 16: 129–144.

Kozubek A. and Tyman J.H.P. 1999. Resorcinolic lipids, the natural non-isoprenic amphiphiles and their biological activity. Chem. Rev. 99: 1–26.

Kuypers M.M.M., Blokker P., Hopmans E.C., Kinkel H., Pancost R.D., Schouten S. and Sinninghe Damsté J.S. 2002. Archaeal remains dominate marine organic matter from the early Albian oceanic anoxic event 1b. Palaeogeogr. Palaeoclimatol. Palaeoecol. 185: 211–234.

Leppig U. and Montenari M. 2000. Organic-walled microfossils of possible dinoflagellate affinity from the Lower Permian of Sonora (northwest Mexico). Mar. Micropaleontol. 40: 1–8.

Lewis J., Harris A.S.D., Jones K.J. and Edmonds R.L. 1999. Long-term survival of marine planktonic diatoms and dinoflagellates in stored sediment samples. J. Plankton Res. 21: 343–354.

Lille Ü. 2003. Current knowledge on the origin and structure of Estonian Kukersite kerogen. Oil Shale 20: 253–263.

Marchant H.J. 1977. Cell division and colony formation in the green alga *Coelastrum* (Chlorococcales). J. Phycol. 13: 102–110.

Marret F. 1993. Les effets de l'acétolyse sur les assemblages des kystes de dinoflagellés. Palynoscience 2: 267–272.

McKinney D.E., Bortiatynski J.M., Carson D.M., Clifford D.J., de Leeuw J.W. and Hatcher P.G. 1996. Tetramethylammonium hydroxide (TMAH) thermochemolysis of the aliphatic biopolymer cutan: insights into the chemical structure. Org. Geochem. 24: 641–650.

Metzger P. and Largeau C. 1994. A new type of ether lipid comprising phenolic moieties in *Botryococcus braunii*. Chemical structure and abundance, and geochemical implications. Org. Geochem. 22: 801–814.

Metzger P. and Largeau C. 2002. Natural polyacetals. In: Matsumura S. and Steinbüchel A. (eds.), Miscellaneous Biopolymers and Biodegradation of Synthetic Polymers, Wiley-VCH, Weinheim, pp. 113–127.

Moldowan J.M., Dahl J., Jacobson S.R., Huizinga B.J., Fago F.J., Shetty R., Watt D.S. and Peters K.E. 1996. Chemostratigraphic reconstruction of biofacies: molecular evidence linking cyst-forming dinoflagellates with pre-Triassic ancestors. Geology 24: 159–162.

Montsant A., Zarka A. and Boussiba S. 2001. Presence of a nonhydrolysable biopolymer in the cell wall of vegetative cells and astaxanthin-rich cysts of *Haematococcus pluvialis* (Chlorophyceae). Mar. Biotechnol. 3: 515–521.

Morrill L.C. and Loeblich A.R. III 1981. The dinoflagellate pellicular wall layer and its occurrence in the division Pyrrophyta. J. Phycol. 17: 315–323.

van Mourik A. 2000. Chemical characterization of fossil palynomorphs; gonyaulacoid, peridinioid dinoflagellate cysts and sporomorphs. Masters Thesis Department of Geochemistry, Utrecht University, 49 pp.

Mösle B., Finch P., Collinson M.E. and Scott A.C. 1997. Comparison of modern and fossil plant cuticles by selective chemical extraction monitored by flash pyrolysis-gas chromatography-mass spectroscopy and electron microscopy. J. Anal. Appl. Pyrolysis 40–41: 585–597.

Mulder M.M., van der Hage E.R.E. and Boon J.J. 1992. Analytical in source pyrolytic methylation electron impact mass spectrometry of phenolic acids in biological matrices. Phytochem. Anal. 3: 165–172.

Müller P.J., Kirst G., Ruhland G., Von Storch I. and Rosell-Melé A. 1998. Calibration of the alkenone paleotemperature index based on coretops from the eastern South Atlantic and global ocean (60°N-60°S). Geochim. Cosmochim. Acta 62: 1757–1772.

Porcella R.A. and Walne P.L. 1980. Microarchitecture and envelope development in *Dysmorphococcus globosus* (Phacotaceae, Chlorophyceae). J. Phycol. 16: 280–290.

Puel F., Largeau C. and Giraud G. 1987. Occurrence of a resistant biopolymer in the outer walls of the parasitic alga *Prototheca wickerhamii* (Chlorococcales): ultrastructure and chemical studies. J. Phycol. 23: 649–656.

Rascio N., Casadoro G. and Andreoli C. 1979. Ultrastructural features of *Chlorella nana* sp. nov. Bot. Mar. 22: 223–227.

Rodríguez M.C. and Cerezo A.S. 1996. The resistant 'biopolymer' in cell walls of *Coelastrum sphaericum*. Phytochemistry 43: 731–734.

Rodríguez M.C., Noseda M.D. and Cerezo A.S. 1999. The fibrillar polysaccharides and their linkage to algaenan in the trilaminar layer of the cell wall of *Coelastrum sphaericum* (Chlorophyceae). J. Phycol. 35: 1025–1031.

Rozema J., Broekman R.A., Blokker P., Meijkamp B.B., de Bakker N., van de Staaij J., van Beem A., Ariese F. and Kars S.M. 2001. UV-B absorbance and UV-B absorbing compounds (*para*-cumaric acid) in pollen and sporopollenin: the perspective to track historic UV-B levels. J. Photochem. Photobiol. 62: 108–117.

Rozema J., van Geel B., Björn L.O., Lean J. and Madronich S. 2002. Toward solving the UV puzzle. Science 296: 1621–1622.

Rullkötter J. 1993. The thermal alteration of kerogen and the formation of oil. In: Engel M.H. and Macko S.A. (eds.), Organic Geochemistry. Principles and Applications, Plenum Press, New York, pp. 377–96.

Sarjeant W.A. 1978. *Arpylorus antiquus* Calandra, emend., a dinoflagellate cyst from the Upper Silurian. Palynol. 2: 167–179.

Schenck P.A., de Leeuw J.W., van Graas G., Haverkamp J. and Bouman M. 1981. Analysis of recent spores and pollen and of thermally altered sporopollenin by flash pyrolysis-mass spectrometry and flash pyrolysis-gas chromatography-mass spectrometry. In: Brooks J. (ed.)Organic Maturation Studies and Fossil Fuel Exploration, Academic Press, London, pp. 225–237.

Schouten S., Hopmans E.C., Schefuss E. and Sinninghe Damsté J.S. 2002. Distributional variations in marine crenarchaeotal membrane lipids: a new tool for reconstructing ancient sea water temperatures? Earth Planet. Sci. Lett. 204: 265–274.

Schrank E. 1988. Effects of chemical processing on the preservation of peridinioid dinoflagellates: a case from the late cretaceous of NE Africa. Rev. Palaeobot. Palynol. 56: 123–140.

Shaw G. 1971. The chemistry of sporopollenin. In: Brooks J. (ed.), Sporopollenin, Academic Press, London, pp. 305–350.

Simpson A.J., Zang X., Kramer R. and Hatcher P.G. 2003. New insights on the structure of algaenan from *Botryococcus braunii* race A and its hexane insoluble botryals based on multidimensional NMR spectroscopy and electrospray-mass spectrometry techniques. Phytochem. 62: 783–796.

Sinninghe Damsté J.S., Delas Heras F.X.C., de Bergen P.F. and de Leeuw J.W. 1993. Characterization of tertiary catalan lacustrine oil shales: discovery of extremely organic sulphur-rich Type I kerogens. Geochim. Cosmochim. Acta 57: 389–415.

Srivastava A. and Prasad R. 2000. Triglycerides-based diesel fuels. Renew. Sust. Energ. Rev. 4: 111–133.

Staehelin L.A. and Picket-Heaps J.D. 1975. The ultrastructure of Scenedesmus (Chlorophyceae). I. Species with the 'reticulate' or 'warty' type of ornamental layer. J. Phycol. 11: 163–185.

Stankiewicz B.A., Briggs D.E.G., Michels R., Collinson M.E., Flannery M.B. and Evershed R.P. 2000. Alternative origin of aliphatic polymer kerogen. Geology 28: 559–562.

Strother P.K. 1996. Acritarchs. In: Jansonius J. and McGregor D.C. (eds.), Palynology: principles and applications, AASP Foundation, Salt Lake City, pp. 81–106.

Swift E. and Remsen C.C. 1970. The cell wall of *Pyrocystis* spp. (Dinophyceae). J. Phycol. 6: 79–86.

Syrett P.J. and Thomas E.M. 1973. The assay of nitrate reductase in whole cells of *Chlorella*: strain differences and the effect of cell walls. New Phytol. 72: 1307–1310.

Talyzina N.M., Moldowan J.M., Johannisson A. and Fago F.J. 2000. Affinities of Early Cambrian acritarchs studied by using microscopy, fluorescence flow cytometry and biomarkers. Rev. Palaeobot. Palynol. 108: 37–53.

Tegelaar E.W., de Leeuw J.W., Derenne S. and Largeau C. 1989. A reappraisal of kerogen formation. Geochim. Cosmochim. Acta 53: 3103–3106.

Tegelaar E.W., Kerp H., Visscher H., Schenck P.A. and de Leeuw J.W. 1991. Bias of the paleobotanical record as a consequence of variations in the chemical composition of higher vascular plant cuticles. Paleobiology 17: 133–144.

Tiffany L.H. 1924. A physiological study of growth and reproduction among certain algae. Ohio J. Sci. 24: 65–99.

Turon J.-L. 1984. Le palynoplancton dans l'environnement actuel de l'Atlantique nord-oriental. Évolution climatique et hydrologique depuis le dernier maximum glaciaire. Mem. Inst. Geol. Bassin Aquitaine 17: 1–313.

Versteegh G.J.M., Blokker P., Wood G., Collinson M.E., Sinninghe Damsté J.S. and de Leeuw J.W. 2004. Oxidative polymerization of unsaturated fatty acids as a preservation pathway for microalgal organic matter. Org. Geochem. 35: 1129–1139.

Versteegh G.J.M., Jansen J.H.F., de Leeuw J.W. and Schneider R.R. 2000. Mid-chain diols and keto-ols in sediments. A new tool for tracing past sea surface water masses?. Geochim. Cosmochim. Acta 64: 1879–1892.

Versteegh G.J.M. and Zonneveld K.A.F. 2002. Use of selective degradation to separate preservation from productivity. Geology 30: 615–618.

Warnaar J. 2001. Chemical characterization of fossil organic walled dinoflagellate cysts. Masters Thesis, Department of Geochemistry, Utrecht University. 26 pp.

Wehling K., Niester C., Boon J.J., Willemse M.T.M. and Wiermann R. 1989. *p*-Coumaric acid – a monomer in the sporopollenin skeleton. Planta 179: 376–380.

Wicander R., Foster C.B. and Reed J.D. 1996. Green and blue-green algae. *Gloeocapsomorpha*. In: Jansonius J. and McGregor D.C. (eds.), Palynology: Principles and Applications, AASP Foundation, Salt Lake City, pp. 215–225.

Wurdack M.E. 1923. Chemical composition of the walls of certain algae. Ohio J. Sci. 23: 181–191.

Yule B.L., Roberts S. and Marchall J.E.A. 2000. The thermal evolution of sporopollenin. Org. Geochem. 31: 859–870.

Zárský V., Kalina T. and Sulek J. 1985. Notes on the sexual reproduction of *Chlamydomonas geitleri* Ettl. Arch. Protistenk. 130: 343–353.

Zelibor J.L. Jr, Romankiw L., Hatcher P.G. and Colwell R.R. 1988. Comparitive analysis of the chemical composition of mixed and pure cultures of green algae and their decomposed residues by ^{13}C nuclear magnetic resonance spectroscopy. Appl. Env. Microbiol. 54: 1051–1060.

Zhang E., Hatcher P.G. and Davis A. 1993. Chemical composition of pseudo-phlobaphinite precursors: implications for the presence of aliphatic biopolymers in vitrinite from coal. Org. Geochem. 20: 721–734.

Zonneveld K.A.F., Versteegh G.J.M., de Lange and G. J. 1997. Preservation of organic walled dinoflagellate cysts in different oxygen regimes: a 10,000 years natural experiment. Mar. Micropaleontol. 29: 393–405.

Zonneveld K.A.F., Versteegh G.J.M. and de Lange G.J. 2001. Palaeoproductivity and post-depositional aerobic organic matter decay reflected by dinoflagellate cyst assemblages of the Eastern Mediterranean S1 sapropel. Mar. Geol. 172: 181–195.

SUBJECT INDEX VOLUME 182 2006

SPECIES INDEX VOLUME 182 2006

AUTHOR INDEX VOLUME 182 2006

Tasks for vegetation science

1. E.O. Box: *Macroclimate and plants forms.* An introduction to predictive modelling in Photography. 1981 ISBN 90-6193-941-0
2. D.N Sen and K.S. Rajpurohit (eds): *Contributions to the ecology of halophytes.* 1982 ISBN 90-6193-942-9
3. J.Ross: *The radiation regime and architecture of plant stands.* 1981 ISBN 90-6193-942-9
4. N.S. Margaris and H.A.Moony(eds): *Components of productivity of mediterranean-cliamte regions.* Basic and applied aspects. 1981 ISBN 90-6193-944-5
5. M.J.Muller: *Selected climate data for a global set of standard stations for vegetation science.* 1982 ISBN 90-6193-945-3
6. I.Roth: *Stratification of tropical forests as seen in leaf structure [Part 1].* 1984 ISBN 90-6193-946-1 For Part 2, see Volume 21
7. L.Steubing and H.-J.Jager (eds.): *Monitoring of air pollutants by plants. Methods and problems.* 1982 ISBN 90-6193-947-X
8. H.J. Teas (ed.): *Biology and ecology of mangroves.* 1983 ISBN 90-6193-948-8
9. H.J. Teas (ed.): *Physiology and management of mangroves.* 1984 ISBN 90-6193-949-6
10. E.Feoli, M.Langonegro and L.Orlci: *Information analysis of vegetation data.* 1984 ISBN 90-6193-950-X
11. Z.Svestak (ed): *Photosynthesis during leaf development.* 1985 ISBN 90-6193-951-8
12. E.Medina, H.A.Mooney and C.Vzquenz-Ynes(eds.): *Physiological ecology of plants of the wet tropics.* 1984 ISBN 90-6193-953-4
13. N.S. Margaris, M.Arianoustous-Faraggitaki and W.C.Oechel (eds.): *Being alive on land.* 1984 ISBN 90-6193-953-4
14. D.O.Hall, N.Myers and N.S.Margaris (eds.): *Economics of ecosystem management.* 1985 ISBN 90-6193-505-9
15. A.Estrada and Th.H.Fleming (eds.): *Frugivores and seed dispersal.* 1986 ISBN 90-6193-543-1
16. B. Dell, A.J.M.Hopkins and B.B.Lamont (eds.): *Resilience in Mediterraneam-type ecosystems.* 1986 ISBN 90-6193-579-2
17. I.Roth: *Stratification of a tropical as seen in dispersal types.* 1987 ISBN 90-6193-613-6
18. H.G.Dassler and S.Bortitz (eds.): *Air pollution and its influence on vegetation.* Causes, Effects, Prophy laxis and Therapy. 1988 ISBN 90-6193-619-5
19. R.L.Specht (ed.): *Mediterranean-type ecosystem.* A data source book. 1988 ISBN 90-6193-652-7
20. L.F.Huenneke and H.A.Mooney (eds.): *Grassland structure and function.* California annual grassland. 1989 ISBN 90-6193-659-4
21. B.Rollet, Ch.Hagermann and I.Roth: *Stratification of tropical forests as seen leaf structure,* Part 2. 1990 ISBN 0-7923-0397-0
22. J.Rozema and J.A.C. Verkleij (eds.): *Ecological response to environment stresses.* 1991 ISBN 0-7923-0762-3
23. S.C. Pandeya and H.Lieth: *Ecology of Cenchrus grass complex.* Environmental conditions and populations differences in Western India. 1993 ISBN 0-7923-0768-2
24. P.L.Nimis and T.J.Crovello (eds.): *Quantitative approaches to phytogeography.* 1991 ISBN 0-7923-0795-X
25. D.F. Whigham, R.E.Good and K.Kvet (eds.): *Wetland ecology and management.* Case studies. 1990 ISBN 0-7923-0893-X
26. K.Flinska: *Plant demography in vegetation succession.* 1991 ISBN 0-7923-1060-8
27. H. Lieth and A.A.Al Masoom (eds.): *Towards the rational use of high salinity tolerant plants,* Vol. 1: Deliberations about high salinity tolerant plants and ecosystems. 1993 ISBN 0-7923-1865-X

Tasks for vegetation science

28. H. Lieth and A.A.Al Masoom (eds.): *Towards the rational use of high salinity tolerant plants,* Vol. 2: Agriculture and forestry under marginal soil water conditions. 1993 ISBN 0-7923-1866-8
29. J.G. Boonman: *East Africa's grasses and folders.* Their ecology and husbandry. 1993 ISBN 0-7923-1867-6
30. H. Lieth and M. Lohmann (eds.): *Restoration of tropical forest ecosystems.* 1993 ISBN 0-7923-1945-1
31. M. Arianoutsou and R.H. Groves (eds.): *Plant-animal interactions in Mediterranean-type ecosystem.* 1994 ISBN 0-7923-2470-6
32. V.R. Squires and A.T. Ayoub (eds.): *Halophytes as a resources for livestock and for rehabilitation of degraded lands.* 1994 ISBN 0-7923-2664-4
33. T. Hirouse and B.H. Walker (eds.): *Global change and terrestrial ecosystems in monsoon Asia.* 1995 ISBN 0-7923-0000-0
34. A. Kratochwil (ed.): Biodiversity in Ecosystems: *Principles and case studies of different complexity levels.* 1999 ISBN 0-7923-5717-5
35. C.A. Burga and A. Kratochwil (ed.): Biomonitoring: *General and applied aspects on regional and global scales.* 2001 ISBN 0-7923-6734-0
36. H.-J. Barth and B. Boer (eds.): *Sabkha Ecosystems.* Volume I: The Arabian Peninsula and Adjacent Countries. 2001 ISBN 1-4020-0504-0
37. R. Ahamed and K.A. Malik (eds.): *Prospects for saline Agriculture.* 2002 ISBN 1-4020-0620-9
38. H. Lieth and M. Mochtchenko (eds.): *Cashcrop Halophytes Recent Studies.* 10 Years after the Al Ain Meeting. 2003 ISBN 1-4020-1202-0
39. M.D. Schwartz (eds.): PHENOLOGY: *An Integrative Environmental Science.* 2003 ISBN 1-4020-1580-1
40. M.A. Khan and D.J. Weber (eds.): *Ecophysiology of High Salinity Tolerant Plants.* 2006 ISBN 1-4020-4017-2
41. J. Rozema, R. Aerts and H. Comelissen (eds.): *Plants and Climate Change.* 2006 ISBN 1-4020-4449-9

MIX
Papier aus verantwortungsvollen Quellen
Paper from responsible sources
FSC® C105338
FSC
www.fsc.org

If you have any concerns about our products,
you can contact us on
ProductSafety@springernature.com

In case Publisher is established outside the EU,
the EU authorized representative is:
Springer Nature Customer Service Center GmbH
Europaplatz 3, 69115 Heidelberg, Germany

Printed by Libri Plureos GmbH
in Hamburg, Germany